CHEMTREK

Small-Scale Experiments for General Chemistry

Stephen Thompson
Colorado State University

PRENTICE HALL, Englewood Cliffs, New Jersey 07632

Thompson, Stephen.
Chemtrek : small-scale experiments for general chemistry / Stephen Thompson.
p. cm.
ISBN 0-205-11913-1
1. Chemistry--Laboratory manuals. I. Title.
QD45.T48 1989
542--dc20

89-6852
CIP

Signing representative: Robert G. Cohen
Cover designer: Richard Hannus
Cover administrator: Linda Dickinson
Production administrator: Lorraine Perrotta

Acknowledgements
Figures on pages 87-90: John Fluke Manufacuring Company, Inc.
Figure on page 91: Amperex Corporation.
Tables on pages 92-93: Amperex Corporation.
Figure on page 149: Mallinckrodt Inc.
Figure 9.1: R.C. Teitelbaum, S.L. Ruby, and T.J. Marks, *Journal of the American Chemical Society* 102, 1980, p. 33222.
Figures on page 221 and 232: From *The Lively Membranes* by Rutherford Robertson. Published by Cambridge University Press.
Figures 16.1 and 16.2: From *Carbon-Nitrogen-Sulfur* by V. Smil. Published by Plenum Press.
Figure on page 346: R.C. Teitelbaum, S.L. Ruby, and T.J. Marks, *J. Amer. Chem. Soc.* 102, 1980, p.3322.

Published by Prentice-Hall, Inc.
A Paramount Communications Company
Englewood Cliffs, New Jersey 07632

Printed in the United States of America

20 19 18 17 16 15 14 13 12 11

ISBN 0-205-11913-1

Prentice-Hall International (UK) Limited, *London*
Prentice-Hall of Australia Pty. Limited, *Sydney*
Prentice-Hall Canada Inc., *Toronto*
Prentice-Hall Hispanoamericana, S.A., *Mexico*
Prentice-Hall of India Private Limited, *New Delhi*
Prentice-Hall of Japan, Inc., *Tokyo*
Simon & Schuster Asia Pte. Ltd., *Singapore*
Editora Prentice-Hall do Brasil, Ltda., *Rio de Janeiro*

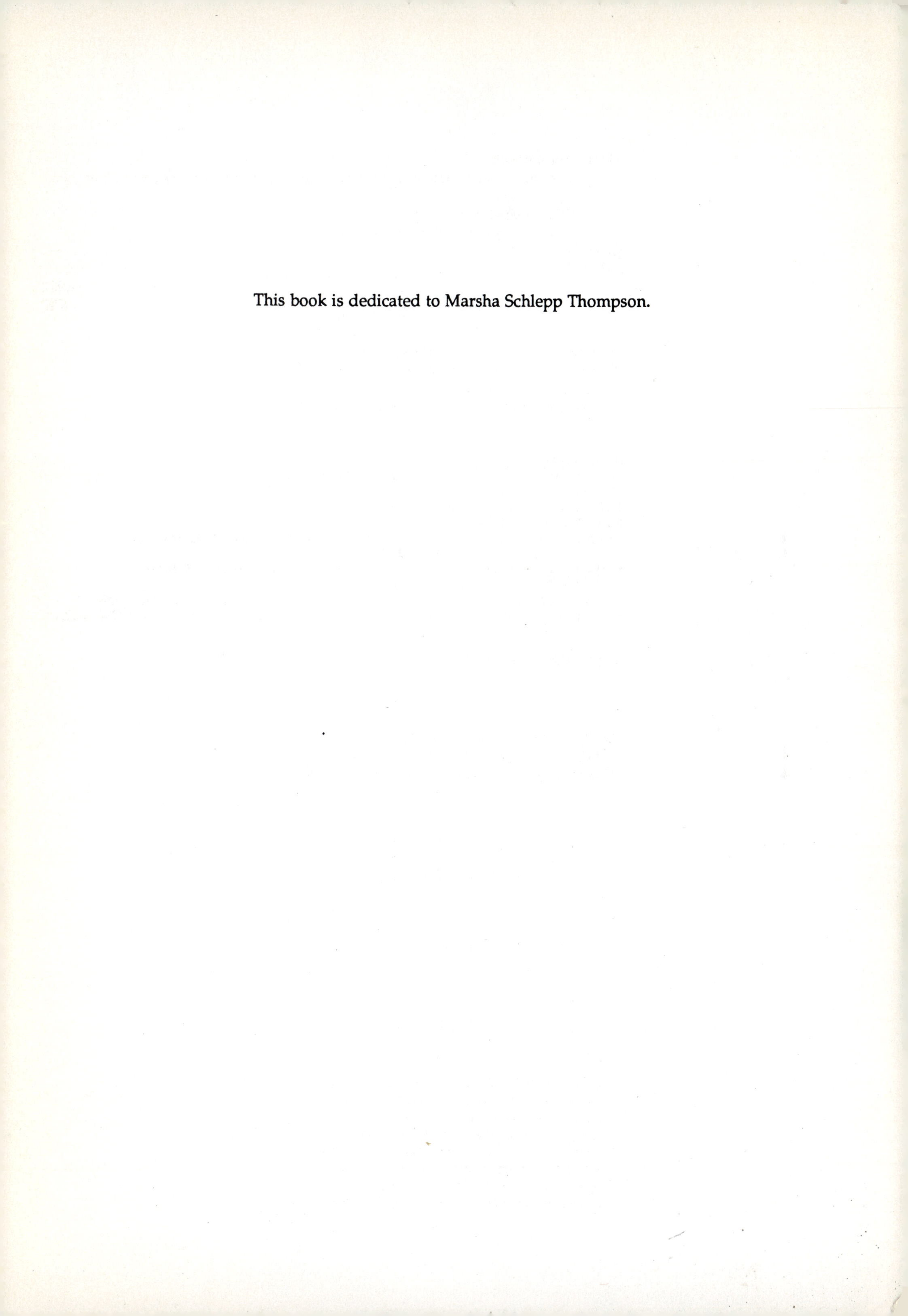

This book is dedicated to Marsha Schlepp Thompson.

Contents

Contents

Preface

Chemtrek is a laboratory text designed for use by students in a two-semester general chemistry laboratory program for science majors at the college or university level. The chapter topics have been selected to complement areas of chemistry normally presented in an introductory science major's lecture course. The material in the text has been used in the instructional laboratories of a major university over a period of ten years and has thus been thoroughly class tested and revised numerous times.

There are probably dozens of laboratory manuals in print that fit the above description — then why consider another one? The answer is that ***Chemtrek* presents an entirely new, innovative, small-scale, inexpensive, safe, efficient, and extremely effective way to get students involved in the excitement and promise of experimental chemistry.**

Chemtrek is the outcome of nearly two decades of research and development in the use of small-scale methods, equipment, and techniques in instructional contexts. The small-scale approach is based on the use of nontraditional, plastic equipment that was originally developed for use in the fields of clinical chemistry, microbiology, and recombinant DNA research. All of the equipment is inexpensive and readily available (worldwide) at chemical supply houses. *Chemtrek* is extremely easy to implement and requires minimal conversion costs. An instructor's technical manual is available and provides necessary information for ordering supplies, preparing solutions and other laboratory necessities, as well as for organizing the logistics for any number of students in the laboratory course.

The small-scale *Chemtrek* approach provides solutions to most of the multitude of difficult problems associated with teaching the undergraduate chemistry laboratory. *Chemtrek* is inexpensive to implement because only small quantities of chemicals are required. The small amounts and low cost for chemicals have many synergistic effects in other areas of instruction. Solution preparation time and costs are significantly reduced and inventories can be cut to a bare minimum. Faculty and students need not feel inhibited about exploring interesting additional experiments that always seem to surface naturally during laboratory investigations. "Small" means that you can even consider exotic or expensive materials that would be unthinkable in a traditional lab program.

Small-scale experimental methods have proven to be much safer for the student, the instructor, and the institution. The disposal of wastes is cut to an absolute minimum in the *Chemtrek* program. In fact, an attempt has been made to build in environmental safety and awareness in the design stage rather than to try to control things at the execution stage. It is perhaps crucial to point out that the experiments in *Chemtrek* are not just scaled-down or miniaturized versions of the mundane, traditional "cookbook" approach to experiments. By the way, I have strongly resisted attempts to label the *Chemtrek* approach as microscale or microchemistry. It is not. However, it certainly is on a smaller scale than classical macroscopic experimentation.

Long experience in developing instructional laboratories has made me painfully aware of the enormous problems inherent in the use of scientific instrumentation in an instructional context. Most modern instruments are expensive to buy, difficult to maintain, and are regarded by most learners as problematical "black boxes." My approach in *Chemtrek* is to ask students to build their own instruments wherever possible. This approach is based on the use of an inexpensive multimeter as the basic measuring device — and the only black box allowed! The meter can be connected to various, inexpensive sensors to produce simple but effective devices that can be student-built, that are user-friendly, and that reveal the relationship between form and function. Only in this type of context can students begin to gain insight into the hows and whys of instrumental design.

I wrote *Chemtrek* for students. Students actually work from, with, and through laboratory texts. Each chapter represents a particular area of chemistry and is divided in three parts: Introduction, Background Chemistry, and Laboratory Experiments. The Introduction situates the chemistry within a life or application context and tells students why they are being asked to do the experiments. The Background Chemistry section provides more specific details about various chemical aspects of the experiments, without answering the major questions in advance. Students are encouraged to read the Introduction and Background Chemistry before coming to lab. A Pre-Laboratory Quiz is provided so that students can test themselves on the reading material.

Different types of information have been integrated within the Laboratory Experiments sections. Students are encouraged to record real-time observations, answer questions, get involved in peer discussions, design experiments, and interact with their instructor. Every chapter contains qualitative and quantitative aspects of both theory and experiment, and each ends with an application, an unknown sample for identification and analysis, or a research project.

In teaching chemistry laboratory with *Chemtrek*, we have already seen some serendipitous advantages to small-scale methods. Small-scale focuses the students attention on having insights, on realizing concepts, on understanding intelligible processes, and on the value of constantly questioning textual and instructional authority! Small-scale provides a new way to teach creative problem solving, the processes of invention and discovery, analytical thinking, effective writing, and the elements of descriptive chemistry.

"Great problems are solved by being broken down into little problems. The strokes of genius are but the outcome of a continuous habit of inquiry that grasps clearly and distinctly all that is involved in the simple things that anyone can understand."

Bernard J.F. Lonergan, *Insight, A Study of Human Understanding*

Acknowledgments

It has taken a long time to write this book. It would have never happened at all if it were not for the encouragement of my undergraduate and graduate students at Colorado State University. We academics must always remember that the students *are* the university!

I would like to express my appreciation to the U.S. Department of Education Fund for the Improvement of Post-Secondary Education and the Carnegie Corporation for a Mina Shaughnessy Scholar Award that allowed me two summers of time to write and reflect. I have been enriched by the Woodrow Wilson and the Camille and Henry Dreyfus Foundations for giving me the opportunity to meet people with extraordinary energy, enthusiasm, and dedication to teaching. I would also like to thank the faculty and cadets of the Chemistry Department of the United States Air Force Academy for a most delightful and productive year as a visiting professor.

I would like to thank Randy Matsushima, John Schmidt, Bob Breuer, John Murray, Marilyn Bain Ackerman, K. Vasudevan, Louise Saddoris, Pete Markow, Steve Sadlowski, Alan Kopelove, Ed Brehm, Shawn Fujita, Sri Bringi, Carl Schanbacher, John Stringer, John Haase, Rene Elles, and Don Dick for their insights, suggestions, hard work, and faith.

I can never repay the debt of friendship to Rob Cohen, Bill Cook, Ted Kuwana, Thad Mauney, Ed Waterman, and Mike Braydich. These friends never failed to shine light during my own dark struggle with the flight from understanding.

Without doubt, no project like this could ever be successful without the total commitment from dedicated technical staff. Mrs. Jackie Resseguie was there at the beginning, providing continuous encouragement throughout the project, and solved all of the enormous logistical problems in the actual implementation of *Chemtrek*. Thanks also go to all her staff, particularly Lynne Judish, who helped enormously in the development work. I would also like to thank the staff at Technical Texts for their patience and help in creating the final product.

Finally, my heartfelt thanks go to my wife Marsha Schlepp Thompson, who produced this camera-ready text and chemical structures (on a MacPlus) from my handwritten draft and endless revisions. She coordinated the whole project with equanimity, good humor, grace, and love!

Introduction

Small is Beautiful:

A Discussion with the Student about the Advantages of Using Small-Scale Equipment and Techniques in the Study of Science

THE NEED FOR A NEW APPROACH IN THE LABORATORY

I recognize that most of you are thinking seriously about a career in science and that only a few of you wish to be chemists. However, I would like to try to explain why the chemistry laboratory is one place where you can actually do science and start becoming a scientist. The laboratory, unlike the lecture room, is one place where you can and should question and test all those accepted facts about how nature works. Doing science, rather than listening to someone else talk about it, always generates new perspectives and new ways of synthesizing knowledge. Professors and textbooks (including this one) should only serve as a guide. I have every confidence that you can make original observations, design new types of experiments, and seek out new knowledge — even though you may only be a freshman!

I am sure that you are asking the question, "Why should I have to take a *chemistry* laboratory, of all things?" One answer is that a knowledge of chemistry and of the experimental methods by which chemical information is obtained has proved to be useful in almost all areas of science and technology. Chemistry is the study of matter and its transformations. Chemists have invented and refined a unique and powerful ideography (pictorial way of representing matter), that can be used to rationalize and explain the results of over two centuries of experimentation with matter. Matter is described by three-dimensional structures consisting of atoms held together by a variety of electronic forces. These structural representations have proven to be crucial concepts in fields as diverse as designing new plastics and genetic engineering. Transformations of matter are pictured in the form of dynamic chemical reactions in which old bonds are broken and new arrangements produced. What is truly remarkable about this view on matter is that it was constructed out of the information from chemical experiments carried out on real gases, liquids, and solids — long before anyone had actually "seen" a molecule. What is perhaps even more astonishing is that the molecular images produced by the scanning tunneling microscope, a new microscope capable of resolving individual atomic dimensions, appear to confirm and support the chemical ideography. The congruence of scientific imagination and technological image clearly demonstrates that chemistry is an information science of great beauty, utility, and predictive power.

The development of the creative scientific approach to solving problems has come at a time when the scale of human interference in natural processes has reached dangerous levels. Life in the global village is becoming increasingly complex and subject to technological disintegration. We are faced simultaneously with devastating population increases, severe soil erosion, environmental degradation, and potentially disastrous global climate changes. Energy production from fossil fuels becomes increasingly costly, both in terms of the waste of nonrenewable resources and in the aquatic and atmospheric degradation caused by combustion products. The technological backlash is inevitable and will occur worldwide. The disasters at Minamata (Japan), Three Mile Island (United States), Bhopal (India), Chernobyl (Soviet Union), and Valdez in Prince William Sound (United States) attest to the destructive potential.

The headlines become more strident everyday: "Can dry armpits cause a world crisis?," "Is the 1988 drought a result of the greenhouse effect?," "There is a hole in the ozone layer again!," and "Exxon-Valdez — America's Chernobyl."

It is now more important than ever to remain skeptical about the claims made about new "breakthrough" technologies, especially those in the areas of energy production, biotechnology, and agriculture. The thoughtful application of human scientific wisdom and an ongoing commitment to conservation are the keys to solving these very complex problems.

SMALL — AND THE BEST OF ALL WORLDS!

The chemistry laboratory is a good place in which to begin to understand the scientific principles inherent in some of these complex issues. One of the strengths of the scientific approach to problem solving is the experimental testing of concepts. The concepts derive from human imagination and are found to be true or false by the test of their behavior. The constant tension and interplay between insight, imagination, thought, and experimentation constitute the foundation of all method, including science.

I am sure that most of you have not had the opportunity (or the inclination or the time, for that matter) to compare various texts for chemistry laboratory. I have, and I would like to explain how this book is different and why I believe that the difference is important. Earlier in this introduction, I pointed out that chemistry is the study of matter and its transformations. It would seem practical and sensible to study small amounts of matter wherever possible. After all, unless you want large amounts for some reason, 10 milligrams of matter is as representative of matter as, say, 100 grams. The reason is obvious. Molecules are so small — 10^{20} molecules (~ 10 mg) is as good as 10^{24} molecules (~ 100 g)! All the chemical investigations included in this text have been designed to require only small amounts of matter, usually in the tens of milligrams and less than one milliliter range. There are some tremendous advantages to working with small amounts of matter:

- Small-scale experiments are *much safer* for you.
- The use of small amounts of chemicals means that there is *less waste* to dispose of.
- The use of small amounts naturally leads to *conservation* of valuable resources.
- Small-scale systems tend to be *user friendly* and facilitate scientific comparisons, even in complex environments.
- Small-scale means *less expensive,* for you, the student, and for the institution.

Experimental chemistry carried out on a small scale requires apparatus and instruments that are different from the traditional, large-scale glass and metal ware that most of us usually associate with chemistry laboratories (beakers, test tubes, ring stands, Bunsen burners, etc.). Fortunately, there have recently been tremendous advances in the development of sophisticated small-scale equipment for use in the fields of genetic engineering, clinical chemistry, and recombinant DNA. Much of the nontraditional small-scale equipment is made out of plastic, and most of it was originally designed to be disposable. We are going to reuse it and recycle it wherever possible. The plastics are polystyrene (e.g., microreaction trays), polyethylene (e.g., pipets and microburets), and polypropylene (e.g., straws) — yes, the very kind you use to drink your sodas. All three plastics have chemical properties that can be exploited in the innovative design and use of tools for science. All have surfaces that are nonwetted by aqueous solutions. Nonwetted surfaces make the storage, transfer, and delivery of aqueous solutions much easier than, say, glass surfaces. The high surface tension of water means that in the presence of a plastic, a drop can be its own container and a reproducible volume increment at the same time. The ease of production of small drops from plastic tubes leads naturally to digital methods in volumetric work. Counting drops can circumvent the problems inherent in the analog methods of meniscus reading. The low softening and melting point of these materials, together with the ease with which they can be stuck, bent, cut, and otherwise mutilated naturally leads to innovative construction. Of course, in an experimental science such as chemistry, there is occasionally a need for glass or metal materials. When this need arises, we will not hesitate to use small quantities and small tools.

The modern scientific research laboratory also contains many types of sophisticated and expensive instrumentation — such as gas chromatographs, nuclear magnetic resonance spectrometers, pH meters, and electronic balances. These instruments are often interfaced with dedicated microcomputers and are therefore referred to as "smart" instruments. Unfortunately, these "smart" instruments are "black boxes" that may be operated by anyone who knows which button to push — hence the cliché, "smart instruments, dumb students." In this laboratory course you will have the opportunity to build many of the instruments needed for quantitative chemical studies. The instrument design has deliberately been kept as simple as possible to enable you to focus on the fundamental operational principles. When you build your own scientific instrument,

- *You* know how it works.
- *You* can fix it when it breaks.
- *You* can play with it whenever you wish to.
- *You* begin to understand the connection between form and function.
- *You* begin to understand the real nature of experimental error and approximation.
- *You* can usually take on the more sophisticated versions with confidence.

HOW TO USE THIS BOOK

Each chapter of Chemtrek contains three major divisions: Introduction, Background Chemistry, and Laboratory Experiments. The Introduction gives you an overview of a particular area of chemistry and tells you why you are being asked to carry out the laboratory experiments. The Background Chemistry contains more information about the chemical principles that are important in the laboratory experiments. You should read the Introduction and Background Chemistry sections before you go to the laboratory. Find out for yourself whether you have grasped the main ideas by taking the Pre-Laboratory Quiz. Go back to the Introduction and Background Chemistry, and I am sure that you will be able to find the answers fairly easily.

Each section of the Laboratory Experiments will usually include questions and requests to do calculations which will be in bulleted (•) format. You should respond to these items in your laboratory report. Your instructor will give you information as to how you will be expected to write your laboratory reports. One of the most important skills a scientist must acquire is the ability to write as thoughts occur and as experiments evolve. Learning to write as you work in the laboratory is not easy. However, you will find that the more you write, the better you will become. If you keep an ongoing scientific diary, you will be amazed at how your communication skills will develop and mature. The actual format of your laboratory report that is required by your instructor will be given to you during the first laboratory session.

THE RELATIONSHIP BETWEEN LABORATORY AND LECTURE

It is virtually impossible for the laboratory course to be completely coordinated with the lecture course — they represent two entirely different methods of learning. However, your instructor has planned the sequence of laboratory work very carefully so that you will not be confronted with major new concepts until you have encountered them in lecture. Occasionally, you may find that you have not had much time to assimilate lecture concepts before you start using them in laboratory. Don't panic. Read the Introduction and Background Chemistry and don't hesitate to ask your instructors (in both laboratory and lecture) for help. This is an educational institution, and we are here to help you!

LABORATORY SAFETY

Experimentation in a chemistry laboratory always has an element of danger and risk associated with it. This is particularly true if the surroundings, tools, and techniques are new and unfamiliar. In the process of writing this text, I have performed all the experimental procedures dozens of times (some of them hundred of times). I did this in order to critically assess the risk factors inherent in each step. It is obviously impossible, in any situation, to be sure that there is zero risk. However, the risk can be kept to an absolute minimum by using a small-scale approach (as in this text) and by adhering to the following precepts:

- Recognize in advance, before you come to laboratory, which operations are likely to involve more risk than usual. Your instructor will give you information about these operations at least one week ahead of the laboratory.
- Try to be safety conscious when you work in the laboratory. Be aware not only of your personal safety, but also of your peer group safety.
- Recognize that even a safe task could possibly be dangerous if not carried out according to instructions.
- Be aware of the general safety features appropriate to the room, laboratory, and building that you are working in. Again, your instructor will familiarize you with the location of fire extinguishers, safety showers, and the steps to take if the fire alarm goes off.
- Wear your safety goggles.

- Only work in the laboratory when an instructor is present. Experience is invaluable in unusual or unforeseen situations.
- Report any accident, *however slight,* to your instructor. Don't wait; report it immediately.
- Don't eat or smoke in the chemistry laboratory.

I have found from experience that the majority of accidents in undergraduate laboratories are caused by

- The insertion of glass tubing through cork or rubber stoppers
- Cuts caused by broken or chipped glassware
- Burns from touching hot metal or other heated objects

The use of small-scale plastic equipment has been found to eliminate many of these problems. However, it is important to recognize that numerous chemicals and solutions that you will be using are corrosive and can cause burns. Your instructor will provide you with the appropriate cautions and inform you of the steps to take if you do spill chemicals. Most of the chemicals and materials needed for the laboratory experiments will be placed in a convenient central location in the room. Throughout the book this location is called *Reagent Central.*

Chapter 1

The System

"It is a great mistake to think that scientific truths differ essentially from those of every day. They differ only in their greater extension and precision.... The scientist multiplies man's contacts with nature but it is impossible for him to modify in any way the essential character of these contacts. He sees how certain phenomena are produced, though they escape us, but he is inhibited as much as we are from enquiring into why they occur."

Anatole France

"Often that which has come latest seems to have accomplished
the whole matter."

Livy, *History of Rome*

"The System"
will be placed within your control at the start of the laboratory.
Please read this brief introduction
BEFORE
you begin to exercise your authority.
Unlike Dr. Science,
I have a Ph.D.,
and since I am the Original Controller,
I wish to challenge you by setting a few goals
and a few rules. Of course, if you don't wish to work towards goals
and if you wish to break rules,
then you may do so. However, since I am
the Originial Controller,
I simply request that you tell me in writing
what motivates you to be so outrageous
and antisocial.

THE GOALS

1. To find out what "The System" will do.

2. To interpret what "The System" does.

3. To tell the Original Controller how you found out.

THE RULES

1. You may do anything you wish to "The System," but remember that whatever you do will change it and you.

2. You cannot remove the cap until the Original Controller is convinced that you know what you are doing.

➥ One further request. Since this institution purports to be one of Higher Education, the Original Controller begs you to keep a real-time record of your day with "The System." After all, this is the beginning of your life as a true thinker and natural scientist.

Are you sitting comfortably? Then we shall begin.

★ "The System" is in front of you. Don't touch — yet. Why don't you record its birthdate (and time). (Then at least you will know when it gets old.)

★ Don't worry, the Original Controller has an English (some call it weird) sense of humor, as well as a Ph.D.!

★ Don't touch — yet.Why not try to describe "The System" as accurately as you can — for example,

What is it standing on?
Are the lights switched on?
What is the stuff you can see through?
How do you know that? What visual cues
tell you that something is not solid?
What shape is it?
What are its dimensions?
Can you find that out without touching it?

★ Make a simple drawing — sometimes pictures are worth more than a million words (Confucius, not Controller).

★ Are you writing all this down? Or are you bored?

★ Now write down the goals in the box below:

1.
2.
3.

★ You have the goals and a lot of facts. Try to interpret some of these observations with the goals in mind — e.g., what's happening to all that energy pouring into "The System" in the form of light? What's the ratio of space to liquid?

- ☆ Don't you find it tough to really see "The System" well when it's on a black background? Before you move it, think about that . If it were milk, it would be easy to see, right?

- ☆ Why not move it to a white background? *Hold it!* How are you going to do that? With your fingers — then which part to touch? What temperature are fingers? For that matter, what temperature is the system? What makes you think that it's at room temperature? I would like to know! Note how you do it.

- ☆ If it's on a white background now, you should make a note in your book that it now has about 50 times more light going into (through) it.

- ☆ Anything peculiar happen when you moved it?

- ☆ Are you still sitting in the same place? Perhaps if you moved and looked, it might change your point of view! Places where two things meet — interfaces — are often revealing.

- ☆ OK, let's get radical — better note the time. If "The System" self-destructs in a blue flash, at least you will know when it died. Pick it up and give it a small motion — *small*! What happened? The reason I am suggesting small is that small perturbations can be increased. Large perturbations could finish it.

- ☆ Write down how you picked it up.

- ☆ Shake "The System" gently. Watch (you have a hand lens). Now describe exactly what you saw and exactly how you shook it. Your shake is unique. This is important because nobody shakes like you do. Did the stuff at the bottom touch the cap?

- ☆ Try that again, this time a *little* harder. How much kinetic energy do you think you gave it? No, seriously — how does one measure human shakes?

- ☆ Did you create any bubbles? Did you watch them? Perhaps bubble power is the secret.

- ☆ There must be an explanation for that change. For example, here are three that seem reasonable:

  1. There is something on the cap that catalyzes the change.

  2. Kinetic energy from the motion makes the change happen.

  3. The change won't happen in the dark.

- ☆ What's your interpretation? Then you must design simple experiments to prove yours and disprove mine (stated below). Remember the K.I.S.S. principle!

[I think that there are really 2 liquids!]

★ Back to bubbles for a minute. What is a bubble in "The System?" Draw one. What happens when one bursts?

★ Perhaps you need to design a standard shake so that you can reproduce it. The Original Controller's standard shake involves holding the cap firmly between the thumb and the second finger with the index finger on top to steady it. (I am right handed.) Hold the left hand out at the side of the system and at the height of the cap, palm down. Lift "The System" up about 1 cm. Now sharply shake in a *vertical* direction so that the bottom of "The System" reaches the height of the left hand and then descends back down to about 1 cm above the table. Then set it down.

★ Try it again. Describe the dynamics, and don't forget to keep a record of "System" shakes.

★ What are those strange *structures* inside "The System?" Are they reproducible? Describe, or better still, draw what happens spatially — i.e., the bottom half versus the top half.

★ Perhaps it's time to get real scientific! Since radical things happen when you shake "The System", then a graph of number of shakes versus time might be most enlightening. Time for what? Time for the bottom part to clear. You will need a clock or watch with a second hand to do this. Try up to about 6 standard shakes all at once. At the same time, try to assess the intensity of color in "The System."

★ On the lined paper in your book, make a rough graph of the number of standard shakes versus the time to change back. On the same graph plot number of standard shakes versus intensity of color (approximately).

★ How old is "The System" now?

★ Do you like music? (The Original Controller (O.C.) is a Grateful Dead freak — you know, the only good freak is a dead freak!) Try this one. Tap the system several times with a pen or pencil and listen to the sound. Now shake the system and *immediately* tap again. Notice any difference? It's almost an octave! By the way, this is one way to beat "The System."

★ How about a total review of the facts? Why not do what chemists do? Choose some simple symbols for things — e.g., C could mean colorless, → could mean change, G could be gas. Now organize the symbols into a scheme of things that matches what the system does. What you are doing now is trying to create a *model*.

★ Does your scheme fit *all* the facts (observations)?

★ Do you think "The System" involves physical change or chemical change, or both? What observations support your opinion? Let's see if you can make a prediction (in writing). Describe what you think will actually happen if "The System" is cooled or heated. Think this through with your scheme in mind.

☆ Why not see if you are right? Cool the system first. This is less dangerous than heating it. Make an ice bath (about 50% ice and 50% water) in a styrofoam coffee cup. If you cool things remember that the process takes time. Why? What experiment(s) are you going to do to test your prediction? OK, do it (them).

☆ Don't forget — now it's old *and* cold.

☆ Were you right?

☆ Here's a non sequitur. Do you think that light has had any effect on "The System?" Make a brief list of what light can do to things.

☆ If you ask pleasantly, the Original Controller's assistant will show you another "System" untouched by human mind or hand. Make a comparison of the two "Systems" — yours and the untouched one. Remember this significant fact — the untouched "System" has been standing in the room light since the start of the laboratory.

☆ Now how about some risky business? What will happen if "The System" is heated? Bang! *Before* you send "System" shrapnel whistling through your fellow scientist (O.C. is behind the wall), explain in your notes why heating things like "The System" will sometimes produce explosions.

☆☆☆ Perhaps it is time to gain access to "The System" — i.e., remove the cap. ***Hold it!***! Before you undo anything, think about what could happen. How you take the cap off may tell you a lot about "The System." Let's look at some possibilities. Well,

1. The cap won't come off.
2. There is a potential vacuum in "The System" and ordinary air will rush in.
3. "The System" contains a gas (or gases) that is/are denser than air and nothing will happen. Or it might contain a gas (or gases) less dense than air and the gas (or gases) will fly away never to be smelled again.
4. "The System" might be toxic.

☆ How about undoing the cap carefully and slowly so that you can watch what happens in that ring of liquid that's always in the area of the cap thread? Listen, smell and watch for changes — but don't snort it! Use a microtowel to wipe the cap and outside top of "The System" dry.

CAUTION: If you get system stuff on your fingers, wash your hands in cold water.

★ Now what? Well, you now have access to the inside of "The System," which will be referred to as *a* "system", since it actually is only one of any number of "systems" in the universe. The question is, Do you have to repeat everything you already did, or can you make some assumptions about the relationship between "The System" and "a system"? Try. What are your assumptions? (Hint — "The System" became "a system" when the cap was removed.)

★ Do you think that the space above the liquid in the original vial was air?

★ Now you can clone "The System" and make little "systems." "Small is beautiful" is a motto to be revered. Perhaps it is worth sacrificing a little one in the cause of science. Use a plastic pipet to transfer a small "system" to a small vial. Make sure there is no cap on the small vial. Light a match and hold it in one hand while you grasp the vial containing the small "system" with tweezers in your other hand . Now bring the match under the vial and heat it for five seconds.

CAUTION: Do not boil the liquid or it will shoot out of the vial. Do not point the vial at yourself or your neighbor.

★ I know the vial gets black. Explain why it gets black. Wipe the black away. Now what happened? Will the little "system" still change?

★ OK. Let's take a break. During the break, please introduce yourself to your neighbors so that you can communicate on a first-name basis. Compare your scheme with theirs. You can incorporate any features of their scheme(s) into yours if you like. However you need to make sure that the original vials ("The System") of your neighbors are very like yours, and you must get their permission in writing. An acknowledgment in your notes might also be courteous. They must do the same for you. Anyway, it's always nice to meet such good-looking people in the same course!

★ Let's get back to what happened to the little "system." Make a brief list of what happens when things are heated. What is heat anyway? Add an extra symbol and step to your scheme on the basis of what happened with heat.

★ Slight problem, however. What color was the little "system" when you put a match to it? See what I mean?

★ Back to the original vial — "The System". Put the cap on and stand it in some warm water in a styrofoam cup for a few minutes. Now what color is it? Give it one standard shake and time how long it takes to go back to its original color. Compare this time with the previous one carried out at room temperature.

★ Now you can do an experiment that has some very interesting practical applications. Remove the cap and dry the glass and cap with a microtowel. Use a plastic pipet to suck up some fresh cold water (you will find it in a beaker at Reagent Central). Add the water to "The System" liquid *without* making bubbles. Stir gently. Record what happens. Why not try that again? Is the effect reproducible? Now, why does "The System" do that?

☆ You have added enough water to roughly double the original volume. Warm it with your hands to make sure it's back to about room temperature. Now put the cap on and give it one standard shake. Time how long it takes to go back to its original color. Interpret your result. Does your explanation fit into the scheme you proposed earlier?

☆ Happiness. We are going to divide the class up into four discussion groups. You get to meet more good looking scientists! Once you are in your group, introduce yourselves and discuss "The System." Make a *collective* decision on the following two issues:

1. What general scheme of symbols best describes what "The System" will do?
2. How many components are there inside "The System?"

☆ Write the group decisions in your notes. Elect a spokesperson.

☆ Your instructor, the assistant controller, will now organize a minimeeting in order to astound you with demonstrations of such bravery and skill that you will be truly thankful to be in this chemistry course. Your spokesperson will be asked to render public the collective wisdom of your group.

* * *Meeting* * *

☆ You are now promoted to scientist first-class. At Reagent Central there are, in beautifully labelled containers, all of the components of

"The System"

☆ The goal is for you to decide within your collective group and by experiments

which combination of components is responsible for the phenomenon that you have observed while interacting with "The System."

☆ You
are
confused.

OK, let me give an example. What combination (apart from the Grateful Dead and good food) makes the blues go away?

Conclusion

"The System" will be revealed to you in all its sophisticated detail at the end of the laboratory. Any questions you have will be answered fully to the best of our ability. We will also give you a glimpse of the incredible diversity of practical applications of "The System." Components can be found in Japanese food stores, in sewage treatment plants, in large-animal veterinary clinics, in recombinant DNA laboratories, in cancer treatment programs, and in erasable laser holograms. We do have one request of you before you leave the laboratory. This is a large chemistry course, and I am sure that you feel, as we do, that to reveal any information of what transpires in this laboratory to other students would truly deprive them of a scientific experience almost equal to a personal appearance on NOVA. The Original Controller begs you to sign the affidavit on the following page. ⇨

I, (print name)____________________ Student No.__________

Major____________________ Star Sign__________

Musical preference____________________

Will not reveal anything (verbally or in writing) of what transpired in the laboratory concerning "The System" to any other student, Controller, or friend until after the laboratory has been completed by all students in the course.

Signature____________________ Date__________

Please tear this out and give it to your instructor. It will be kept on record for at least 17 years.

Chapter 2

Spectroscopy:

The Interaction of Light and Matter

"We all *know* what light is; but it is not easy to *tell* what it is."

Samuel Johnson

Introduction

In certain parts of South America, there exists a most beautiful and remarkable species called the railroad worm (*Phrixothrix*). Yellowish-green luminous spots are arranged in eleven pairs on the worm's abdominal segments, and on its head there are two additional red luminous spots. When the worm is disturbed, all the spots give off light, and as it undulates through the night, it looks like a miniature railroad train. Of course, this biological midnight express emits an amazingly small amount of energy in the form of visible radiation (light) compared to the many other natural and man-made sources. Our world is literally inundated with all kinds of radiation, much of it coming from our own sun. Daylight that comes from the sun is a very small part of the spectrum of electromagnetic radiation. From cosmic rays to radiowaves, the total electromagnetic spectrum includes all the known types of radiation (see Table 2.1).

Radiation may be described in one of two ways: either as a stream of energy pulses (photons) or as energy waves sent out from a source (worms, suns, etc.) at the speed of light. Scientists dodge the bullet by using whichever interpretation works best to explain an experiment involving radiation. The photon and wave theories are linked by Planck's law:

$$E = h\nu$$

where E is the photon energy in joules (J), ν is the frequency of the radiation (Hz or s^{-1}) and h is Planck's constant (6.63×10^{-34} J s). Wavelength and frequency are related by

$$c = \lambda\nu$$

where c is the speed of light (3×10^{8} m s^{-1}), λ is the wavelength of the radiation (often reported in nm), and ν the frequency. It is useful to note that energy and frequency are directly proportional to each other, whereas energy is inversely proportional to wavelength.

Sources of radiation are extraordinarily diverse, ranging from complex nulear fusion reactions in supernovae to burning zirconium foil in photographic flash cubes. In spite of this diversity, the amount and type of emitted radiation are intimately related to the chemistry of the processes that produce the radiation. It is this link that forms the basis for the science of spectroscopy. *Spectroscopy* is the study of the interaction of electromagnetic radiation with matter. When matter is energized (excited) by the application of thermal, electrical, nuclear, or radiant energy, electromagnetic radiation is often emitted as the matter relaxes back to its original (ground) state. The spectrum of radiation emitted by a substance that has absorbed energy is called *an emission spectrum*, and the science is appropriately called *emission spectroscopy*.

Another approach often used to study the interaction of electromagnetic radiation with matter is one whereby a continuous range of radiation (e.g., white light) is allowed to fall on a substance; then the frequencies absorbed by the substance are examined. The resulting spectrum from the substance contains the original range of radiation with dark spaces that correspond to the missing, or absorbed, frequencies. This type of spectrum is called an *absorption spectrum*. In spectroscopy the emitted or absorbed radiation is usually analyzed, i.e., separated into the various frequency components, and the intensity is measured by means of an instrument called a *spectrometer*.

The resultant spectrum is mainly a graph of intensity of emitted or absorbed radiation versus wavelength or frequency. There are in general three types of spectra: continuous, line, and band. The sun and heated solids produce *continuous spectra* in which the emitted radiation contains *all* frequencies within a region of the electromagnetic spectrum. A rainbow and light from a light bulb are examples of continuous spectra. *Line spectra* (illustrated in Figure 2.1) are produced by excited *atoms* in the gas phase and contain only certain frequencies, all other frequencies being absent. Each chemical element of the periodic chart has a unique and, therefore, characteristic line

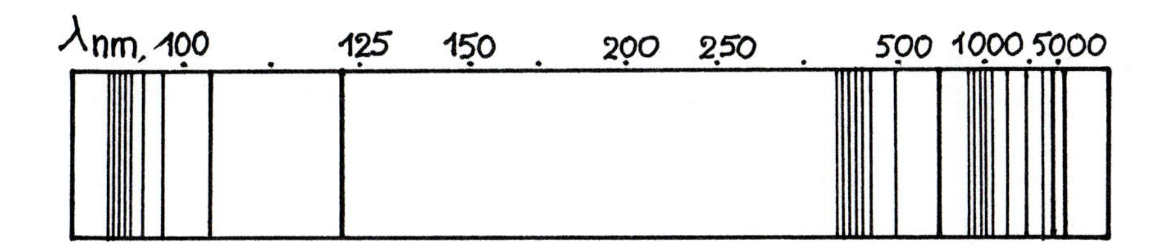

Figure 2.1 Line Spectrum of Hydrogen

spectrum. *Band spectra* are produced by excited *molecules* emitting radiation in groups of closely spaced lines that merge to form bands.

These categories of emission and absorption spectra contain tremendous amounts of useful information about the structure and composition of matter. Spectroscopy is a powerful and sensitive form of chemical analysis, as well as a method of probing electronic and nuclear structure and chemical bonding. The key to interpreting this spectral information is the knowledge that certain atomic and molecular processes involve only specific energy ranges. Table 2.1 shows the regions of the electromagnetic spectrum and the associated energy transitions that occur in atomic and molecular processes.

A simple example of the energy changes involved in particular transitions and the resulting spectrum is the hydrogen emission spectrum. This spectrum is especially interesting for historical, theoretical, and practical reasons. Over a period of 40 years, from 1885 to 1925, all of the lines in the emission spectrum in the ultraviolet, visible, and infrared regions were found experimentally and were identified with various electron transitions. Spectroscopic experiments like the above gave the major experimental evidence for the Bohr theory of the atom and eventually for the modern quantum theory.

Much of the scientific knowledge of the structure of the universe, from stars to atoms, is derived from interpretations of the interaction of radiation with matter. One example of the power of these techniques is the determination of the composition, the velocities, and the evolutionary dynamics of stars. The source of the incredible amount of energy produced by the sun is nuclear fusion reactions going on within the hot interior (temperature 40×10^6 K). Two fusion cycles, the carbon cycle and the proton cycle, convert hydrogen nuclei into helium nuclei via heavier nuclei, such as carbon 12 and nitrogen 14. The enormous radiation of energy from the hot core seethes outwards by convection. This radiation consists of the entire electromagnetic spectrum as a continuous spectrum. Towards the surface of the sun (the photosphere), the temperatures are much lower, and the cooler atoms of different elements all absorb at their characteristic frequencies. The radiation that shoots into space towards the earth is a continuous emission spectrum with about 22,000 dark absorption lines present in it (Fraunhofer lines), of which about 70% have been identified. These absorption lines — i.e., missing frequencies — prove that more than 60 terrestrial elements are certainly present in the sun.

Stellar spectroscopy via satellites has shown unequivocally that all the known elements of the periodic chart are made in complex sequences of fusion reactions in stars. The absorption lines also have different intensities at different temperatures, and the spectra are therefore excellent indicators of the temperature of a given stellar atmosphere. Absorption lines of stellar spectra can also be used to measure star

Type of Radiation (i.e. Spectral Region)	Energy Range (E, joule)	Frequency Range (ν, Hz)	Wavelength Range (λ)	Energy Transition
γ – ray	4.0×10^{-14}	6.0×10^{19}	$< 5 \times 10^{-3}$ nm	Nuclear
X – ray	$4.0 \times 10^{-14} - 2.0 \times 10^{-17}$	$6.0 \times 10^{19} - 3.0 \times 10^{16}$	$5 \times 10^{-3} - 10$ nm	Inner-shell electrons
Vacuum UV	$2.0 \times 10^{-17} - 1.1 \times 10^{-18}$	$3.0 \times 10^{16} - 1.7 \times 10^{15}$	10 – 180 nm	Middle-shell electrons
Near UV	$1.1 \times 10^{-18} - 5.7 \times 10^{-19}$	$1.7 \times 10^{15} - 8.6 \times 10^{14}$	180 – 350 nm	Valence electrons
Visible	$5.7 \times 10^{-19} - 2.6 \times 10^{-19}$	$8.6 \times 10^{14} - 3.9 \times 10^{14}$	350 – 770 nm	Valence electrons
Infrared	$2.6 \times 10^{-19} - 4.0 \times 10^{-21}$	$3.9 \times 10^{14} - 6.0 \times 10^{12}$	770 nm – 50 μm	Molecular vibrations
Far Infrared	$4.0 \times 10^{-21} - 2.0 \times 10^{-22}$	$6.0 \times 10^{12} - 3.0 \times 10^{11}$	50 – 1000 μm	Molecular rotations
Microwave	$2.0 \times 10^{-22} - 6.6 \times 10^{-25}$	$3.0 \times 10^{11} - 1.0 \times 10^{9}$	0.1 – 30 cm	Molecular rotations
Radiowave	$< 6.6 \times 10^{-25}$	$< 1.0 \times 10^{9}$	> 30 cm	Nuclear and electron spin

Table 2.1 Spectral Regions in the Electromagnetic Spectrum

velocities (relative to the earth) by measuring the Doppler effect. When a source sends out light of a specific frequency, the frequency remains the same only if the distance between the source and receiver stays the same. If the source is moving towards the earth, the light will appear to be of a higher frequency — i.e., bluer — than the light of a similar source in the laboratory. Conversely, the light of a receding source appears to be of lower frequency and, thus, redder (see Figure 2.2). The Doppler effect is easily measured by comparing the stellar spectrum with the spectrum of matter made luminous in the laboratory.

The science of spectroscopy can take us into the subatomic world or out into the farthest distance of space. The development of lasers is now causing a revolution in many areas of spectroscopy. These coherent, almost monochromatic, sources can be used to study extraordinarily fast chemical processes, as well as such previously inaccessible processes as flames and combustion. Recent research has shown that *all* living objects emit low-intensity radiation that may be detected and analyzed with modern photoelectronic techniques.

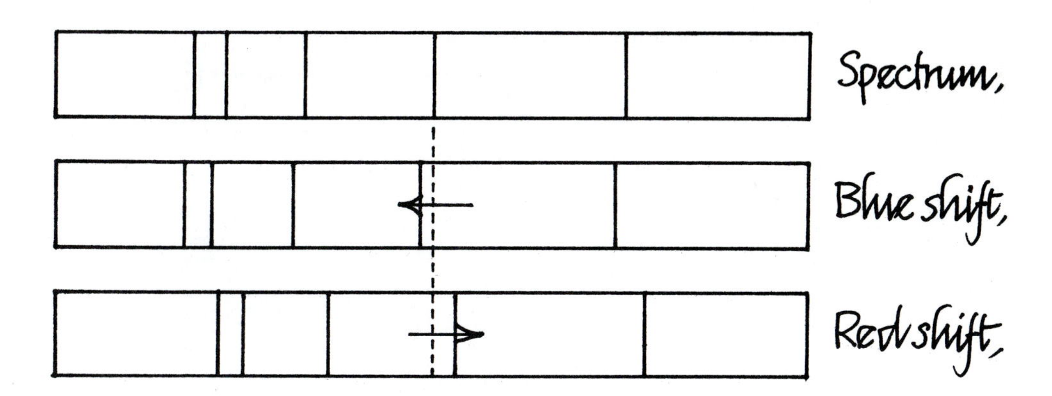

Figure 2.2 Spectral Shifts Due to the Doppler Effect

Background Chemistry

"The most directly compelling evidence for the quantization of energy comes from the observation of the frequencies of light absorbed and emitted by atoms and molecules."
P.W. Atkins

Atoms, ions, and molecules contain electrons that occupy discrete energy levels. The actual energy of each state (level) is dependent upon several factors: the nuclear charge, the distance of the electron from the nucleus, and the number of electrons between the nucleus and the electron in question. The *transition* of an electron from one level to another must be accompanied by the emission or absorption of a discrete amount of energy. The magnitude of this energy depends on the energy of each of the levels between which the transition occurs.

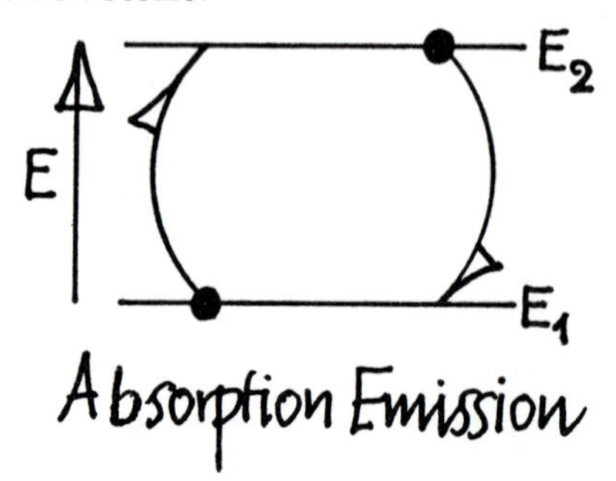

Figure 2.3 Energy Level Transitions

In the instance illustrated above in Figure 2.3, the energy involved in the electron transition (ΔE) will equal the energy difference between E_2 and E_1:

$$\Delta E = E_2 - E_1$$

For *energy emission* to occur, electrons must first be given energy from some external process — e.g., an electrical discharge, a combustion reaction, or a heated wire. They are then said to move from the *ground state* to an *excited state*. The excited electrons then "relax" back to their original levels, and energy is emitted. The number and type of these transitions depend on the particular structure of the energy levels in a given chemical species and on various quantum selection rules. These properties are unique to each individual species and give rise to an emission of energies that characterizes that species. If the value of ΔE lies within the visible region of the total electromagnetic spectrum, then the frequency corresponds to visible light, and the emission can be seen by the eye. From a practical point of view, since each electron can undergo many transitions and since many species have many electrons, emission spectra usually consist of a very large number of discrete frequencies.

The wave theory of radiation is particularly useful in providing models for interpreting the behavior of light emitted from atoms. Radiation is a form of energy consisting of oscillating electric and magnetic fields that move the direction of propagation at the speed of light. The wave motion, which is illustrated in Figure 2.4, is described in terms of some fundamental properties such as amplitude A, wavelength λ, and frequency ν.

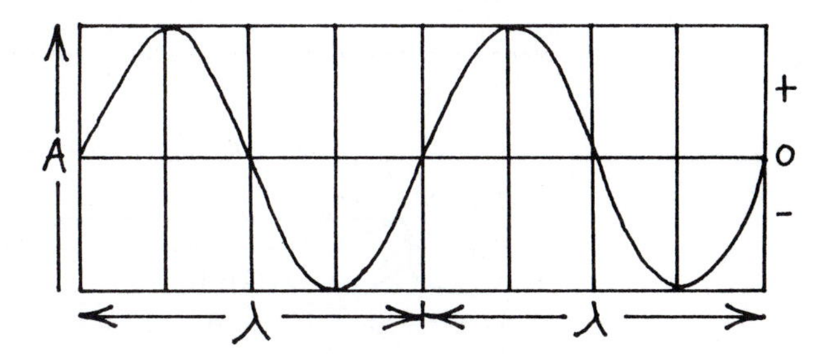

Figure 2.4 A Wave

The frequency is defined as the number of waves (of wavelength λ) passing a point per second. The relationship that links λ and ν is

$$c = \lambda\nu$$

The human eye is sensitive only to a tiny band of radiation of the total electromagnetic spectrum. This band, called *white light,* ranges from about 400 nm to about 800 nm in wavelength and is made up of the colors of the rainbow. Table 2.2 shows the wavelength ranges within the spectrum of white light:

Color	Wavelength (nm)
Violet	400 – 430
Blue	430 – 490
Green	490 – 570
Yellow	570 – 590
Orange	590 – 640
Red	640 – 750

Table 2.2 Wavelength Ranges of Colors in White Light

The energy of the various colors of light can be easily calculated using Planck's law,

$$E = h\nu$$

where E is the energy (in joules), h is Planck's constant (6.63×10^{-34} J s), and ν is the frequency of the radiation.

The separation of light into its spectral components can be done by *refraction* or *diffraction*. In this series of experiments, the separation of light into its component colors is accomplished by diffraction in a device called a spectroscope. A *spectroscope* is simply a box, with a slit at one end (to let in light) and a light-separating device at the other end. The separating device you will be using is called a *transmission diffraction grating,* and it consists of a sheet of transparent plastic that has thousands of tiny grooves ruled on it. The way in which the grating works to separate light into colors is by *wave interference.* Imagine that light of one color (monochromatic) is shone on the grating in Figure 2.5.

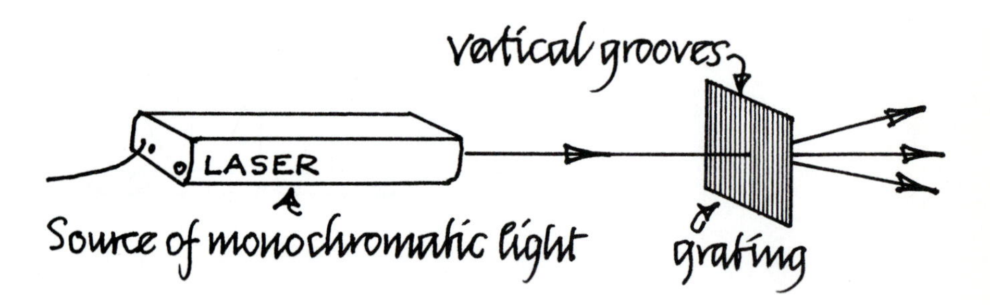

Figure 2.5 Diffraction of Laser Light

The grating device in Figure 2.6 illustrates wave interference. Viewed from above, the light waves can be seen as a series of straight lines representing the crests of the waves and the grooves look like tiny openings in the plastic. As these waves hit the grating, ripples of light come from each hole.

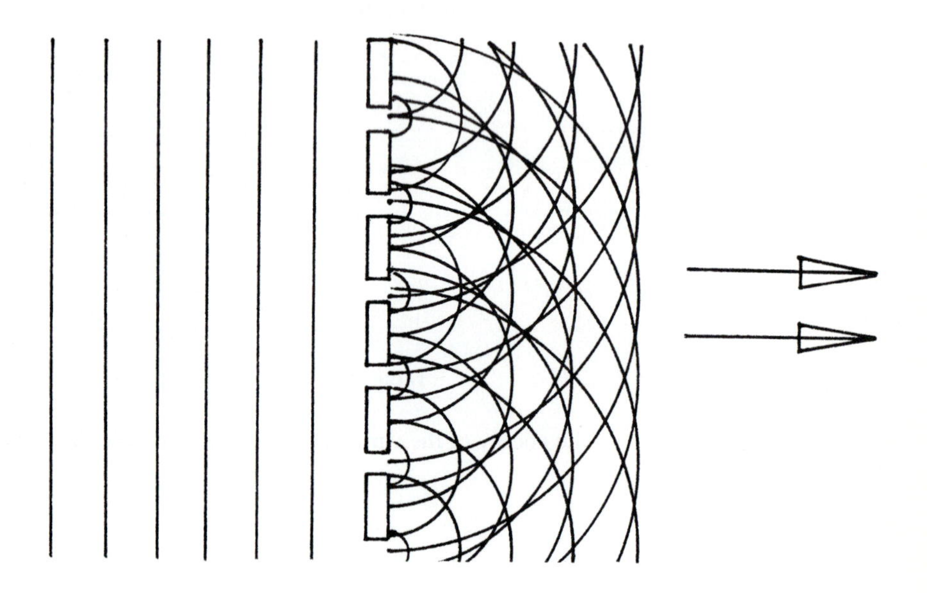

Figure 2.6 Light Falling on the Grooves of a Grating

The ripples move outward and interfere with each other. Where the crests are in the same place (crest lines cross), the waves will actually *add* to give a higher wave and a bright spot of light, which is called *constructive interference. Destructive interference* occurs when crests and valleys meet; the waves will almost cancel each other, resulting in darkness. Figure 2.7 illustrates these differences.

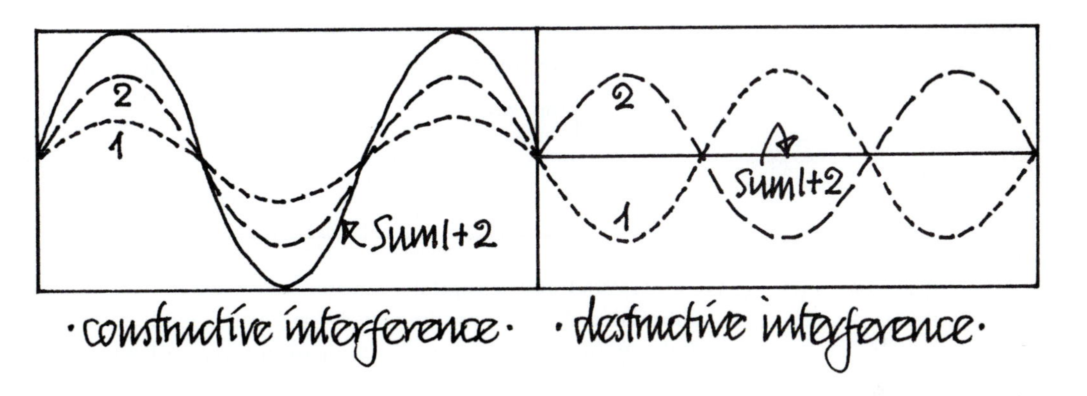

Figure 2.7 Constructive and Destructive Interference

The overall cumulative effect is that the light is bent at an angle to the incident beam. This phenomenon is called *diffraction* and the angle is called the *angle of diffraction* θ. A quantitative relationship can be derived from the geometry as shown in Figure 2.8.

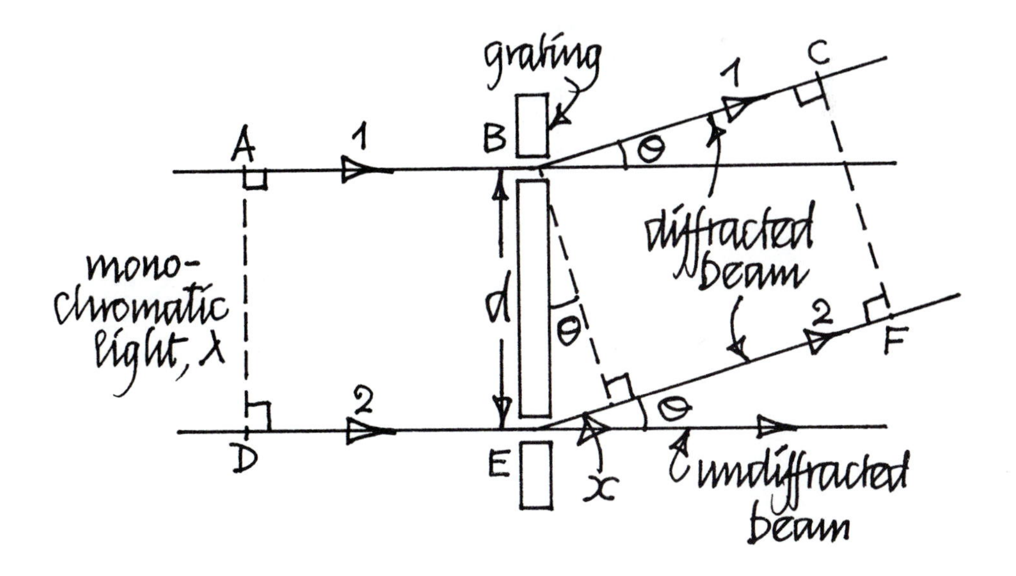

Figure 2.8 Geometry of Diffraction

where **d** is the distance between two adjacent grooves, θ is the angle of diffraction and **1** and **2** are two rays of monochromatic light of wavelength λ. Now both rays, **1** and **2**, are travelling at the same speed (the speed of light), and after hitting the grating, both are bent at an angle θ. Ray **2**, however, has to travel a little bit farther (distance **x**) than ray **1** in the same amount of time. This path length difference causes a delay in ray **2**, and interference can occur. If the path difference is an integral number of wavelengths, then constructive interference will occur. From simple geometry,

For constructive interference, $x = n\lambda$ (where n = 1, 2, 3...)

and, since $\frac{x}{d} = \text{sine } \theta$,

then $x = d \text{ sine } \theta$,

thus $n\lambda = d \text{ sine } \theta$

For a given number of grooves per cm and when n = 1 (said to be first order), the angle of diffraction depends on the wavelength of the light. White light, or any other kind of light composed of a combination of colors, will be separated into the individual component colors.

Three common terrestrial sources of visible radiation are the electrical discharges in gases, thermal energy from combustion, and heated metals. A convenient source in the laboratory is the *electrical discharge tube* — i.e., a glass tube that contains metal electrodes at each end and is filled with a gas, such as hydrogen, helium, or mercury, at a low pressure. A high voltage is placed across the electrodes, and when the current is switched on, a stream of fast-moving electrons shoots through the gas from the cathode to the anode. Energy is transferred from the electrons to the gas atoms, and the electrons in the gas atoms are excited to higher energy levels. The return of excited electrons to the ground state results in the emission of light, which may be analyzed with a spectroscope. A simple discharge tube is pictured in Figure 2.9.

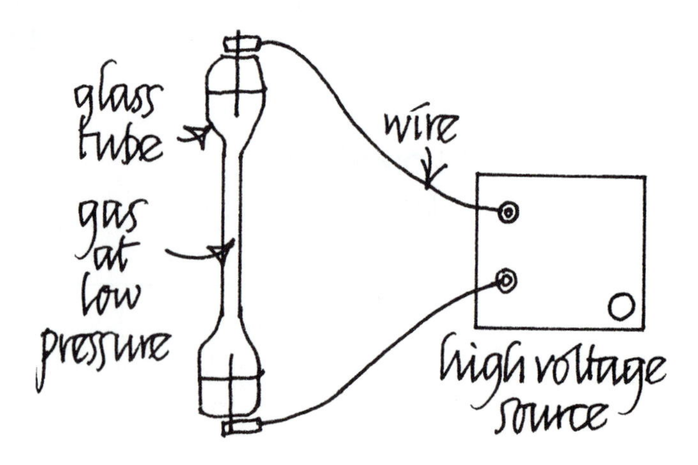

Figure 2.9 A Simple Discharge Tube

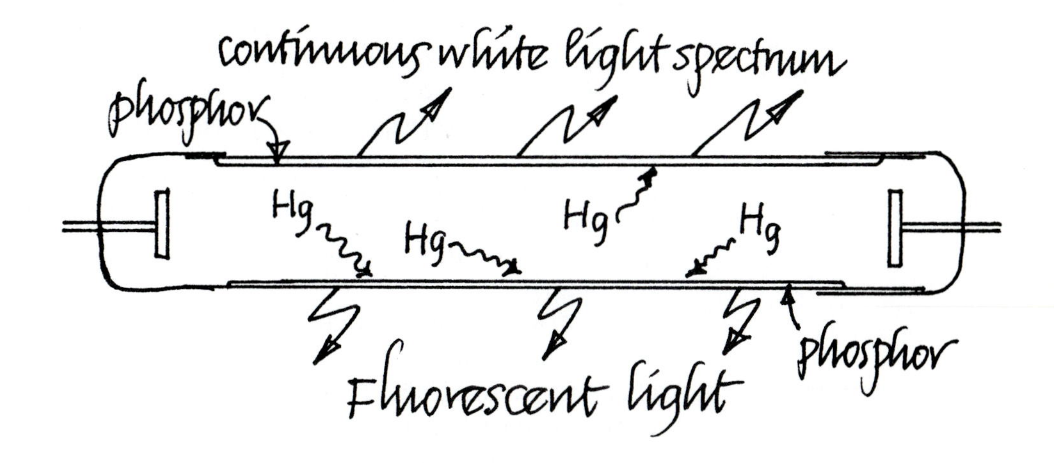

Figure 2.10 A Fluorescent Lamp

Some practical devices are the common fluorescent light, neon signs, and street lights. A fluorescent lamp (such as the one pictured in Figure 2.10) is a discharge tube that is filled with mercury vapor at low pressure. The inside walls of the tube are coated with a phosphor of calcium halophosphate ($Ca_5(PO_4)_3F_{1-x}Cl_x$) doped with Mn^{2+} and Sb^{3+}. Mercury atoms are excited by the process discussed above and emit their characteristic spectrum, which is in both the visible and in the ultraviolet spectrum (mostly UV, at 254 and 185 nm). This uv radiation is absorbed by the Sb^{3+} and passed on to the Mn^{2+}. The Sb^{3+} dopant gives off blue light, and the Mn^{2+} gives off orange light; the combination appears as white light.

Many solid substances — e.g., sodium chloride (NaCl) — may be excited by the thermal energy (heat) of flames. The heat comes from the exothermic combustion reactions occurring in the flame, e.g.,

$$CH_4 + 2O_2 \rightarrow CO_2 + 2H_2O + \text{heat}$$

Solid sodium chloride consists of sodium ions (Na^+) bonded to chloride ions (Cl^-). In a bunsen burner flame, these ions are dissociated, and the separated Na^+ ions combine with free electrons in the flame to form sodium atoms:

$$Na^+ + e^- \rightarrow Na$$

Electrons within the neutral sodium atoms can then be excited to higher energy levels by the heat of the flame:

$$Na + \text{heat} \rightarrow Na^* \text{ (an excited Na atom)}$$

The excited atoms then "relax" back to the ground state and emit light:

$$Na^* \rightarrow Na + h\nu \text{ (light)}$$

The emitted light is characteristic of all of the electronic transitions that occur with energy differences corresponding to visible radiation. Analysis of the light by means of a spectroscope gives the atomic emission spectrum of sodium. In this series of experiments, you will be able to measure wavelengths of the emitted light, but *not* intensity. This is because you will be looking at the spectra by eye. Various types of detectors are available for measuring radiative intensities, and if they were used, a quantitative analysis for the amount or concentration of a chemical element could be accomplished.

Continuous emission spectra contain so many emitted frequencies of radiation that the lines overlap and the light looks like a rainbow. Common terrestrial sources of continuous emission are fluorescent lamps, ordinary incandescent light bulbs, heated metals in general, and flames containing soot. Notice that all these sources are solids — e.g., the wire filament in a light bulb and soot particles in a candle flame. The atoms and molecules in heated solids are continuously bumping against each other because they are so close. This dynamic contact results in much of the energy being transferred as kinetic energy which is not quantized. As the temperature of a solid is raised, more and more of the radiation is emitted at shorter wavelengths.

ADDITIONAL READING

1. Henderson, S.T., *Daylight and Its Spectrum,* John Wiley and Sons, New York, 1977.
   Everything you wanted to know about the sun and its emission spectrum.
2. Kippenhahn, R., *100 Billion Suns,* Basic Books, Inc., New York, 1983.
   A wonderful, readable book about the birth, life, and death of the stars (the nonterrestrial ones!).
3. Herzberg, G., *Atomic Spectra and Atomic Structure,* Prentice-Hall, Inc., New York, 1937.
   A classic and, in my opinion, still the best book on the subject.
4. Feynman, R.P., *QED: The Strange Theory of Light and Matter,* Princeton University Press, Princeton, N.J., 1985.
   Feynman's dedication is beautiful — "what one fool can understand, another can." This book should be read by all faculty and students, but particularly faculty!
5. Ingle, J.D., *Spectrochemical Analysis,* Prentice Hall, Englewood Cliffs, N.J., 1988.
   A more advanced book, but has all the practical details. Encyclopedic.

Pre-Laboratory Quiz

1. Photon and wave theories of radiation are linked by Planck's Law. Give the law and define the symbols. Give typical units for each.

2. Briefly describe emission spectroscopy.

3. Give 2 examples of sources that emit continuous spectra.

4. What process will produce a line spectrum?

5. What types of radiation will be emitted from atoms in which valence electron transitions are occurring?

6. What is the wavelength range of the visible spectrum?

7. List the colors of the spectrum of white light from the highest to the lowest frequency.

8. How do we know that more than 60 terrestrial elements are found in the sun?

9. In the hydrogen emission spectrum, what energy level transitions produce visible light?

10. What is the Doppler effect?

Laboratory Experiments

Flowchart of the Experiments

Section A.	Characterization of a Transmission Diffraction Grating

Section B.	Construction of a Spectroscope

Section C.	Exploring Spectroscope Specifications

Section D.	Wavelength Calibration of the Spectroscope

Section E.	A Comparison of Continuous Emission Spectra

Section F.	Atomic Line Spectra and Electronic Transitions; Electrical Discharge Tubes

Section G.	Flame Emission Spectra

Requires one three-hour class period to complete

Section A. Characterization of a Transmission Diffraction Grating

CAUTION: *Lasers are dangerous. Do not* stare directly into a laser beam, or into a beam reflected from any mirror surface. Although you are using low-power lasers, exercise caution.

Goals:

(1) To use laser light of wavelength 632.8 nm, to determine the number of grooves on a diffraction grating. (2) To be able to calculate the groove distance of a diffraction grating from first and second order diffraction data.

Discussion:

One of the best ways of experimentally determining the number of grooves on a diffraction grating is to use it to diffract light of an exactly known wavelength. This can be done easily if you have access to a Helium-neon laser source. This gas laser is a continuous-wave (CW) source which emits mainly visible radiation at 632.8 nm. The laser is pointed at the grating and the angles of diffraction are calculated form the experimental geometry.

This experiment can be done in any one of three ways depending on how many He-Ne lasers are available in the laboratory. If you have one that you can use, then your instructor will demonstrate the experiment and give you the measurements. If there are none available, you may read through the method and I will give you values that I measured using a Spectraphysics 0.5 milliwatt He-Ne laser. If lasers are available, then work through the following sequence.

Experimental Steps:

1. Set up the diffraction grating (which has been mounted in a 35 mm slide holder). To do this, simply cut an SJ straw through about 1 cm from the end.

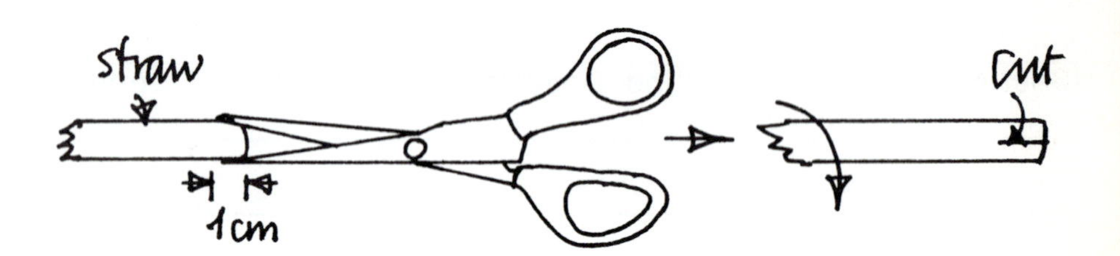

2. Mount the straw into a vertical straw in which a hole is punched 5 cm from the end. Place the vertical straw in a microtray and push the diffraction grating into the cut straw.

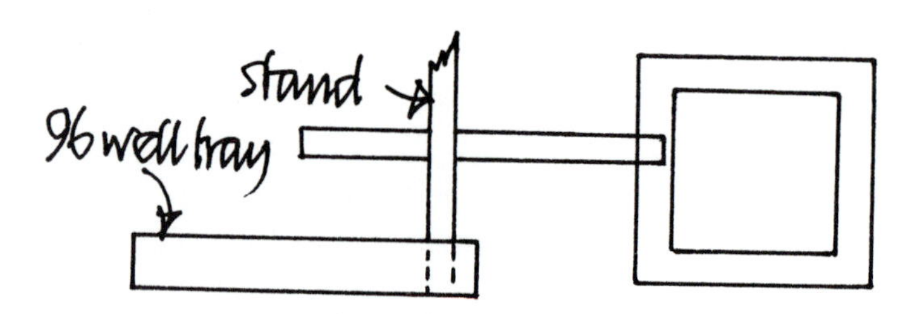

3. The laser should be set up so that it points normal (at 90°) to the grating. Then place a white card to intercept the diffracted light. Turn the laser on.

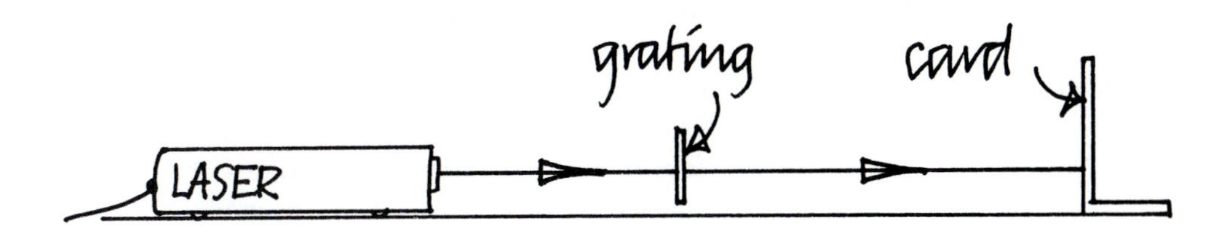

You can see an intense undiffracted beam striking the card and on either side — *if* you have the grating in the right direction — you can see the diffracted spots. If the diffracted spots are in a vertical mode simply rotate the grating 90°.

If the room is dark and the distance between card and grating is not too great then you should be able to see at least two orders of diffraction.

4. Using a ruler or meter stick, measure the appropriate distances to find sin θ for each order of diffraction that you can see.

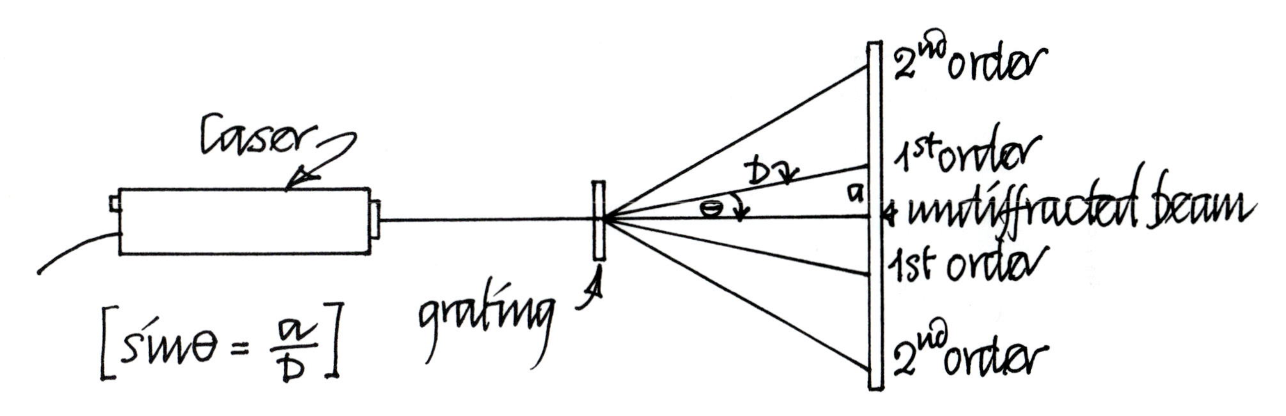

Here are some measurements I made:

For first-order diffraction, a = 15.3 cm, b = 45.8 cm. To obtain second-order diffraction, I moved the card closer to the grating and found a faint second-order spot: a = 8.5 cm, b = 12.7 cm. With the low-intensity beam on my laser, I could not find a third-order spot.

5. Now use the two-dimensional diffraction relationship $n\lambda = d \sin \theta$ to calculate the groove distance d. Be careful with the units — a good tip is to work in meters. You can now easily find the number of grooves per cm (or inch) on the replicate transmission diffraction grating.

Section B. Construction of a Spectroscope

Goal: *To construct a simple but rather accurate spectroscope containing a built-in quantitative calibration system.*

Discussion: All of the materials you need for building the spectroscope are located at Reagent Central. Part 1 and Part 2 provide instructions for building a spectroscope from two different types of boxes. Your instructor will determine which of these sets of instructions should be used.

Section B. Part 1. Construction of Box Type A

Experimental Steps: Diagrams of the completed instrument are shown below.

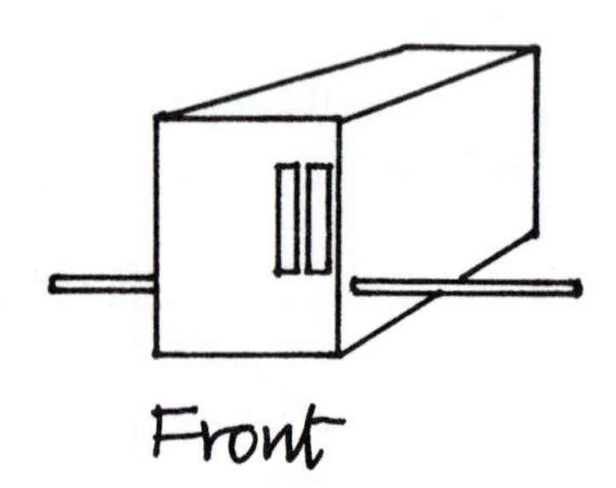

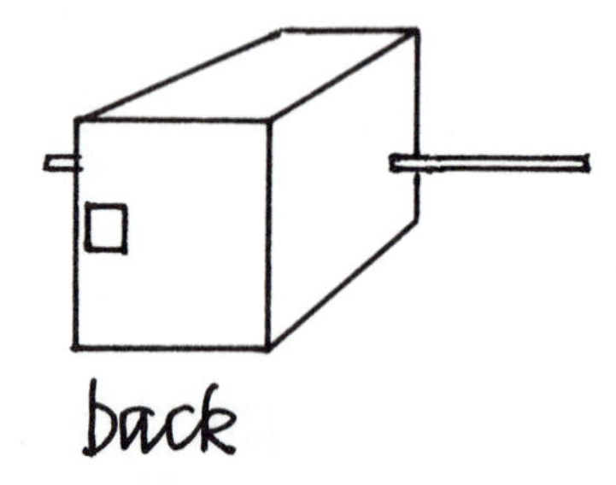

1. Cut off the inside flaps, leaving about 0.5 cm.

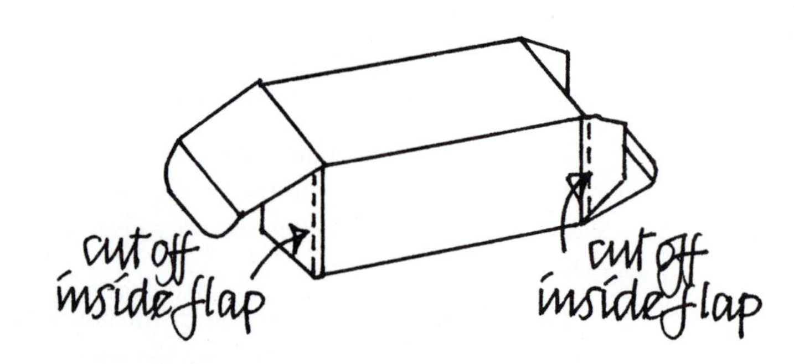

2. Outline the slit area and small square with a pencil before you cut them out. Make sure that these holes are in line before you cut them out.

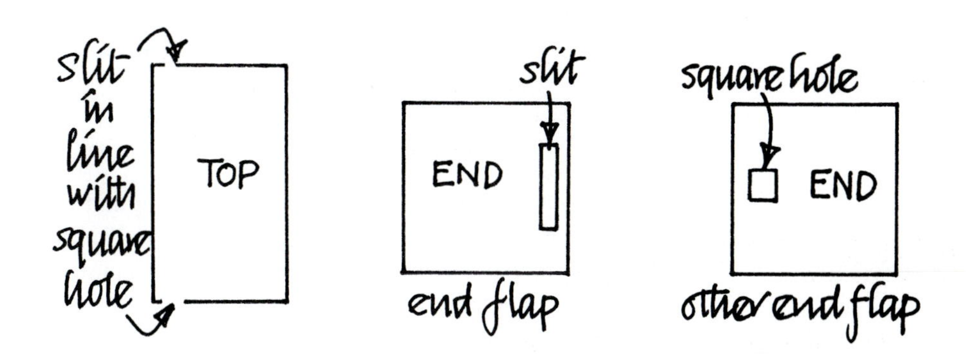

3. Obtain a piece of corrugated cardboard, which will be used as a cutting board.

4. With a razor blade cut out the slit and square.

   CAUTION: Please be careful, these blades are sharp.

5. Now put the box together tightly using masking tape. The only light permitted in should be through the two holes you have cut.

6. Pencil in 2 points, one on each side of the box exactly opposite each other.

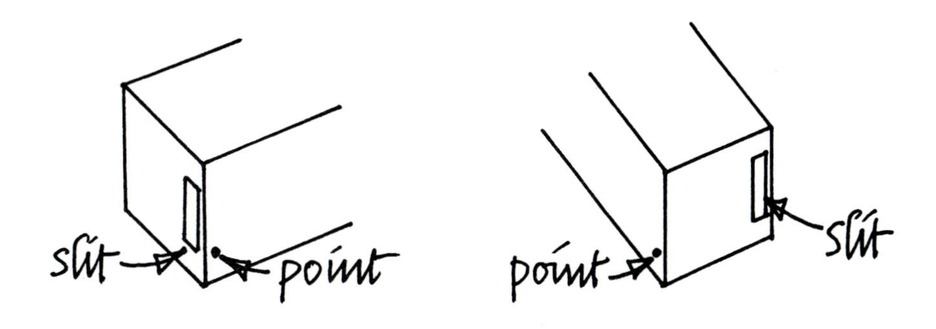

7. Make holes at these points by pushing and rotating a sharp pencil point into the box. The holes should only be large enough to friction-fit the 4 mm glass rod!

8. Lay the glass rod on the table and make a clean scratch in the middle of the rod using a glass scorer or file. Don't cut too deeply or the rod will break.

   CAUTION: Be careful — the ends of the rod are sharp.

9. Now use 2 pieces of black electrical tape to define an approximately 1 mm slit.

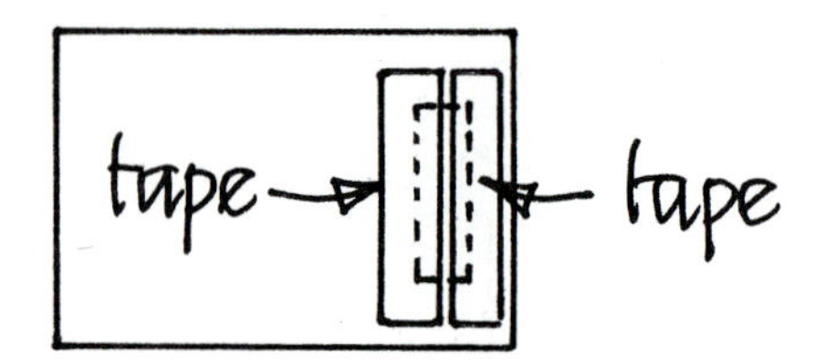

10. Obtain a piece of plastic diffraction grating and orient it correctly on the box *before* taping it permanently with transparent tape. To do this, hold the box so that you can look into it through the square hole, and point it vertically so that light from the fluorescent light enters the slit. Hold the grating over the square hole with your index finger while still looking into the box. You should see a visible spectrum (rainbow) to the right of the slit. If it doesn't look right, rotate the grating 90°.

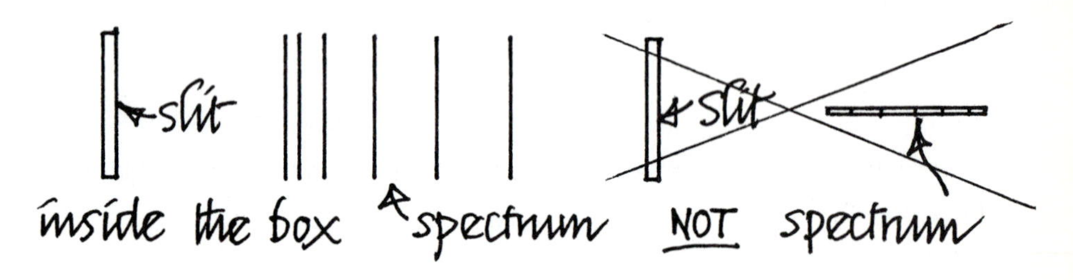

Once you are satisfied, carefully tape the grating to the box (tape around the edges).

11. Hold the box with the slit facing you on the right side of the box end and carefully insert the glass rod into the box as shown below, gently working it so that it is all the way through.

CAUTION: Don't force it, or it will break and cut you!

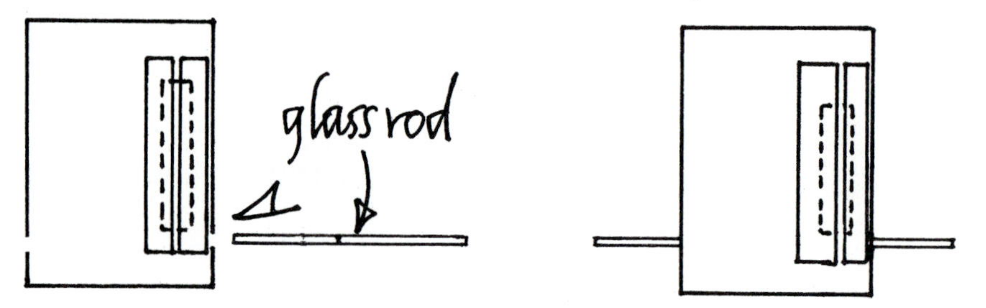

You now have a spectroscope with a calibration device.

12. Point the spectroscope vertically and look at the brightest part of the ceiling light. Inside the box you should see a good spectrum and also an outline of the glass rod. The scratch on the rod should be clearly visible to you. If not, carefully rotate the rod until you see it.

13. Now move the rod further into the box, *slowly*. The scratch will move, allowing you to locate any specific point on the spectrum.

14. The reason you can see the scratch is that light from outside the box is reflected down the rod until it reaches the scratch (as illustrated below). At this point the light is scattered in all directions by the damaged glass. Some of the light will come in your direction and you can see the scratch.

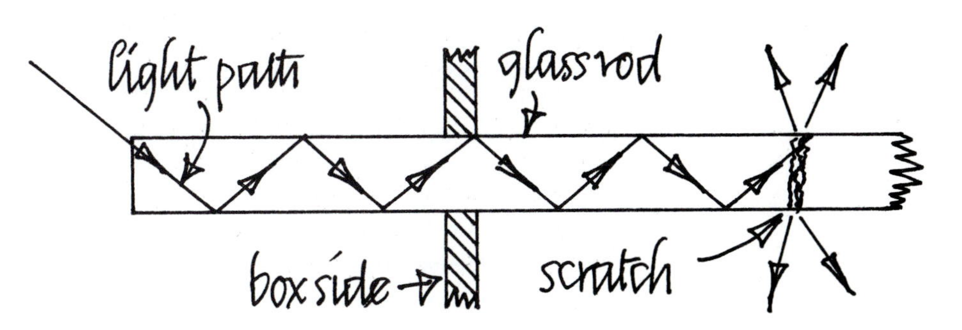

- Do you think that you could see a spectrum without the box?

Section B. Part 2. Construction of Box Type B

Experimental Steps:

1. Obtain a flat bakery box from Reagent Central.
2. Place the box on a piece of corrugated cardboard.

   Steps 3 through 5 are illustrated in the diagram below.

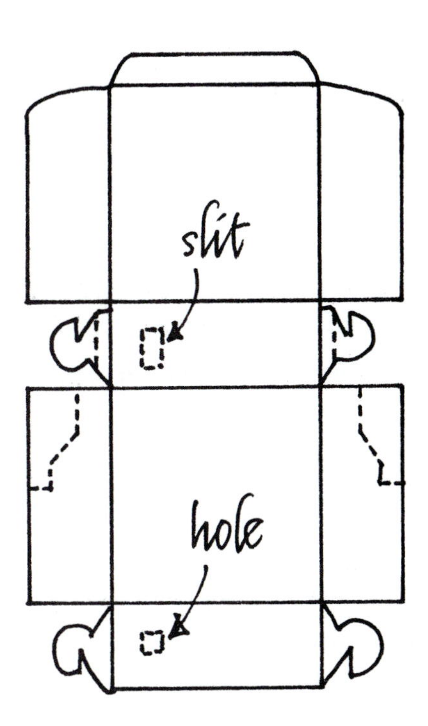

3. Using a razor blade, cut a slit about 2 cm from the fold.
4. Using a pair of scissors, cut off the flaps as indicated by the hatched areas.
5. Using a razor blade, cut a hole 1 cm square. Fold the box on the fold lines and tape all the edges with masking tape.

   Now go back to Step 6 of Part 1 (Section B) to finish constructing the spectroscope and to complete Part 1.

Section C. Exploring Spectroscope Specifications

Goal:

To be able to determine the specifications for your spectroscope.

Discussion:

It is always wise to explore the good points and the limitations of any instrument before launching out and spending time and money (tuition!) using it. Obviously, the spectroscope that you have built has many limitations. However, the nice thing about an inexpensive, simple, self-built machine is that you should have no inhibitions about using it, breaking it, changing it, or fixing it. In fact, you have a unique opportunity to write the specifications for your own spectroscopic instrument.

NOTE: If you wear spectacles, you might find it easier to take them off when looking into the spectroscope.

Experimental Steps:

- Were you confused at any step — and where? How could the instructions for building the spectroscope be improved — e.g., can they be made shorter?
- What are the dimensions of your spectroscope?
- How could you measure the actual dimensions of the spectrum? What are they?
- This may seem like a silly question, but it's not. What is the actual location of the spectrum? Is it really inside the box?
- Predict what differences in the spectrum you would see if the spectroscope were made out of a giant pizza box.

1. Now let's change some things. Vary the *slit width* by unpeeling one of the pieces of electrical tape and repositioning it onto the box.
   - If you make the slit larger or smaller, what happens to the fluorescent light spectrum?
   - Draw a few simple pictures to illustrate what you see.
   - Does the slit have to be a long, thin rectangle?
   - What is the relationship between the distance of the source from the spectroscope and the characteristics of the spectrum?
   - What changes would you have to make in your spectroscope in order to see a second-order spectrum?

- Give your spectroscope a rating (1 – 10) for

  a) craftspersonship
  b) aesthetics
  c) spectroscopic efficiency

Section D. Wavelength Calibration of the Spectroscope

Goal: *To carry out a calibration of your spectroscope by using known spectral lines in the emission spectrum of a fluorescent light.*

Discussion: All scientific instruments must be calibrated before any quantitative measurements can be made. Commercial instruments are almost always calibrated by the manufacturer before they are actually used in the factory or laboratory. The glass rod in your spectroscope is a suitable measurement device, but the only way to calibrate it is to be able to view a visible spectrum that has emission lines of exactly known wavelength. The light emitted from a typical fluorescent light is such a spectrum.

Experimental Steps:

1. Make sure that the slit width is about optimum (1 mm), and point your spectroscope directly up at the brightest part of a fluorescent light. It helps if you shut one eye!

   The spectrum that you can see is actually composed of two types of spectra: a continuous and a line emission spectrum, *superimposed* on each other. This superimposition produces a rainbow with 3 fairly prominent lines of light in it.

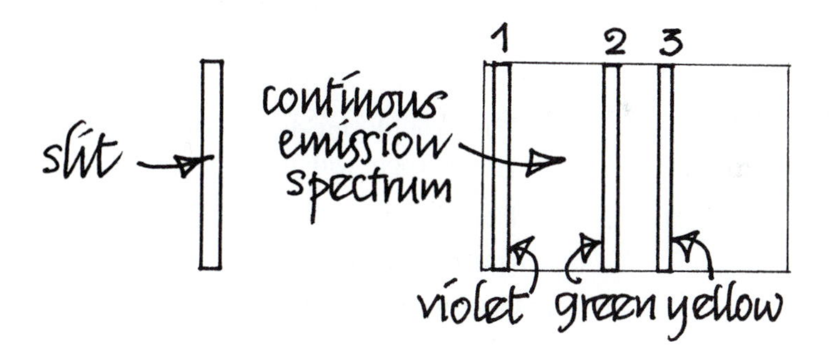

View of Spectrum Inside Box

The three lines 1, 2, and 3, are violet, green, and yellow and are the strong emissions from Hg vapor in the fluorescent tube. These lines have well-known wavelengths and we can use them to calibrate the spectroscope:

Line 1, violet	436 nm
Line 2, green	546 nm
Line 3, yellow	580 nm

2. Keep looking at the spectrum and slowly push the rod until the scratch mark is lined up just underneath and exactly in line with the violet line (line 1).

3. Without moving the rod at all, use a ruler to measure the distance from one end of the rod to the side of the box (outside the box). Measure to an estimated 0.1 mm!

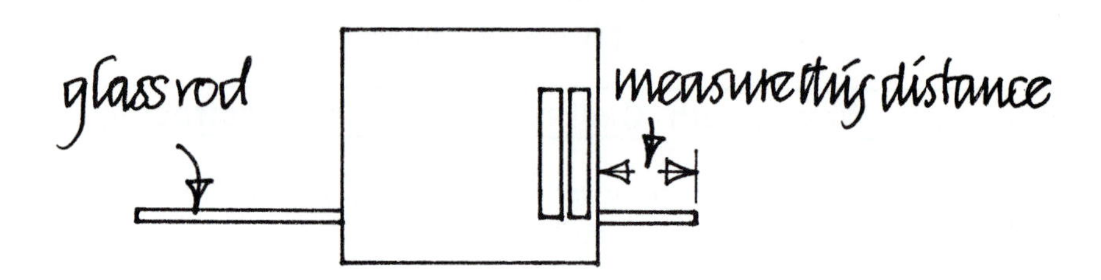

4. Look at the spectrum again and push the rod so that the scratch mark is now lined up with the green line (line 2). Measure the rod distance again (on the same side of the box as before).
5. Repeat the procedure for the yellow line (line 3) and measure the distance.
6. Now plot a graph of wavelength of lines 1, 2, and 3 versus rod distance for each line. Use a wavelength axis from 400 nm to 650 nm. Draw the smoothest curve you can through all 3 points.

   NOTE: Graph paper is bound into the back of this book. Simply tear it out and tape it into your laboratory record.

   This graph is the spectroscope calibration line which will enable you to find the wavelengths of unknown spectral lines, bands, etc., in any emission or absorption spectrum.

   - Would this calibration line work for your neighbor's spectroscope?
   - What could you do to ensure that the calibration line was the same for any spectroscope (with the same size box)?
   - How does the slit width affect the calibration line?

   NOTE: The next three experiments (E, F, and G) can be done in any order. The sources for these experiments are in various places in the room. Check for accessibility and go ahead!

Section E. A Comparison of Continuous Emission Spectra

Goal: *To use the spectroscope to compare various continuous emission sources.*

Discussion: Most people are aware that white light from different sources can look very different. Colors that match in artificial light look very different in daylight. In this experiment you can use the spectroscope to investigate the difference in spectral characteristics of common sources of white light. Use your spectroscope to analyze the light from (a) fluorescent light, (b) incandescent light (i.e., 60 watt bulb), (c) candlelight, and (d) daylight. Several methods are described below; try each method!

Experimental Steps:

Method 1

1. Look at one source and immediately go to another and back to the first. Differences in the two spectra can then be easily seen.
   - Draw pictures to illustrate these differences.

Method 2

1. Determine the total spectral range and the width of each band of color (red, for example) by measuring with the glass rod.
2. Line up the scratch on the edge of the blue side of the spectrum, measuring the rod distance, and then moving the scratch all the way over to the very end of the red side and measuring. By consulting your calibration line you can obtain the spectral range for that light.
3. To obtain a *daylight* spectrum, point your spectroscope at a bright sky.

CAUTION: Do not look directly at the sun.

Method 3

Studying the candle flame requires a different observational approach.

1. Look carefully at the flame.
   - Draw a simple picture.

Even though the spectroscope shows one bright continuous spectrum, close inspection reveals several zones which are different intensities or colors. At the base of the flame, the blue light is emitted by the combustion reaction products molecular carbon (C_2) and a hydrocarbon (CH). About halfway up the flame, the temperature reaches about 1400 °C. Most of the candlelight (and, therefore, the emission spectrum) comes from tiny soot particles that are about 50 nm in diameter. It is the heating of these particles that produces the yellow glow of the flame.

Discussion:

The chemical processes that produce the light emitted from the sun, a candle, a 60 watt bulb, and a fluorescent light are all quite different. However, all these sources give a *continuous* emission spectrum.

- Why is this?

You may take your spectroscope with you at the end of the laboratory.

CAUTION: Remove the rod when you are carrying it.

- As a homework project, look at and describe the spectral characteristics of 4 sources different from the ones you studied in lab. Some suggestions are white light from a TV screen, street lights (there are many types), and neon signs.

Section F. Atomic Line Spectra and Electronic Transitions; Electrical Discharge Tubes

Goals:

(1) To obtain the atomic line spectra of light emitted from electrical discharge tubes. (2) To be able to calculate photon wavelengths, frequencies, and energies from line spectra data.

Experimental Steps:

1. The electrical discharge tubes are along one side of the room and are connected to a high-voltage supply. Ask your instructor to switch them on for you.

   CAUTION: Be careful not to touch the tubes or metal strands or you will get an electrical shock.

2. View several discharge tubes before making any measurements so that you can get a good idea what line spectra look like. The brighter tubes can be viewed from several feet away.

   NOTE: You might find it advantageous to increase the slit width to about 1.5 to 2 mm.

   - Why?

3. Now view the tube containing hydrogen gas. This is not very bright so you must view it quite closely, perhaps about 10 to 20 cm away.

   - Use the glass rod measurement technique and your spectroscope calibration line to determine the wavelength of the spectral lines that are prominent.
   - Be careful to record the color of the lines and the overall color of the emitted light.

4. Study the following diagram which shows the energy level transitions for H atom emission.

   - From your data identify the transitions which correlate with the spectral lines you observed.

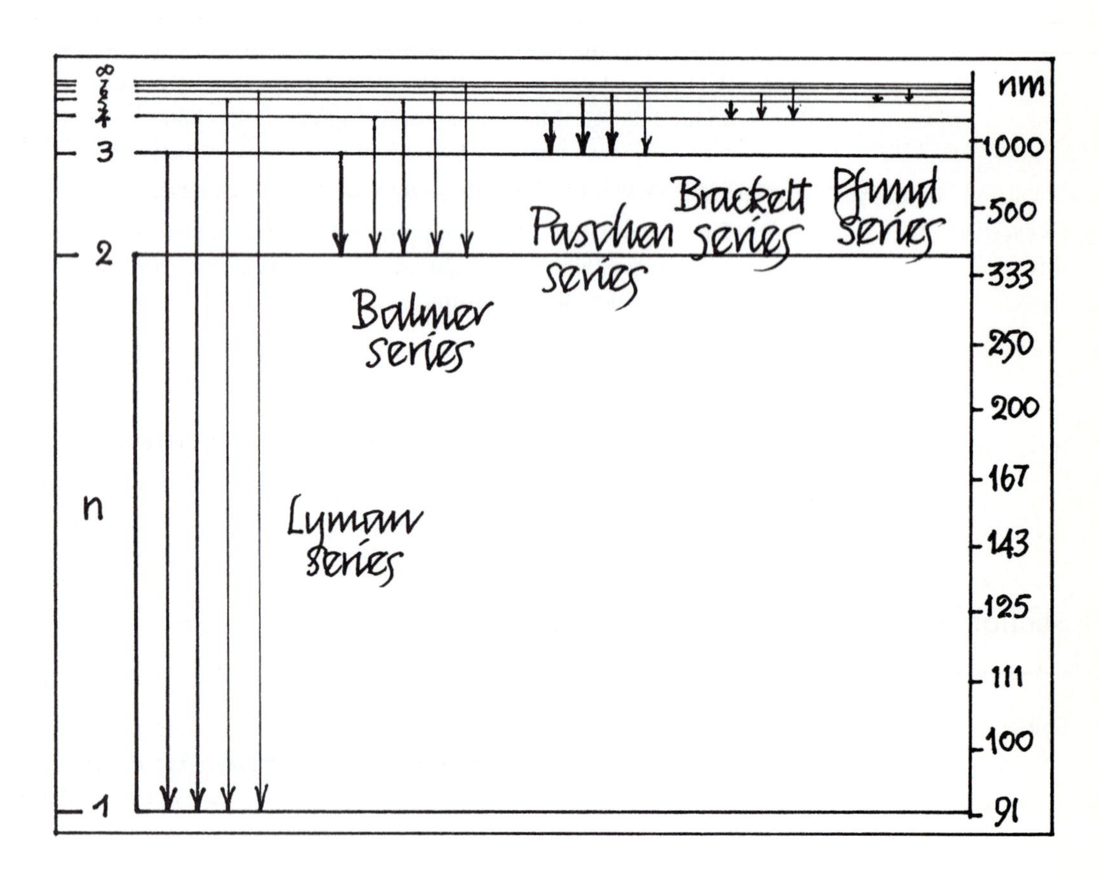

5. Select another discharge tube.

   - Find the wavelength, frequency, and photon energy for the prominent spectral lines.

   Suppose the tube you selected (Step 5) was contaminated with hydrogen gas. What would the emission spectrum of this mixture look like?

   - Draw the lines and colors on a wavelength grid 450 – 750 nm.

Section G. Flame Emission Spectra

Goal:

To obtain line and band spectra of light emitted from chemical species produced in Bunsen burner flames.

Discussion:

Many substances, particularly alkali metal and alkaline earth salts, produce brightly colored flames when they are heated by a Bunsen burner. Flame emission spectroscopy is a powerful method for qualitative identification and quantitative determination of many elements. One of the most common and useful flame emissions is that from atomic sodium.

Experimental Steps:

1. Find a partner to hold a glass rod in a Bunsen burner flame.
2. Obtain a wire loop that has been stuck into a cork and one of several solid salts ($LiCl$, $NaCl$, KCl, $SrCl_2$, $CaCl_2$, or $CuCl_2$) from Reagent Central.
3. When the flame gets bright, view it through your spectroscope. Get reasonably close to the flame (about 4 inches or so), but try not to set the spectroscope on fire.

   CAUTION: Be very careful to keep the spectroscope at a safe distance from the flame so that it doesn't catch fire!

   - Measure the wavelength in the usual way, calculate the frequency and photon energy, and record in your laboratory record.

4. Adjust the Bunsen burner flame so that it is about 4 to 5 inches high and is a hot flame.
5. Heat the loop in the flame until it is red-hot, which will burn off any impurities. While the loop is still red-hot, quickly dip the loop into the solid. Some of the solid will stick to the wire.
6. Place the loop into the outside part of the flame and note the flame color.
7. Clean the loop by dipping in water and reheating and then obtain the flame colors for all of the salts.

   CAUTION: Please do not contaminate the salts!

8. Find a partner and select one of the salts.
9. Have your partner hold the loop (with sample) in the flame while you view the flame color through your spectroscope. Flame spectra are not very intense, and in order to get

good measurements of line spectra, you will need to (1) have an optimum slit width, (2) do 1 line at a time, and (3) be fairly close to the flame. Try for the most prominent lines.

10. Obtain one of the unknowns from your instructor and carry out a qualitative analysis.
    - Report the result and the reasons for your answer in your laboratory record.

Chapter 3

Small-Scale Techniques and the Absorption of Light

Introduction

Behavioral psychologists tell us that sighted people receive more than 90% of the information about the world through their eyes. The subtleties of line, perspective, shadow, texture, and color form the complex representation of the world we see. These perceptions require three elements: an object, a medium, and an eye-brain system. The medium is light. A source of visible radiation sends out oscillating electromagnetic energy travelling at the speed of light. As the light moves through molecules in the air and impacts on objects, the oscillating energy field interacts with the outer electrons in matter. The light-electron coupling takes less than a picosecond, but in this very short interaction time, a tremendous amount of information is encoded in the light. Eventually, the changed light may pass through the pupil of an eye, travel through the lens and the aqueous humor, and strike the retina. The tiny light detectors, called rods and cones, contain molecular antennas that absorb the light. Rod cells form black and white images in dim light, and cones are responsible for color vision in bright light. The light-absorbing visual pigments consist of a protein, opsin, that is covalently linked to a molecule called 11-cis-retinal. Vision begins when light absorbed by the visual pigments causes the retinal to change its shape, a process called photoisomerization. Shape changes in retinal then trigger a series of changes in the attached protein, which in turn catalyze the conversion of hundreds of secondary messenger molecules into an active state. This conversion is the first step in a cascade of reactions that eventually produces a signal in the brain.

The human eye-brain system is a remarkably flexible and sensitive detector. Recent research in the chemistry of vision has shown that only one photon is required to produce a physiological response in an individual receptor. A person with good color vision can discriminate minute color differences — on the order of wavelength differences of 1 or 2 nm! Of course, there are variations in the response of different humans to light stimuli. About 6% of males are color deficient in some part of the spectrum compared to less than 0.1% for females. However, these variations do not change the fact that the human visual system is very good at deciphering much of the information encoded in light. You have already used your own visual system to investigate the electronic structure of atoms by means of the technique of atomic emission spectroscopy. The encoded information in light emitted from excited atoms is the detailed record of all the electronic energy-level transitions occurring in those atoms. The real key to revealing quantitative information is the use of appropriate standards, calibrations, and comparisons.

Another fundamental way in which valuable information can be obtained from light probes is to measure the *amount* of light energy absorbed or reflected by matter. The amount of light absorbed or reflected depends on the number of interactions that occur between the photons and the absorbing molecules. The absorption is therefore proportional to the amount or concentration of molecules in the chemical system. It is interesting to note that this relationship between the amount of light absorbed and the amount of absorbing species is a general one that holds for *all* types of radiation. Again, the key to a practical measurement system is the use of appropriate visual comparison standards. Small–scale absorption standards allow the eye-brain system to discern rather small differences in light absorption.

Background Chemistry

Chemistry is the study of matter and its transformations. The experimental practice of chemistry has traditionally relied on the use of large-scale metal or glass apparatus. This equipment is used for carrying out chemical reactions and for the storage, transfer, and quantitative delivery of all forms of chemical substances. Recently, there have been tremendous advances in the development of sophisticated small-scale equipment, much of which is now being used in the clinical, microbiological, and recombinant DNA fields. Your chemistry laboratory course has been designed to utilize some of the outstanding advantages of doing chemistry on a small scale. *Small-scale* means that small quantities of chemicals are used. The disposal of small quantities of chemical wastes is relatively simple, inexpensive, and more importantly, much less of a threat to the environment. Doing science on a small scale almost always leads to much greater safety in the laboratory. It is hoped that you will find the equipment user friendly and efficient and that it will allow you to be far more involved in interesting chemistry than is possible in traditional university laboratories.

Much of the nontraditional small-scale equipment is made out of plastic, and most of it was originally designed to be disposable. We are going to reuse it and recycle it wherever possible. The plastics are mainly polystyrene (e.g., microreaction trays), polyethylene (e.g., pipets and microburets), and polypropylene (e.g., straws). All three plastics have chemical properties that can be exploited in the innovative design and use of tools for science. All have surfaces that are nonwetted by aqueous solutions. Nonwetted surfaces make the storage, transfer, and delivery of aqueous solutions much easier than, say, glass surfaces. The high surface tension of water means that in the presence of a plastic, a drop can be its own container and a reproducible volume increment at the same time. The ease of production of small drops from plastic tubes leads naturally to digital methods in volumetric work. Counting drops can circumvent the problems inherent in the analog methods of meniscus reading. The low softening and melting point of these materials, together with the ease with which they can be stuck, bent, cut, and otherwise mutilated naturally leads to innovative construction. Of course, in an experimental science like chemistry, there is occasionally a need for glass or metal materials. When this need arises, we will not hesitate to use small quantities and small tools.

One of the most innovative applications of small-scale plastic equipment has been in the area of colorimetry and spectrophotometry. As you have already seen in the spectroscopy laboratory, the interaction of light with matter provides scientists with a powerful probe for investigating the structure, function, and dynamics of any chemical system. The analysis of light emitted from excited atoms and molecules is only one example of this general approach.

Another technique that is used extensively in many areas of science is that of light absorption by a chemical system. Consider white light falling on a solution of red food dye. The solution looks red because only red light is being transmitted by the solution. All the other colors, mainly blues and greens (490–560 nm), have been absorbed by the molecules of red food dye dissolved in the solution. Bluish-green light is absorbed by these molecules because there are electronic energy-level transitions in the molecule that correspond to the photon energy of bluish-green light. In other words, the food dye molecules become excited. The question is, What happens to the extra energy that the molecules have absorbed? Unlike the excited gas-phase atoms in the spectroscopy experiment, the excited food dye molecules do not emit radiation (although other organic molecules under the right conditions will emit light, e.g., fireflies and railroad worms). The extra energy is removed continuously by solvent molecules bumping billions of times a second against the excited food dye molecules. As a result, there is a small but imperceptible increase in the thermal energy of the solvent. The basic reason for this different mode of energy transfer is that all of the molecules in a solution are much closer to each other than they are in the gas phase. Many, many more collisions occur between solution molecules than gas molecules, each collision removing a small amount of energy. Solutions that have colors other than red undergo the same processes. In general, the observed color is that of the light that is transmitted. All the complementary colors have been absorbed by the solution, as shown in Table 3.1.

Wavelength (nm)	Colors Absorbed	Color Observed
380 – 435	violet	yellowish green
435 – 480	blue	yellow
480 – 490	greenish blue	orange
490 – 560	bluish green	red
500 – 560	green	purple
560 – 580	yellowish green	violet
580 – 595	yellow	blue
595 – 650	orange	greenish blue
650 – 780	red	bluish green

Table 3.1 Absorbed and Observed Colors of Solutions

The color of the light absorbed by a molecule largely depends on the electronic energy levels of the molecule. These levels are determined by the chemical structure of the molecule — i.e., by the number and arrangement of the electrons. Red food dye, for example, has a very different arrangement of electrons than a blue food dye, as shown in their respective structures in Figure 3.1.

A careful analysis of the light absorbed and transmitted by a molecule can be used to provide information about the chemical structure. In this series of experiments, however, we will be more concerned with the *amount* of light absorbed by colored solutions. Light absorption of this type follows a very general law known as Lambert, Bouger, and Beer's law (commonly abbreviated as Beer's law). Beer's law states that if monochromatic radiation is allowed to fall on a solution, then the amount of light absorbed or transmitted is an exponential function of the concentration of absorbing

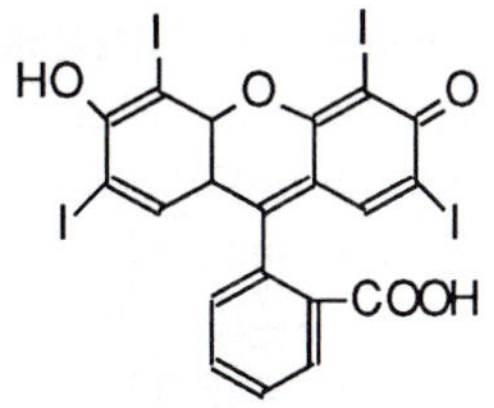

FD and C Red No. 3

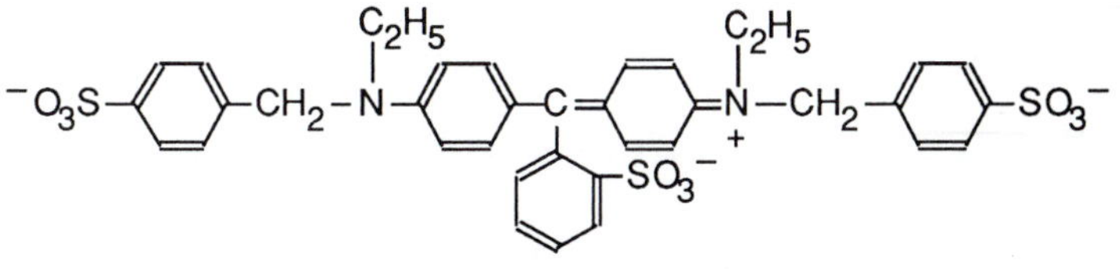

FD and C Blue No. 1

Figure 3.1 The Chemical Structures of Two Food Dyes

substance and of the length of the path of light through the sample. Mathematically, the law is expressed as

$$T = 10^{-abc}$$

or

$$-\log_{10}T = abc$$

where T is called the transmittance, a is a constant that depends on the substance absorbing the light, b is the path length of light through the sample, and c is the concentration of absorbing substance in the solution. The law is often expressed in linear form as

$$A = abc$$

where A, the absorbance of the solution is defined as

$$A = -\log T$$

One of the earliest and most useful applications of light absorption phenomena was in a technique of chemical analysis called colorimetry. In *colorimetry,* a set of solutions of known concentrations of some light-absorbing substance — e.g., food dye — is placed in a series of containers. An unknown concentration solution is then compared with the calibration set, either by eye or with the aid of a simple instrument. Once a match is obtained, the concentration in the unknown solution can be determined. Modern instrumentation has enabled much greater accuracy and sensitivity to be achieved in light absorption analysis. In instruments called *spectrophotometers*, the amount of light absorbed or transmitted is accurately and quantitatively detected and measured by solid state electronic devices. Sophisticated microspectro-photometers are now commercially available in which light absorption of 0.1 mL solution can be automatically measured with great accuracy.

Laboratory Experiments

Flowchart of the Experiments

Section A. An Introduction to Small-Scale Scientific Apparatus

Section B. Making a Microburner

Section C. Making Microburets

Section D. General Microburet Techniques

Section E. Microburet Calibration for Quantitative Volumetric Work

Section F. Quantitative Dilution of Solutions

Section G. Standard Color Solutions and Colorimetry

Section H. The Factors That Govern the Absorption of Light

Section I. Colorimetric Analysis of a Beverage

Section J. Solution Absorption Spectrophotometry

Requires one three-hour class period to complete

Section A. An Introduction to Small-Scale Scientific Apparatus

Goals: *To allow you to become familiar with some of the nontraditional apparatus and techniques you will be using in this laboratory course.*

Discussion: *Equipment List*

1. At the lab area assigned to you is a collection of the small-scale equipment discussed earlier. You should be able to identify the following:

   - 24-well (4 x 6 wells) tray
   - 96-well (8 x 12 wells) tray (round profile wells)

   NOTE: The trays are the same size and are stackable and that the wells are labeled by letters for the horizontal rows and by numbers for the vertical columns.

   - 1 x 12 well (1 x 12 wells) strip (flat profile wells)

   NOTE: The strip wells are numbered 1 through 12. The trays and strips have all been designed so that the bottom of the wells do not touch the table. This is done to avoid scratching the well, thereby maintaining the optical clarity of the well.

   - Plastic thin-stem pipet (pictured below)

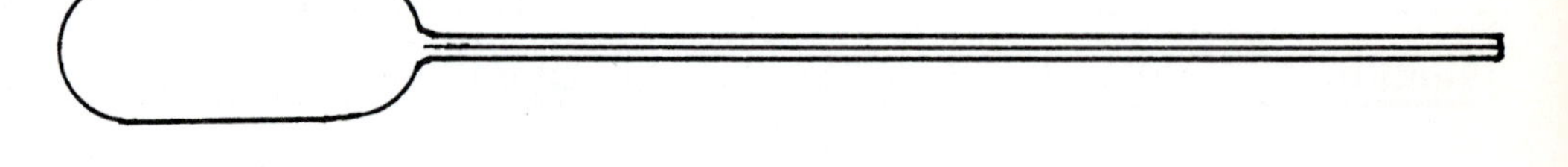

   NOTE: This type of pipet will be used in many ways during the laboratory course. It will be made into a storage container, microscale burets, a microburet cap, a stirrer, a scoop, a spatula, a reaction vessel, a sampling device, and a volumetric flask.

   - Straws Slim and super jumbo. Caps for straws.

   NOTE: There are four standard sizes of straws in the USA: slim, jumbo, super jumbo, and giant. As you will see later, straws are extraordinarily useful devices. You will be using them as tubular integrated containers, stands, racks, clamps, holders, volumetric flasks, pointers, test tubes, reaction vessels, chromatography columns, stirrers, and straws.

   - Plastic reaction surface

   NOTE: The surface is simply an office file protector. However, it will provide a wide variety of backgrounds, templates, and surfaces for chemical reactions.

   - Waste disposal cup

   NOTE: Most chemical investigations generate waste. All the experiments in this laboratory course have been designed from the point of view of using very small amounts of chemicals in safe ways. However, it is extremely important for you, the beginning scientist, to learn to be a responsible investigator. Making mistakes is an essential ingredient in most learning experiences. One of the beauties of small-scale experimentation is that making mistakes will not usually lead to grand disaster. Try to exercise intelligence and promote conservation when you do science, and we will do our part to dispose of wastes and, wherever possible, recycle materials.

- Distilled water wash bottle

NOTE: The wash bottle contains a good supply of distilled water. Your instructor will show you where you can obtain more if you need it.

- safety goggles
- pair of scissors
- pair of tweezers
- 1/4 inch office punch
- hand lens
- ruler

Reagent Central

In many of the experiments, we will arrange materials and solutions, etc., at your place in the laboratory. On other occasions, you will need to obtain supplies from Reagent Central. Your instructor will show you which part of the room is designated as Reagent Central. It is important that you act intelligently and responsibly when obtaining materials from Reagent Central. If it's a mess when you get there — well...!

Section B. Making a Microburner

Goal: *To construct a simple, alcohol-fueled microburner that may be used in molding, melting, and forming plastic apparatus for use in experiments.*

Experimental Steps:

1. Obtain microburner parts from Reagent Central. Carefully thread the wick through the piece of glass tubing until it comes out of the other end.
2. Fluff the wick at one end and pull it out at the other so that it protrudes about 2 or 3 mm.

   NOTE: The size of the flame is governed by the length of exposed wick.
3. Place the burner, fluffed end down, into an outer well of a 96-well tray.
4. Carefully drop 4 or 5 drops of ethanol (alcohol) onto the wick and let it soak in.

   NOTE: Avoid dropping it around the outside. If you do, wipe the excess ethanol up with a paper towel.

   The microburner can be moved around in the tray or outside the tray.

5. Place the microburner against a black background so that you will be able to see the flame and light the burner.

   CAUTION: Ethanol flames are tough to see. Think about why this is so. Be very cautious when placing anything near the lighted wick.

   The flame will go out after about 3 minutes, and at that point the wick will burn. Blow it out before it begins to blacken. It is important not to allow it to burn this far because it will blacken the wick; a blackened wick does not burn smoothly.

   CAUTION: Always let the wick cool before refueling and don't fuel it unless an experiment requires you to use the microburner.

Section C. Making Microburets

Goal: *To learn how to make simple, constant-drop-volume delivery devices.*

Discussion: Throughout this course you will be using several types of microburets for both qualitative and quantitative delivery of solutions.

Experimental Steps:

Thin-Stem Pipet

The *thin-stem pipet* (hereafter simply referred to as *pipet*) can be used directly, without any alterations. However, the long stem length tends to make positioning the tip difficult, so the pipet is used only to carry out qualitative transfer.

Large-Drop Microburet

The *large-drop microburet* is much more useful than the pipet.

1. Cut off the stem with your scissors at a point about 2 cm from the bulb.

   NOTE: Make the cut at right angles!

   Don't throw away the cutoff end. You can make a *microstirrer* out of it! The large-drop microburet will be used most frequently in this text's experiments.

Small-Drop Microburet

You will also need a microburet that delivers much smaller volumes (drops). Your instructor will demonstrate this method before you try.

1. Hold a new pipet firmly between the index finger and thumb of your dominant hand at a position on the thin stem just beyond the bulb.
2. Hold very firmly so that the bulb is in the palm of your hand.
3. Now curl the thin stem between the index finger and second finger of your other hand and grasp tightly. The finger knuckles on both hands are touching.
4. Pull slowly and the stem will stretch. Keep pulling until the stretched part is more than 4 or 5 cm long.

5. Use sharp scissors to cut off the pulled out part about 1 to 1.5 cm from the original thin stem. Save the cutoff part.

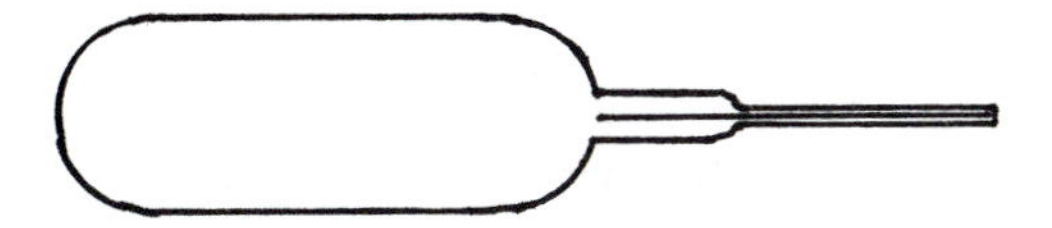

NOTE: Don't be alarmed — only total geniuses and world-class weight lifters can do it the first time around! Please save the disasters. We will recycle them. Try again.

Mini Microburet

Use the *microburner* to make the *mini microburet.*

1. Place the tray and microburner against a black background.
2. Obtain a new pipet.
3. Light the refueled and cooled wick (the microburner that you constructed in Section B).
4. Hold the ends of the pipet in both hands so that it won't sag.
5. Place the thin stem horizontally just above the flame at a place about 1.5 cm from the bulb for about 2 seconds, rotating evenly.

   NOTE: The plastic should go from translucent to clear (transparent).
6. Blow out the microburner flame and very gently pull the pipet apart horizontally for about 4 or 5 cm.
7. Hold the pipet in the horizontal position with slight tension until the plastic becomes cool and translucent again.
8. Once the plastic is rigid, cut at the extended part, leaving at least 1 cm of narrow part left (as shown below).

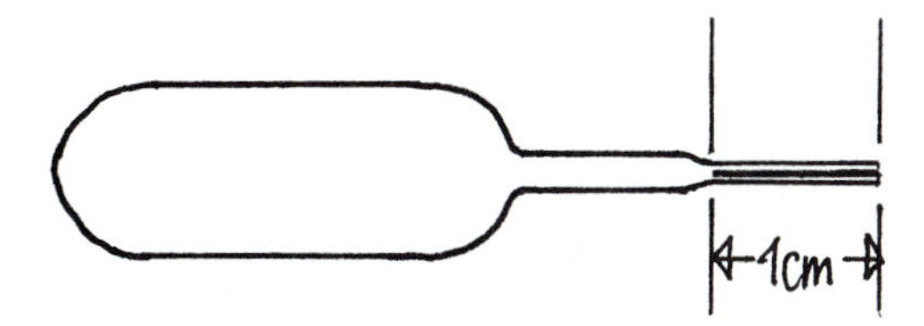

Mini microburets that will deliver even smaller drops can be made by a slight modification of the above method. Your instructor will demonstrate.

Your instructor may also show you how to make cute microcaps for microburets!

You may wish to label the microburets if you find it difficult to differentiate between them.

Section D. General Microburet Techniques

Goals: *(1) To investigate some of the factors that control drop volume. (2) To learn how to handle microscale burets for the quantitative delivery of solutions.*

Experimental Steps:

1. With a hand lens, examine the tips of the microburets you made.
   - Draw a picture of the end cross sections in your notes.
   - Do you notice anything different about the small-drop microburet?
2. Fill 2 wells of a 24-well tray with water that has been colored with green food dye.

   NOTE: The green dye solution is in a plastic bottle which may be passed around or obtained from Reagent Central. The dye is to enable you to see the solution more easily.
3. Fill 2 wells of a 24-well tray with distilled water.

   NOTE: Before you begin any experiment, you have to assume that the bulb of every microburet is dirty and must be cleaned with distilled water.
4. Begin by cleaning the mini microburet. Suck a little distilled water up into the bulb, shake it so that all internal surfaces have been wetted.
5. Holding the mini microburet vertically, expel to the waste cup. Press firmly to get those last drops out.
6. Now suck up a little green food dye, shake to rinse the bulb, and expel to waste cup.
7. Squeeze the bulb and suck up green food dye.

   NOTE: This sequence (steps 4, 5, and 6) is known as *good wash, rinse, and transfer technique.*

   Note how much the bulb fills with one squeeze.
8. Now deliver some drops back to the well. With your hand lens, look at the drops as they leave the tip.
   - Draw pictures in your notebook to show how they form and fall off the tip.
   - What happens if the mini microburet is held at *different angles*?
   - Examine the other microburets.

   The small-drop microburet tip has an oval, somewhat flattened shape. This is due to the fact that the pipets are manufactured with a seam.
9. Practice producing pools of various size on the plastic surface and then suck the pools up completely.

   Sometimes small bubbles will be formed in the stem of a microscale buret and the first drop will not form properly. Simply waste the first drop to the surface or waste cup and then continue and drops will be uniform.

NOTE: For quantitative work it is best to adopt a standard method of making a microburet and to use a standard delivery technique. These practices are discussed in the next section.

Section E. Microburet Calibration for Quantitative Volumetric Work

Goals: *(1) To decide on a standard delivery technique for repetitive delivery of drops of solution at uniform volume. (2) To use that technique to calibrate a microburet.*

Experimental Steps:

1. Wash a 1 x 12 well strip with distilled water and remove as much water as possible by slapping the strip into a paper towel held in your hand. Any remaining water can be removed with a cotton swab.

   NOTE: Step 1 describes the *technique for cleaning trays and strips* before and after use.

2. Select the mini microburet.
   - How can you tell the difference between the microburets?
4. Use good technique to fill it with green dye solution.
5. Hold the mini microburet vertically about 1 cm away from a well and deliver free-falling drops to the well, counting as you go. The final drop will be the one that reaches the rim and forms a slight convex bulge.
6. Repeat the experiment.
   - Report the number of drops.
7. Carry out the same experiment, except this time hold the mini microburet at 45°.
   - Report the number of drops.

After watching several thousand students do these experiments and after doing them myself several thousand times, I found that microburets are best held in a *vertical delivery position*. Let us use the vertical delivery as *standard delivery technique.* (If you prefer the other method, that's OK — nevertheless, your delivery technique must be consistent!)

NOTE: Each well of a 1 x 12 well strip holds 0.40 mL of liquid.

- Assuming that the density of green food dye solution is 1.00 g mL^{-1}, calculate the volume and weight of 1 average drop delivered by the standard vertical technique.

If you wish to obtain a much more accurate calibration and you have access to a balance that will weigh to 1 mg or better, then you can do the following.

(a) Zero the balance.

(b) Place a small piece of plastic or a plastic boat onto the balance pan.

NOTE: A pipet bulb cut in two makes a terrific weighing boat!

(c) Use any type of microburet and standard delivery technique to deliver a known number of drops of distilled water (say 20 to 50) to the boat.

(d) Reweigh.

- Calculate the average drop volume.

If you wish to obtain more statistical information — e.g., a standard deviation, etc. — then you can design and perform the appropriate experiments.

NOTE: You should be aware that the surface tension of the solution being delivered by a microburet also determines the drop volume. A microburet drop calibration for an aqueous solution cannot be used for a solution containing another solvent or a solute that changes the surface tension.

Section F. Quantitative Dilution of Solutions

Goals:

(1) To be able to use a standardized microburet technique to quantitatively dilute a standard solution. (2) To calculate the concentration of any diluted solution.

Experimental Steps:

1. Clean a 1 x 12 well strip. Remove water by slapping the strip in a towel and using a cotton swab.
2. Place the strip on the plastic surface with the small end hole on the right side.

   NOTE: If you look closely with your lens you can then see that each well is numbered starting with 1 at the left side.
3. Fill a well of a 24-well tray with blue food dye.

   NOTE: The solution is in a plastic bottle.

   - Note the *concentration* of the food dye on the label.
4. Use good technique to fill a large-drop microburet with blue food dye solution.
5. Hold the microburet vertically about 0.5 to 1 cm above each well and deliver free drops in the following sequence: 10 drops to the first well, 9 drops to the second well, 8 drops to the third, and so on. Save any unused dye in the 24-well tray.
6. Wash the microburet and fill it with distilled water.
7. Use standard technique to deliver 0 drops of water to the first well, 1 drop to the second, 2 drops to the third, and so on, until you have added 10 drops of water to the eleventh well.
8. Stir by gently swirling a microstirrer in each well.

   Touch the stirrer to a clean towel between each well. Be careful not to spill the liquid.

   NOTE: The process you have just completed is called a *serial dilution* of the blue food dye.

   Now let's calculate the concentration of blue food dye in the solution in each well. The solution in the first well is undiluted and has the concentration stated on the label of the original container. In any sample dilution process, the final concentration of the diluted solution can be calculated from

$$C_i \times V_i = C_f \times V_f$$

where

C_i = initial concentration V_i = initial volume

C_f = final concentration V_f = final volume

If the drops are of the same volume, then we can calculate the food dye concentration in the second well by

$$C_i \times 9\text{ drops} = C_f \times 10\text{ drops}$$

because 9 drops of dye at C_i was diluted to a total final volume of 10 drops (9 drops dye + 1 drop water). If C_i is known, then C_f can be calculated.

- Calculate and report the concentration of food dye in each well. Don't forget to report the units.

NOTE: Keep the 1 × 12 well strip with the solution in it for the next two sections. This strip is called strip 1 to distinguish it from the other strips you will be using later in the laboratory.

Section G. Standard Color Solutions and Colorimetry

Goal:

To use a color standard system as a calibration in the determination of an unknown concentration solution.

Experimental Steps:

NOTE: The serial dilution of blue food dye solution carried out in the previous section can now be used as color standards of known concentration.

1. Obtain a sample of the commercial product that contains an unknown concentration of blue food dye and fill a well in the 24-well tray.
   - Make a note of the well number!
2. Clean the large-drop microburet used earlier and use good transfer technique to suck up some of the product.
3. Clean another 1 × 12 well strip and use standard technique to deliver 10 drops of the product to well 7.
4. Gently pick up strip 1 and the strip with the product, put one on top of the other, and view at an angle that allows you to look through the *sides* of both strips simultaneously.

5. Slide them relative to each other and compare the "intensity" of blue in the product with the color standards.

6. Find a match for the product. If the match is in between two wells estimate where in between.

   - Calculate the concentration of food dye in the product and report it in your record.

   NOTE: The type of analysis you have just completed is a simple but powerful and widely used method of chemical analysis called *colorimetry*. Colorimetry is based on the principle that the amount of light absorbed by a sample is proportional to the number of absorbing molecules interposed in the path of the light. In your experiment, the light source is the fluorescent light reflected by a white surface. As the light passes through the solution, red light is absorbed, leaving blue light to enter your eye-brain detection system.

7. Keep strip 1 (the color standards) and clean the strip with the product in it.

Section H. The Factors that Govern the Absorption of Light

Goals:

(1) To investigate the factors that control the absorption of light by homogeneous, transparent, colored solutions. (2) To examine the quantitative laws of the absorption of light by solutions.

Experimental Steps:

1. Place strip 1 (from the previous experiment) on the plastic surface with the first well on the left.

2. Clean a second strip and call it strip 2.

3. Place it on the plastic surface with the first well (which is on the left.)

4. Repeat Steps 3 through 5 of Section F, using the same large-drop microburet.

   - What is the concentration of blue food dye in wells 1 through 11 of strip 2?
   - Which well of strip 1 has the same blue food dye concentration as in strip 2?

5. Place strip 1 above strip 2 and lift them both up together at an angle at which you can see through the *sides* of both strips.

   NOTE: Make sure you have a white background.

6. Compare the "blueness" or "intensity" of blue in the wells.

   - Do any wells match? Did you predict that in Step 4?

7. Look closely again.

   - Can you see the strange variation in color "intensity" across *each* well?

   Until recently, most colorimetric methods of chemical analysis were carried out in just this manner. When you look through the sides of a well, the path length of solution that the light travels through is the diameter of the well (at least, it is for the middle of the well!).

- What is the diameter of a well? Record it.

As discussed earlier, the general law that governs the absorption of light by solutions is called Lambert, Bouger, and Beer's law (commonly abbreviated to Beer's law). This law states that the amount of light absorbed or transmitted by a solution or medium is an exponential function of the *concentration* of absorbing substance present and of the length of the path of light through the sample. Strictly speaking, the law is only obeyed for monochromatic light, not white light. The mathematical form is written

$$-\log T = abc$$

where T is called the transmittance, a is a constant for a particular substance — e.g., blue food dye — b is the path length, and c is the concentration of absorbing substance, — e.g., concentration of blue food dye.

8. Place both strips next to each other and lift them up together at an angle that allows you to look directly through the tops of the wells to the bottom (against a white background).
9. Holding both with strips with both hands, slide them against each other and compare the light absorption of the solutions.
   - What conclusion do you come to about how they match?
   - Is there any strange variation like what you saw in Step 7?
   - What is the path length of light in the solutions in strip 1?
10. Now look simultaneously through the *side* of strip 1 and the *top* of strip 2 and compare the light absorption.
    - What conclusion can you come to?
    - Explain your answer in your notes.

There are many advantages to a colorimetric system in which the light travels into the bottom and out of the top of the sample. This type of light absorption system is called a *variable path length system.* The trays and strips you have been working with are designed to be operated in a variable path length manner.

A different form of Beer's law applies to variable path length light absorption. For wells of a constant cross sectional area,

$$-\log T = a \times \text{moles of absorbing substance in well}$$

11. Keep the two strips. Do not wash them out yet.

Section I. Colorimetric Analysis of a Beverage

Goal: *To design and execute a colorimetric analysis of a beverage.*

Experimental Steps:

You are presented with the following problem. A lime drink is artificially colored with two FD and C food dyes: a blue and a yellow dye. You are provided with a standard solution of yellow food dye and a sample of the drink.

- Design and execute a colorimetric analysis which will allow you to analyze the soft drink.
- Find the ratio of the two dyes present in the beverage.

Section J. Solution Absorption Spectrophotometry

Goal: *To become familiar with an instrumental spectrometer.*

Discussion:

The colorimetry you carried out earlier was based on a comparison, made by eye, of an unknown with a series of solutions of known concentration. Much more information, greater accuracy, and far more sensitivity can be obtained with the use of an instrument called a spectrophotometer. In this instrument white light from a suitable source is separated into a spectrum by means of a diffraction grating (just as was done in the spectroscope). A narrow wavelength range of the spectrum is selected and allowed to pass through a sample solution contained in an optically clear well, cell, or tube. The amount of light absorbed by the sample is measured by a sensitive solid state device that converts photons into an electric current. The light absorption is read from a logarithmic scale (or stored in a computer) on the instrument. Lambert, Bouger, and Beer's law is generally obeyed, and the light absorption is recorded in absorbance units where

$$-\log T = A = abc$$

In this equation, A is called the absorbance of the solution, and the other terms are as they were defined earlier. Your instructor will introduce you to the instrument and discuss its use in chemical analysis.

Chapter 4

The Use and Abuse of Aluminum and its Compounds

Introduction

What do solid rocket fuel, rubies and sapphires, most RVs, antiperspirants, London buses, fireworks, and 50% of writing paper have in common? All of these products contain aluminum metal or aluminum compounds of various kinds. Aluminum is the most abundant metallic element and the third most abundant element in the earth's crust. The metal itself has not been found in nature because it is very reactive. The major naturally occurring compounds are oxides, e.g., bauxite ($Al_2O_3 \cdot 2H_2O$) and silicates, e.g., many clays.

The basic process for the manufacture of aluminum metal was invented independently (in 1886) by two 22-year-olds: Charles Hall, a student at Oberlin College, Ohio, and Paul Héroult, a Frenchman. The modern electrolytic method of production is based on the same process: the reduction of bauxite (dissolved in cryolite) with an electric current. This invention has brought aluminum, in little more than a century, from a chemical curiosity costing over $1000 per pound to the world's second most commonly used metal. The Hall-Héroult process consumes almost 5% of the electricity output of the United States!

There is now a wide variety of aluminum alloys used for a multitude of purposes, from thin foils in the food industry, through every engineering industry, to high technological applications in aeronautics, space exploration, and electronics. Unfortunately, metallic aluminum is rather expensive to make, primarily for two reasons. Most of the high-grade bauxite deposits occur outside of the United States, and the Hall-Héroult process is extremely energy intensive. Changes in the international situation and the depletion of these high-grade deposits are forcing the aluminum industry to use lower grade ores, with correspondingly higher prices (in spite of much research into new methods).

At the same time as these economic pressures are being felt, there is also considerable environmental concern about the broadcast of aluminum beverage cans in the environment. In an ironic twist, the properties of aluminum that are responsible for its widespread use in modern society are also those that are cause for concern in the environment. Aluminum does not corrode in the same way as iron and steel. The aluminum surface reacts rapidly with oxygen in the air to form a tenacious, thin film that effectively stops further corrosion. The discarded aluminum can has become almost immortal! It is estimated that the can has an average "lifetime" in the environment of more than 100 years — unless you pick it up and recycle it!

The recycling of solid wastes, particularly metals, is a societal issue of great importance. In our so-called technologically advanced society, we still try to dispose of the ever-increasing mountains of solid waste by the most primitive of methods: burying or burning. Landfill disposal is rapidly becoming expensive, difficult, and dangerous. Urban areas with high population densities are running out of suitable sites for garbage dumps, and old sites are causing severe local and groundwater pollution. Municipal incineration of garbage, which typically includes a mixture of plastics, paper products, metals, garden wastes, etc., is expensive and causes air pollution due to emission of a wide variety of toxic combustion products including NO_xs and SO_xs. Conservation of the valuable resources that are currently being squandered, together with recycling programs to reduce energy and material costs, are the *only* solutions to these solid waste problems.

Interestingly, it is in the area of recycling of metals, in particular aluminum, where the most progress has been made. Recycling of large-scale aluminum structural materials has been going on for a long time, and over the last ten years, a number of recycling programs for aluminum cans have proved to be successful. The problem of recycling cans that finish up in municipal garbage is still not solved, however. Only in the arena of industrial scrap aluminum, such as construction materials, do we see recycling. The aluminum is usually shredded, melted down, cast, and then made into another aluminum product.

One of the most widely used group of aluminum compounds is the alums. A true alum is a double salt combination of aluminum sulfate and a group IA or ammonium sulfate, e.g., $KAl(SO_4)_2 \cdot 12H_2O$. The pulp and paper industry alone consumes more than 50% of the one million tons of alum produced annually in the United States. In order to make writing paper, etc., the open spaces in the cellulose fibers must be "filled" with substances that stick strongly to the fibers and stop ink from "bleeding." This is accomplished by adding clay and pine or other rosins (size) to the wet pulp. To fix the rosin strongly to the cellulose fibers, "papermakers" alum is added. The positive aluminum ion "neutralizes" the negative charge on the rosin and allows the rosin to chemically bind to the fibers. The slurry of treated pulp is poured onto screens of a paper-making machine, and the resulting mat is washed, dried, and rolled to produce a smooth sheet of paper like this one.

Unfortunately, this 100-year-old method of sizing paper has been found to have a serious flaw. The aluminum salts present in the paper, together with moisture, produce an acidic reaction that results in a pH of 4.8 or less. The acid catalyzes the breakdown of the cellulose, destroying its strength and suppleness and eventually making the paper very brittle and brown. The extent of the problem is typified by an inventory of the 13.5 million volumes at the Library of Congress. Of these, 3 million are too brittle to handle, and each year about 70,000 more volumes are added to this group. Research has shown that other methods of sizing paper — using calcium carbonate, for example — can produce acid-free paper. The paper industry is moving more and more in this direction, but because of the costs involved, about 50% of paper is still sized with alum in the traditional manner.

Several methods are now being used to save some of the more rare and important books that are disintegrating. Photography can produce excellent master copies, but the cost is high: up to $100 per book. A new large-scale plant that is being built will be able to treat about 9000 volumes every week. The process involves vacuum pumping the books until they are dry and then treating them with diethylzinc ($Zn(C_2H_5)_2$) to neutralize acid and react with moisture. The cost per volume is about $3.

Alum compounds are also used extensively as flocculating agents and phosphate removal agents in water and waste treatment plants. Other uses include soap, greases, fire extinguisher compounds, textiles, drugs, cosmetics, plastics, and pickles (look on your jar)! Aluminum oxides and hydroxides are very commonly manufactured compounds. The oxides are used in ceramics and catalysts, and the pharmaceutical grades of aluminum hydroxide are used in commercial antacid medicines.

Background Chemistry

The average American consumes more than 250 cans of carbonated beverages every year. More than 800 billion cans have been manufactured since 1950. With this kind of market, it is obvious that the beverage container industry is extremely competitive, with an increasing demand for a wide variety of packaging materials. The major container materials are refillable and nonrefillable glass bottles, steel cans, aluminum cans, and plastic bottles. Aluminum cans have about a 25% market share for packaged soft drinks and about a 50% share for the packaged beer market.

The modern all-aluminum can is, in a sense, the evolutionary descendant from the all-steel can, which was first test-marketed in 1936 and which became successful in the early 1960s. Much of the success of the aluminum can package is due to the many technical advances that have been made. One of the first major convenience improvements was an aluminum top for the steel can. Although the new top made the cans lighter and easier to open, the consumer still had to use a punch-type opener. In 1962, ALCOA introduced the easy-open, pull-tab opener for aluminum can ends, which provided for easier gripping and safer openings. The nonremovable opener — such as Reynold's "Stay-on-Tab" and ALCOA's tabless "Easy-Open" end — were developed (1974) as a response to safety and environmental complaints. Other design improvements, which utilized the chemical properties of aluminum, were the drawn-and-ironed two-piece construction and the use of lighter-weight alloys. All these advances have made possible high-speed production (>1000 cans per minute) with maximum use of material (containers weighing less than 26 pounds per thousand).

Many factors play a part in the selection of a particular material for can construction. Manufacturing costs, recyclability, breakage, product purity, and consumer satisfaction are all important. All beverage container materials interact chemically with the solutions contained by them. Steel cans must be electroplated in order to reduce the corrosion that occurs with acidic soft drinks. Chemicals from plastic containers slowly dissolve in the beverage and can cause taste and health problems. Aluminum cans are also susceptible to corrosion by acids and bases and must be coated on the inside walls by a lacquer, resin, or plastic coating. The aluminum alloy (in sheet form) that is used for cans contains about 1 to 2% magnesium, which gives optimum performance at the lowest cost.

Aluminum is a very reactive metal. A clean metal surface reacts instantaneously with dioxygen to produce a thin, transparent, and tough layer of aluminum oxide (Al_2O_3) that protects the metal from further corrosion. This oxide layer is *amphoteric*, which means that it is dissolved by both acids and bases. Bases — e.g., potassium hydroxide (KOH) — dissolve the oxide layer quickly:

$$Al_2O_{3(s)} + 2KOH \rightarrow 2KAlO_2 + H_2O$$

The very soluble potassium aluminate ($KAlO_2$) is thus produced. Once the oxide layer has gone, potassium hydroxide can then directly attack the metal:

$$2Al_{(s)} + 2KOH + 2H_2O \rightarrow 2KAlO_2 + 3H_{2(g)}$$

where the subscripts (s) and (g) mean solid and gas, respectively. The products of this reaction are soluble potassium aluminate and dihydrogen gas. It is interesting to note that this is the reason why basic (alkaline) products like detergents, cleaners, and drain openers are never stored in an aluminum container. The aluminum container would slowly disappear!

Once the aluminum has been dissolved and is in solution, a variety of very useful products can be produced by carrying out further chemical reactions. In this experimental series, an alum, potassium aluminum sulfate dodecahydrate, will be synthesized. The addition of sulfuric acid (H_2SO_4) will cause two sequential chemical reactions to occur. Initially, before the addition of all the acid, the potassium aluminate is neutralized by the acid to give a thick gelatinous precipitate of aluminum hydroxide, ($Al(OH)_{3(s)}$):

$$2H_2O + 2KAlO_2 + H_2SO_4 \rightarrow 2Al(OH)_{3(s)} + K_2SO_4$$

As more sulfuric acid is added, the precipitate of aluminum hydroxide dissolves:

$$2Al(OH)_{3(s)} + 3H_2SO_4 \rightarrow Al_2(SO_4)_3 + 6H_2O$$

Both of these reactions are very *exothermic* (give off heat). The solution now contains dissolved aluminum sulfate ($Al_2(SO_4)_3$), potassium sulfate (K_2SO_4), and a slight excess of sulfuric acid (H_2SO_4). The solution can now be cooled, and potassium aluminum dodecahydrate ($KAl(SO_4)_2 \bullet 12H_2O$) will slowly crystallize out of the solution:

$$Al_2(SO_4)_3 + K_2SO_4 + 24H_2O \rightarrow 2KAl(SO_4)_2 \bullet 12H_2O_{(s)}$$

In the experiment the crystallization process is speeded up by providing a small "seed crystal" of alum for the newly forming crystals to grow on. Cooling is needed because alum crystals are soluble in water at room temperature. *Alum* is a generic name for a variety of aluminum compounds that are combinations of aluminum sulfate and a group IA metal sulfate.

The success of a synthesis, like that of alum, is judged by the yield of the compound of interest and by its purity. In the industrial or laboratory manufacture of chemicals, the quality-control evaluation usually involves both a qualitative and a quantitative analysis. *Qualitative analysis* is finding out what elements or substances are present in the sample, and *quantitative analysis* is the determination of how much of each component is in the sample. The synthesis of almost pharmaceutical grade alum in this laboratory provides you with an opportunity to carry out some simple qualitative analysis tests for the components of potassium aluminum sulfate dodecahydrate.

ADDITIONAL READING

1. King, F., *Aluminum and Its Alloys*, John Wiley and Sons, New York, 1987.
A good basic source of technical facts and figures. The only place where I could find a reference to the alloy used to make beverage cans!

2. Kaplan, S. (ed.), "100 Year History 1882–1982 and Future Probe," *Beverage World* , New York, 1982.
The 100th Anniversary Issue of the trade journal *Beverage World.* A remarkable collection of "inside" information and history on the packaging, marketing, production, distribution, franchising, ingredients, and future of beverages.

3. Abelson, P. H., "Brittle Books and Journals," (editorial), *Science* **238**, 30 October, 1987.
A short and to-the-point editorial.

4. Swanson, J. W., "Internal Sizing of Paper and Paperboard," TAPPI Monograph Series No. 33, New York, 1971.
Where I found out about sizing — fascinating.

Pre-Laboratory Quiz

1. Give two reasons why aluminum metal is expensive to produce.

2. Explain the difference between the corrosion of iron and the corrosion of aluminum.

3. What are the two most common methods of solid waste disposal?

4. Give two uses of alum.

5. How many cans of carbonated beverages does the average American consume in one year?

6. Give the chemical formula for

a) Aluminum oxide ______________________________

b) Potassium hydroxide ______________________________

7. What is the name of an aluminum compound that is used in many antacid medicines?

____
____
____
____

8. What is an alum?

____
____
____
____

9. What does dodecahydrate mean?

____
____
____
____

10. Do you recycle aluminum cans? How?

____
____
____
____

Laboratory Experiments

Flowchart of the Experiments

Section A.	The Aluminum Can

Section B.	Recycling Aluminum: The Synthesis of Alum (Potassium Aluminum Sulfate)

Section C.	Qualitative Analysis of an Alum Sample

Section D.	A Practical Use for the Reaction of Aluminum with a Base

Requires one three-hour class period to complete

CAUTION: Several chemicals used in this experiment are dangerous. Potassium hydroxide, sulfuric acid, and drain cleaner are all corrosive. Eye protection must be worn. If you get any of these chemicals on your hands, etc., wash well with cold water and inform your instructor.

Section A. The Aluminum Can

Goals:

(1) To analyze various factors that are important in the design of that ubiquitous container — the aluminum can. (2) To see whether paint and plastic protect aluminum metal from corrosion by highly concentrated strong acid and base solutions.

Experimental Steps:

1. Obtain an aluminum can. Look at your can, which should have a lot of information printed on the outside. Make a list of information that is perhaps important to can design.

Discussion:

Is the information actually painted (printed) onto the can? What are the ingredients that the can contains? Can you read the bar code? What is the can capacity? Do you think the top, sides, and bottom of the can are the same metal and thickness? Is there a seam? Does the can have to withstand pressure?

Consider some design criteria that might be related to chemistry. Obviously, beverage producers must design a container that has the following (minimum) requirements: (a) The can must maintain the total integrity of the beverage through shipping, storage (shelf life), and on to the consumer; (b) the can should not in any way change the taste, color, odor, or any other characteristic properties of the beverage; and (c) the can must be inexpensive, lightweight (to save on transportation costs), easily accessed, and recyclable.

- If you consider these criteria, what advantages and disadvantages do you think aluminum has as a container material?

NOTE: Many beverages contain acids and other ingredients that slowly dissolve aluminum. In order to stop this internal corrosion, most cans have a thin plastic coating on the inside.

2. Cut out a small section (~ 3 cm x 5 cm) from the side of the can.

CAUTION: Be Careful! The aluminum is very thin and sharp!

3. Straighten the piece out and cut it in two. Place the two pieces onto a microtowel on the table. Use the sharp end of the scissors to scratch some of the paint away from 2 small areas on the outside of one piece and some plastic away from 2 small areas on the inside of the other piece.
4. Place the pieces onto the plastic surface.
5. Go to Reagent Central and use good transfer technique to obtain 1/2 full buret of 1.4 M KOH (potassium hydroxide) and 1/3 full buret of 9 M H_2SO_4 (sulfuric acid).

6. Drop 1 drop of the reagents on each of the scratched surfaces as shown:

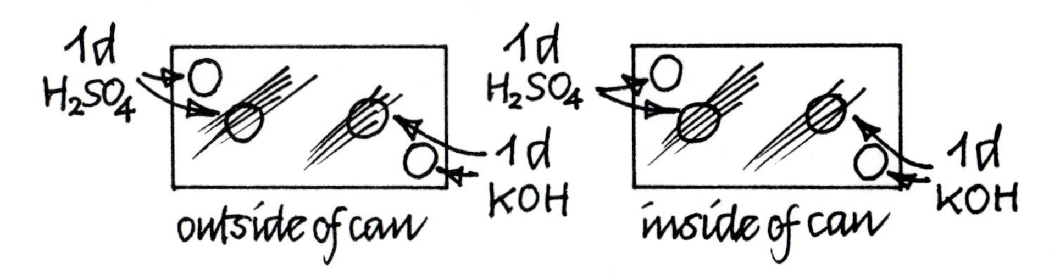

7. Place a Petri dish top over the can pieces to prevent excessive evaporation and examine the surfaces carefully with your hand lens after about 30 minutes. Record what you see.
   - Why are carbonated beverages generally quite acidic? Give the chemical reactions.
   - What is the acid that gives most cola drinks their characteristic taste?
   - What other products besides beverages are contained in aluminum cans?
   - How could you tell the difference between an aluminum can and a steel can?

Section B. Recycling Aluminum: The Synthesis of Alum (Potassium Aluminum Sulfate)

Goal: *To make an excellent grade of alum (potassium aluminum sulfate, $KAl(SO_4)_2 \cdot 12H_2O$) from scrap aluminum foil.*

Discussion: You will find that aluminum metal is quite reactive and can be used as starting material for a wide variety of commercially important chemicals.

Experimental Steps:

1. Cut out a piece of regular aluminum foil (~ 1" × 2"), and if you have access to an analytical balance, weigh it accurately. If not, check with your instructor for the average weight of a 1" × 2" piece.
2. Cut the foil into small strips and put all the strips into a small plastic cup.
3. Fill a styrofoam coffee cup with hot water from the tap to about 2 cm from the brim.
4. Transfer about 1/2 of a microburet of 1.4 M KOH to the aluminum in the small cup.
5. Place the small cup into the hot water in the styrofoam cup. Swirl to wet the metal strips.
   - Record what you see and interpret it by means of a chemical reaction.
6. Swirl gently every few minutes until all of the aluminum dissolves.
7. While the dissolution is going on, prepare an apparatus for microfiltration.

(a) Construct an all-purpose clamp by cutting twice (about 8 cm) down a straw, peeling one half back and cutting it off, as shown below.

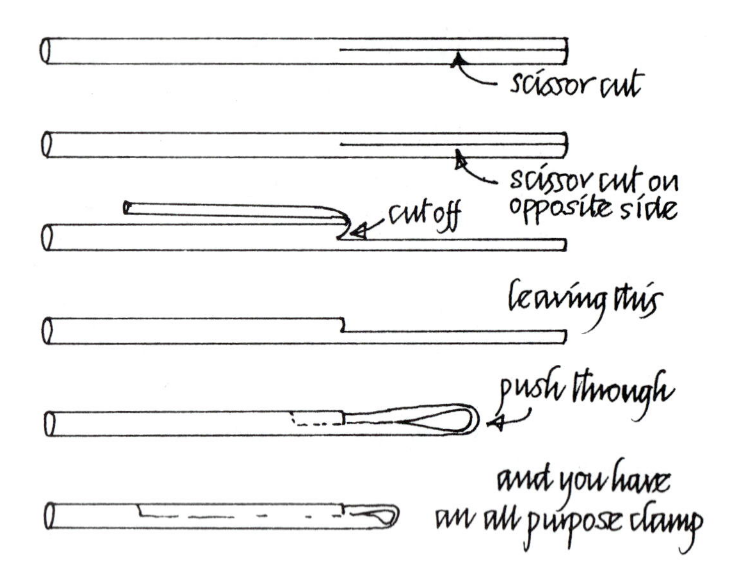

(b) Now make a stand by punching a hole about 5 cm from the end of another straw.

(c) Place the stand in a microreaction tray (round profile wells).

(d) Cut about 8 cm off the end of the stand to make the filtration column.

(e) Push a filter paper circle (already punched out for you) down the filtration column with a slim straw.

(f) Insert a column cap (red plastic cap) onto the end and gently push the paper circle up against the cap.

The microfiltration apparatus is pictured below:

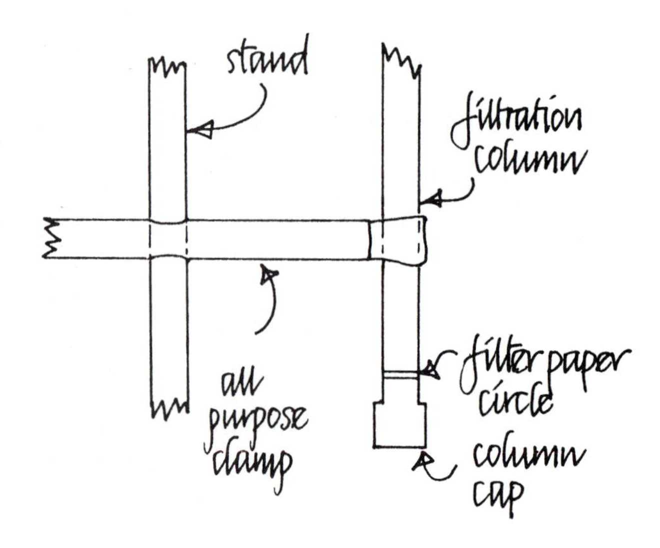

NOTE: Straws can conveniently be threaded through holes and clamps by cutting the end at an angle. This point can always be cut off later if necessary.

8. The aluminum should have dissolved by now. Suck up all the liquid, etc., from the cup with a clean microburet.
9. Wash the cup out with 3 rinses of distilled water from a wash bottle. Wipe dry with a microtowel.
10. Place the clean cup under the filtration column.
11. Transfer the liquid into the filtration column.
12. Filter by gravity filtration until all the liquid has come through. Remove the filtration column and gently shake any remaining liquid into the cup.
13. Remove the cup and liquid and put it aside for a moment.
14. Put the filtration column back into the clamp and place a waste cup under it. Leave the column set up until you can get back to it in about 15 minutes.
15. Fill a microburet about 1/3 full with 9 M H_2SO_4.

    CAUTION: Use care. This solution is very corrosive.

16. Using a straw as a stirrer, add the H_2SO_4 quickly to the liquid in the cup. Stir.
    - Take a moment to record what you saw and write 2 chemical reactions that describe what happened.
    - Why did the mixture get hot?
17. Empty the water from the stryofoam coffee cup and fill it with ice (2/3) and water (1/3) to make an ice bath.

    NOTE: You *must* use 2/3 ice in order to ensure the bath will be cold enough for the length of time required. If the bath is not cold enough, the crystals *will* dissolve and you will lose your product. Hence, the term *ice* bath.

18. Place the small cup with liquid into the ice bath to cool it. Cool for at least 15–20 minutes.
    - While it's cooling, bring your record up to date.
19. Back to the ice bath. A good crop of crystals should now be forming. If crystals are not forming, add a seed crystal or two to start the crystallization.
20. After 20 minutes of cooling, remove the cup from the ice bath.
21. Now you have the difficult problems of removing the liquid (which is very acidic), washing the crystals, and drying them. Steps 22 through 29 describe one method.

22. Tilt the cup gently so that the liquid goes slightly away from the crystals.
23. Suck the liquid up with a clean microburet. The key here is gentle skill! Remove as much liquid in this way as you can. Discard the liquid to a waste cup. Wash the microburet.
24. With the clean microburet, suck up at least 1/2 a buret of ice cold water from the ice bath and add to the crystals.
25. Swirl the cup gently and remove the liquid as before (Step 23).
26. Use a straw to dislodge the crystals of potassium aluminum sulfate, forming a flat pile on the plastic surface.
27. Fold a microtowel so that it is 4 layers thick and place it on the crystals.

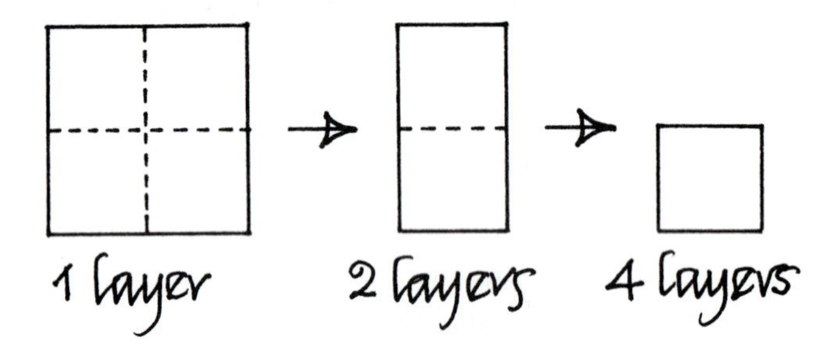

28. Place the small cup, or any other object with a flat bottom, onto the folded towel and press to make the towel absorb any remaining liquid.

    CAUTION: You will probably want to use your hand to press against the crystals — but this is not a good idea since the product is very acidic.
29. Remove the towel and dislodge any remaining crystals back into the pile.
30. Look at the crystals through your hand lens.
    - Can you see any definite shape to them?
    - If you can, draw a picture of it in your notebook.
31. If you have access to an analytical balance, transfer the crystals to a preweighed, clean weighing boat and weigh them. Save them because you are going to use them in other experiments. If you did weigh them, continue through Step 28.

    The overall equation for the synthesis of alum is

$$2Al_{(s)} + 2KOH + 4H_2SO_4 + 22H_2O \rightarrow 2KAl(SO_4)_2{\cdot}12H_2O + 3H_2$$

32. Calculate and record the theoretical yield and the % experimental yield.

$$\%\ \text{experimental yield} = \frac{\text{grams of alum obtained in experiment}}{\text{grams of alum theoretically produced}} \times 100$$

The atomic weights of K, Al, S, O, and H can be found in Appendix 2.

Section C. Qualitative Analysis of an Alum Sample

Goals:

(1) To carry out a qualitative analysis of the potassium aluminum sulfate ($KAl(SO_4)_2{\cdot}12H_2O$) that you synthesized in Section B. (2) To confirm the presence of SO_4^{2-}, K^+, and Al^{3+}.

NOTE: Use your plastic sheet as a reaction surface and do this experiment against a *black* background.

Experimental Steps:

1. Use a straw spatula to transfer a few crystals (about 5 mg) to the plastic. Add 3 drops of water to the crystals. Stir gently until the crystals dissolve.
2. Use a small piece of indicator paper to see whether the solution is acidic.
   - Interpret the result of the last step (see the Background Chemistry section) and record.
3. Now add 1 drop of 0.5 M $BaCl_2$ (barium chloride) to the solution.

   The change is the result of the following chemical reaction:

$$KAl(SO_4)_2 + 2BaCl_2 \xrightarrow{H_2O} 2BaSO_4\downarrow + KAlCl_4$$

   One of the products of this reaction ($BaSO_4$) is not soluble in water and precipitates as a solid. This is a good test for the sulfate ion SO_4^{2-}.

4. A really good test for potassium is a flame test. Push a stainless steel pin into a small cork. Hold the pin head in the flame of a Bunsen burner (or a Thompson burner) to volatilize impurities from your fingers.
   - What kind of impurities do you have on your fingers?
5. This step is tricky: Get the pin head red hot, remove it, and quickly touch it to a small cluster of crystals. Several should stick. Don't melt any plastic!
6. Slowly bring the pin (plus crystals) toward the flame and watch carefully. Finally, hold the crystals in the flame for at least 5 seconds (until the solid glows).
7. Remove and hold the cork until the pin cools.
8. Place the pin, with adhering solid, onto the plastic sheet (against a black background).
   - Record what colors you saw in the flame as the crystals on the pin were heated by the flame.

   NOTE: The flame color from the hot solid is characteristic of the group IA element, potassium. You may want to refer to the chemistry of flame colors discussed in the chapter on spectroscopy (Chapter 2).

   At the high temperature of the flame (about 1000 °C), potassium is volatilized and gives a blueish color in the flame. After a few seconds sulfur dioxide is driven off, and the remaining oxides start to glow.

10. Add 2 drops of water to the solid adhering to the pin and stir.
11. Now add 1 drop of 0.5 M $BaCl_2$.

- Why is the result of the reaction with $BaCl_2$ in Step 10 different from the reaction in Step 3?
- What experimental observation would be evidence that the crystals contain water in the structure?

12. Repeat Step 1.
13. Dilute 1 drop of 1.4 M KOH tenfold using the methods developed in the Small-Scale Technique laboratory (Chapter 3).
14. While stirring, add 1 drop of 0.14 M KOH to the dissolved alum.

    NOTE: The wispy gelatinous precipitate is aluminum hydroxide, which is used extensively as an antacid in various pharmaceuticals.
15. Add 1 drop of 1.4 M KOH to the precipitate. Stir.
    - Give a chemical reaction for what happens.

Section D. A Practical Use for the Reaction of Aluminum with a Base

Goal: *To examine the composition and investigate the chemical action of a commercially available drain-opening product. NOTE: This experiment is optional. Check with your instructor.*

Discussion: Most of the products sold commercially as drain openers contain strong bases that can react with and dissolve away the combination of insoluble grease, soap, and hair that causes blockage of sink, bath, and shower drains. One of these products, Crystal Drano®, is very interesting because it is formulated with aluminum metal, solid sodium hydroxide, bleach, and an iron salt.

CAUTION: Drano is very caustic and corrosive, and extreme care must be used when handling it — *never* allow it to touch the skin.

Experimental Steps:

1. Use a *dry* clean straw spatula to obtain a representative sample of Crystal Drano®. Place the sample on your plastic surface.
2. Carefully observe the sample with your hand lens.
    - How many individual morphological forms (shapes) can you see?
    - Make a rough drawing.
    - Do you see anything changing with time?
3. Add a few drops of water from a microburet.
    - Make careful observations for the next few minutes.
    - Give some reasons why you think that this combination of ingredients is particularly effective in removing blockages.
    - Explain the color changes.
4. Read the side directions-for-use panel on the can of Crystal Drano®. The directions caution to remove any standing water and to make sure water in drain is cool.
    - Why would following the above directions be important when using this product?
    - From what type of material do you think the Drano® container is made?

Chapter
5

Instruments:

What They Do and What They Don't

Introduction

"Instrument, *n.* [L. *instrumentum,* a tool or tools, implement, stock in trade, furniture, dress, from *instruere,* to furnish, equip; *in,* in, and *struere,* to pile up, arrange.]
1. (a) a thing by means of which something is done; means; (b) a person used by another to bring something about.
2. a tool or implement, especially one used for delicate work or for scientific or artistic purposes."

Webster's New Universal Unabridged Dictionary

Much of the research carried out in modern science is done by means of instruments. This is as true in chemistry as it is in high-energy physics. Many of these instruments are expensive, sophisticated, extremely accurate, and require a Ph.D. (and a team of dedicated technicians) to operate. The advent of microelectronics, lasers, and computers has revolutionized the design, application, and accuracy of instruments of all types.

Unfortunately, many of these developments have tended to obscure the link between the scientific principles of the experiment and the output from the instrument. For the well-experienced scientist or engineer, this is generally not a problem, however; for a student who is just getting started in science, the instrument can become an intimidating "black box," spewing forth a mass of meaningless output signals. In an attempt to circumvent some of these problems, we have adopted a different approach to the use of instrumentation in your chemistry course. The general idea is to design, and wherever possible allow you to construct, simple and small working versions of the expensive, commercially available machines. You will find that these simple versions are in the original research papers in the literature. It is perhaps important at this point to note that simple and small does not mean trivial. If you build your own scientific instrument,

- You know how it works.
- You know when it isn't working.
- You can fix it when it breaks.
- You can play with it however you please.
- You know what it can do and what it can't.
- You don't need a Ph.D. to do it!

Eventually, as you progress through the scientific world, you will surely be introduced to the sophisticated versions. When this happens, remember that you really know how "it" works. Read the instructions manual carefully so that you can find out where the buttons are.

The basic principle of most scientific instruments is that there is a sensor of some type that can probe the system under investigation and that is connected to a measuring device of some type. A simple, classical example is the instrument that has been used for many centuries to measure the amount of "hotness" or "coldness" of a system — i.e., the thermometer. The sensor in this instrument is usually a liquid like mercury or alcohol that expands or contracts as it is heated or cooled by the environment. The key to the thermometer being a meter is, of course, the choice of some type of measuring scale to be attached to it. Unfortunately, throughout the more than three hundred years of development, many different scales have been attached to many different designs of the thermometer. Even today there are at least four or five scales commonly used throughout the world. It is proving an extraordinarily difficult task to choose one as a standardized system (hence, the temperature conversion questions on your freshman chemistry test!). It is worth emphasizing that, even if one particular scale were chosen, the choice would be arbitrary. Of course, every thermometer that is manufactured (no matter what the scale) must somehow be calibrated against some better thermometer, and so on. In this series of experiments, you will have the opportunity to investigate several very different instruments that are used to measure that strange property called temperature.

Background Chemistry

All matter has electrical characteristics. Atoms are composed of negatively charged electrons and positively charged protons. Electrons are responsible for the chemical properties of solids that form the basis for modern electronics. Solids can be divided into three types: conductors, semiconductors, and insulators, depending on how well the material conducts electricity. Metals are excellent *conductors* of electricity because there are many free electrons that can be easily pushed through the macroscopic structure. Most nonmetallic elements and compounds contain strongly bound electrons that are fixed in bonding positions close to the original atoms and are not free to move. These substances do not conduct electricity and are called *insulators*. *Semiconductor* substances have an intermediate electrical conductivity that can be controlled to any desired level during the manufacture of the substance. Combinations of all of these types of materials in extraordinarily thin microlayers and regions can be arranged in carefully designed circuits to control the flow of charge in microelectronic devices.

The flow of charge through a substance is called an *electric current*. Current may flow in only one direction — in which case it is called *direct current* (DC) — or in both directions, which is termed *alternating current* (AC). The fundamental law that quantitatively describes the flow of direct current is called Ohm's law,

$$V = IR$$

where V is the voltage, I is the current, and R is the resistance. *Voltage* (V) (sometimes called *potential*) is the electrical pressure or force that pushes charge through the conductor. A good analogy to describe current is to think of it as water flowing through a pipe. The voltage is then the water pressure that is pushing the water through the pipe. The *current* (I) is the quantity of electrons passing a point in one second and is measured in ampere units. One ampere is 6.25×10^{18} electrons passing a point in one second. All conductors tend to resist the flow of charge to some extent. A conductor has a resistance (R) of 1 ohm (Ω) if a voltage of 1 volt (V) will force a current of 1 ampere (A) through it. Another useful term is that of power (P), which is the work performed by all electrical current measured in watt (W) units. The power of a direct current is voltage multiplied by current:

$$P = VI$$

An *electronic circuit* is any arrangement of electrical materials that will allow a current to flow. The most basic circuit for direct current flow has a power source (e.g., a battery), a resistor, and conducting connectors, and is represented in the circuit diagram of Figure 5.1. Resistors play a very important role in the design of circuits for scientific instruments because they limit the flow of current and because they exhibit a temperature dependence. This latter property can be extremely useful in designing

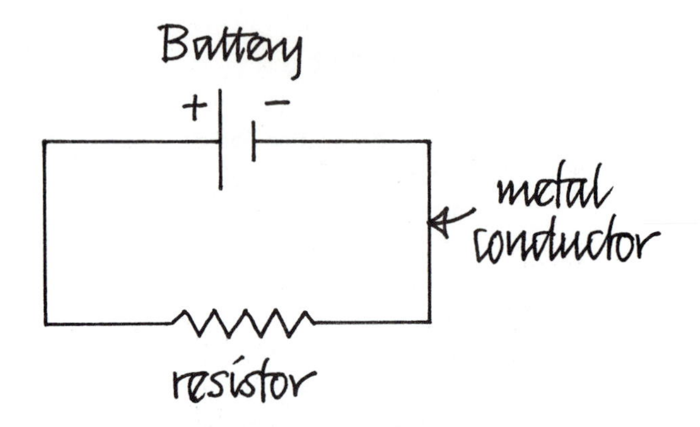

Figure 5.1 A Basic DC Circuit

sensitive temperature sensors. Resistors come in all shapes, sizes, and types, several of which you will be using in this series of experiments. One very common type is the *carbon film resistor*, which is made by depositing a carbon film on a small ceramic cylinder. The length of the carbon film between the leads controls the resistance of this device. Resistors, such as the one pictured in Figure 5.2, are often color coded to indicate the value of resistance, as shown in Table 5.1. The color bands numbered in Figure 5.2 correspond to the band numbers in Table 5.1.

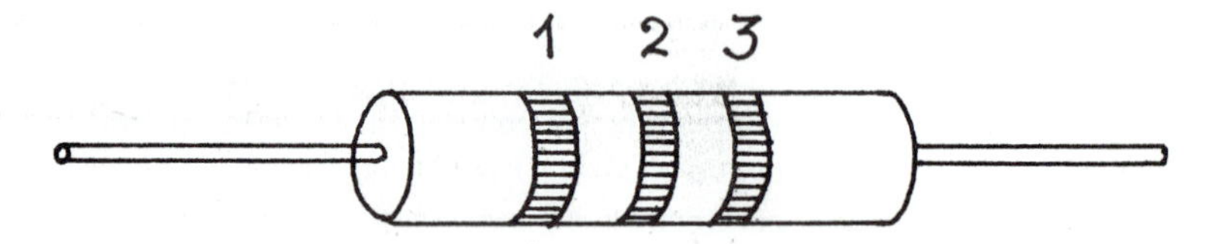

Figure 5.2 Resistor with Color Bands

Band Color	Band 1	Band 2	Band 3 (Multiplier)
Black	0	0	1
Brown	1	1	10^1
Red	2	2	10^2
Orange	3	3	10^3
Yellow	4	4	10^4
Green	5	5	10^5
Blue	6	6	10^6
Violet	7	7	10^7
Grey	8	8	10^8
White	9	9	none

Table 5.1 Resistance Values (Ω) from Color Bands

Sometimes there is a fourth band that gives the accuracy of the resistance value — e.g., gold is ± 5%, silver ± 10%, and none is ± 20%.

You will also be using a relatively uncommon resistor that is a semiconductor silicon temperature sensor. This sensor, shown in Figure 5.3, is a microdevice manufactured from n-type silicon with a doping level of about 10^{15} cm^{-3}.

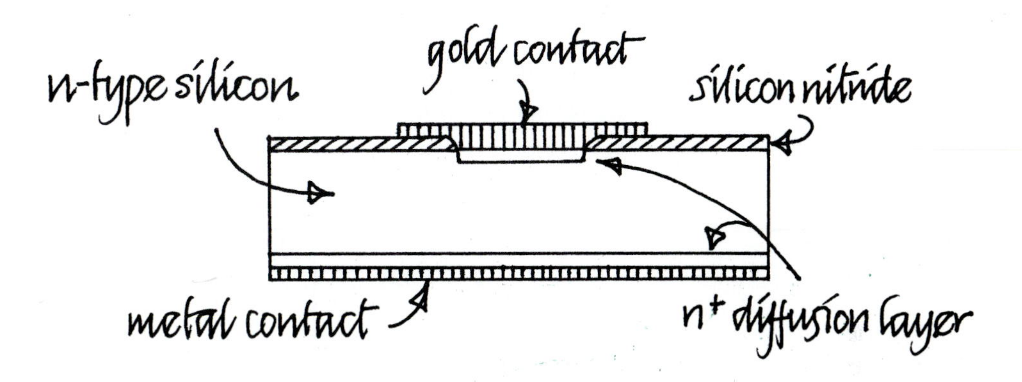

Figure 5.3 Semiconductor Temperature Sensor

Between about –50 °C and 150 °C, this device has a positive temperature coefficient of resistance due to a fall in the charge carrier mobility with rising temperature. Solid state electronic temperature sensors of this type are rapidly displacing the traditional mercury-in-glass thermometers in many scientific instruments.

Quantitative measurement of voltage, current, and resistance is most conveniently made with an instrument called a *digital multimeter*. You will be using this basic electronic instrument in conjunction with a number of sensors or probes that can be used to monitor physical and chemical properties of a system during an experiment. One of the key features of these modern multimeters is that they have a huge (10 MΩ) built-in resistance, which means that the meter requires only a very small flow of charge to make a measurement. A measurement can therefore be made *without* perturbing the chemistry of the experiment. Digital multimeters are themselves rather sophisticated examples of microelectronic circuitry; it is not necessary for you to understand the intricacies of this instrument. However, if you wish to examine the insides or obtain more information about how the meter works, check with your instructor who will organize it for you.

One of the main objectives of this series of experiments is to allow you to explore the nature of applications and limitations of radically different types of instrumentation for the measurement of temperature. One of the first instruments for sensing different degrees of hot and cold was called an air thermoscope. The device was simply a bulb (containing air) that was attached to a long, thin stem. The stem was partially filled with liquid and was dipped into liquid in a container. Expansion or contraction of the air in the bulb due to temperature changes led to a rise or fall in the liquid level in the stem of the thermoscope. Eventually, it was realized that by marking the stem with a graduated scale and by adopting uniform methods of construction, the thermoscope could be a quantitative instrument of temperature measurement. The basis for the gas thermometer described above is the principle that the volume of a gas is directly proportional to the temperature of the gas at constant pressure and amount of gas. This principle is known as Charles's law:

$$V = \text{constant} \times T$$

For 1 mole of an ideal gas at 1 atmosphere, the plot of volume versus temperature is shown in Figure 5.4.

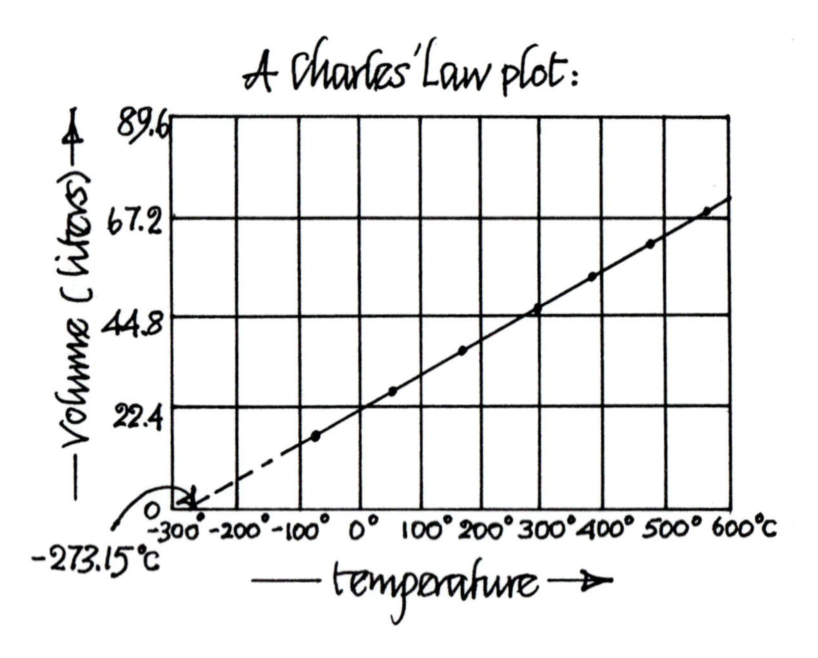

Figure 5.4 A Graph of Charles' Law

Note that the volume of the gas becomes zero at −273.15 °C. This temperature is called *absolute zero,* the lowest possible temperature. The British physicist Lord Kelvin realized that it would be very useful to define a temperature scale that has absolute zero as the zero point. The Kelvin temperature scale is related to the Celsius scale in that zero K is − 273.15 °C. You will have the opportunity to construct and calibrate a rather novel polyethylene gas thermometer in this sequence of experiments.

ADDITIONAL READING

1. Mims, F.M., III., *Getting Started in Electronics,* Radio Shack®, 1987.
   Written for, and obtainable at, Radio Shack® , it is without doubt the best introduction written for the beginner in electronics. Clear, interesting, and funny.
2. Middleton, W.E.K., *A History of the Thermometer and Its Use in Meterology,* The John Hopkins Press, Baltimore, Md., 1966.
3. *Silicon Temperature Sensors,* Amperex Electronic Corporation, Technical Sheet, 1988.
4. *Instruction Manual for Fluke 75 and 77 Multimeter,* John Fluke Mfg. Co., Inc., Everett, Wash., 1983.

Laboratory Experiments

Flowchart of the Experiments

Section A. Basic Electronics

Section B. The Operation of a Digital/Analog Multimeter as a Measurement Instrument

Section C. A Semiconductor Silicon Temperature Sensor System

Section D. The Construction and Calibration of a Gas Thermometer

Section E. An Extended Range Gas Thermometer

Section F. A Novel Pressure Gauge Balance

Requires one three-hour class period to complete

Section A. Basic Electronics

Goal:

To use Ohm's law to calculate current, resistance, and voltage in simple electronic circuits.

Discussion:

First, an important word about the direction of an electrical current. If you look at a battery — the 9 V battery in front of you — you will see that one terminal is marked positive (+) and the other, negative (–). Engineers and solid state designers generally say that *positive current* flows from the positive to the negative terminal. This is the traditional convention that was developed by Benjamin Franklin. It is now known that electrical current is the flow of electrons through a conductor or semiconductor. Since electrons are negatively charged, they move from a negatively charged region to a positively charged region. Throughout these and subsequent experiments, "current flow" means "electron flow."

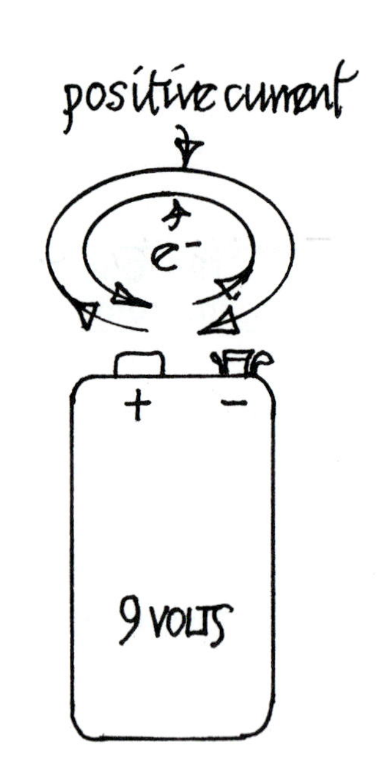

- An interesting question for you. You need to jump start your car or truck. You are given a set of jumper cables and you've found a second vehicle with a good battery. The two batteries are pictured on the right. How do you connect them in order to start your car rather than destroying both batteries?

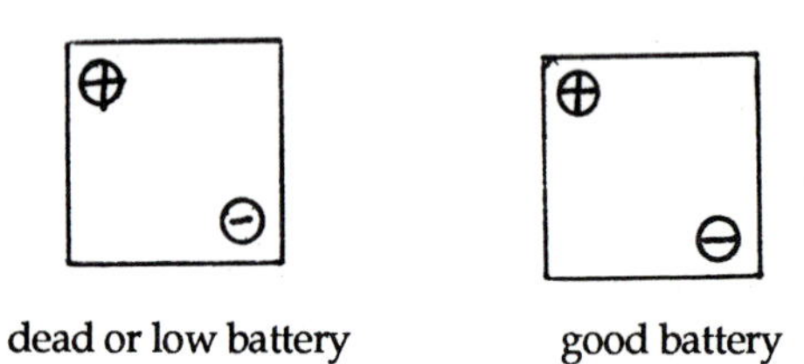

dead or low battery   good battery

- Which way will electrons flow in the following circuit?

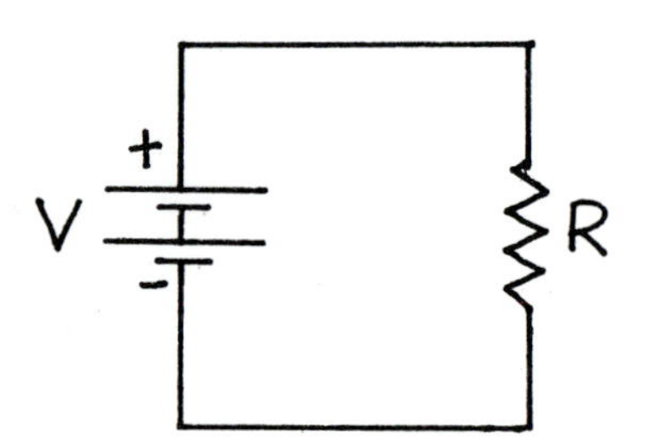

- Draw these circuits and write the answers in your laboratory record.
- For the above circuit calculate the following:

  a) What is the resistance if the voltage is 9.26 V and the current is 0.5 A?

  b) How many mA of current will flow if the resistance is 10 kΩ and the voltage 1.5 V?

  c) What will happen to the direction of electron flow if the resistor is reversed?

  d) Is the above circuit a DC or an AC circuit?

Section B. The Operation of a Digital/Analog Multimeter as a Measurement Instrument

Goal: *To operate a digital multimeter (DMM) as a measurement instrument in electronic circuits.*

Discussion: There are a large number of digital/analog multimeters (DMM) on the market. For this series of experiments and all the others in this book the instrumental requirements are very simple. The meter should have a high internal impedance, be rugged and portable, and have suitable range selection controls. You will be working with an excellent instrument — the Fluke 75 multimeter. The face of the meter is shown below.

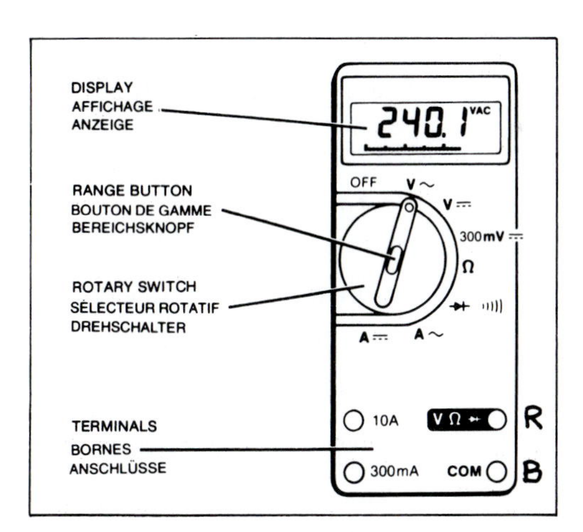

Experimental Steps:

1. Turn the instrument on by rotating the switch from the off position to the function you wish to measure.

   The meter will display all segments, do a self-test, chirp, and is then ready to use. Function symbols are:

   V~ voltage, alternating

V ⎓	voltage, direct
300 mV ⎓	voltage, in mV, for below 300 mV
Ω	resistance
⟶|⟶	diode test and continuity beeper
A ~	current, alternating
A ⎓	current, direct

Voltage Measurements

1. Measure the *voltage* as follows (refer to the illustration below.)
2. Connect the test leads as shown. *R* represents *red lead*, *B* represents *black lead.*

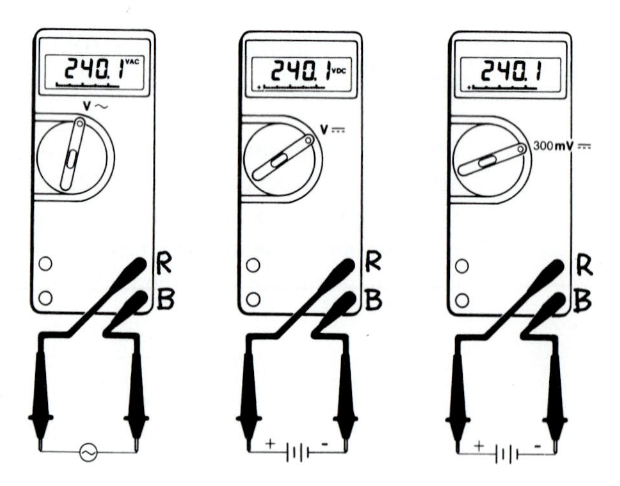

3. Measure the voltage of the battery that is assigned to you, using needle probes.

   NOTE: The red position (and lead) is the positive terminal of the DMM and the black position is the negative terminal of the DMM.

   This means that electrons are coming out of the black lead.

   - Note what the meter reads and which lead is on which battery terminal.
4. Switch the leads and note any change in reading.
5. Look at the analog scale that is just below the digital reading and relate the analog reading to the digital reading.
6. Press the range button that is in the center of the rotary switch.
7. Remeasure the battery voltage.
8. Press the range button again.

- Explain what happens when you press the range button.

NOTE: Autorange is automatically the most sensitive range for all measurements. To get back to autogrange, either switch the meter off and then switch it back on, or press the range button for 1 second; the meter will chirp and go to autorange.

Resistance Measurements

Resistance measurement is made by switching the rotary switch to the Ω symbol. The units of *resistance,* indicated by Ω, kΩ, or MΩ, are shown in the display.

9. Obtain a carbon filament resistor from Reagent Central and connect the test leads as shown (refer to illustration below.)

Remember, *R* represents *red lead, B* represents *black lead.*

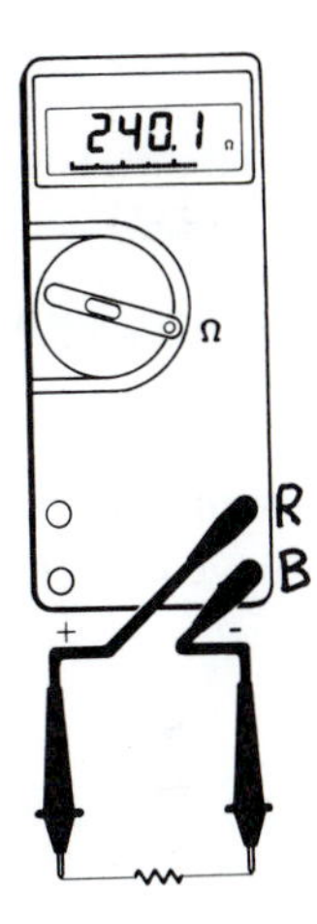

10. From your knowledge of resistor color coding, find the resistance of the resistor.
    - Record the resistance value in correct units.
11. Connect the DMM to the resistor and measure the resistance using the meter in autorange.
    - Is the resistor within tolerance?
    - Do you think that the resistance of a resistor is temperature dependent?

NOTE: If you have a high-value resistor say, greater than 50 kΩ, you might try holding it between your fingers to raise the temperature while measuring the resistance.

Making a Resistor

NOTE: One simple way of making a resistor is by drawing a line with a soft lead (2B) pencil.

12. In your notebook draw a line about 12 cm long with a pencil and ruler.

13. Draw over the line about 10 times, pressing firmly and evenly.

    You have now probably laid down enough carbon to have a planar resistor.

14. Mark the line off in 1 cm intervals.

15. Set up the meter for resistance measurement (on autorange). Press one needle probe firmly at one end of the line and the other probe exactly 1 cm away.

    - Record the resistance.

16. Repeat the measurement at 1 cm intervals up to about 9 cm.

    NOTE: It is important to keep the initial probe position the same each time.

    - Make a graph of resistance versus distance.
    - What does this tell you about resistors which are placed *in series* with each other?
    - Draw a circuit diagram in which each cm of pencil line is represented by a resistor symbol (⩘).
    - What factors play a part in controlling the resistance of the line?

Current Measurement

NOTE: The measurement of current is carried out with the meter in series and with the red lead plugged into the current input.

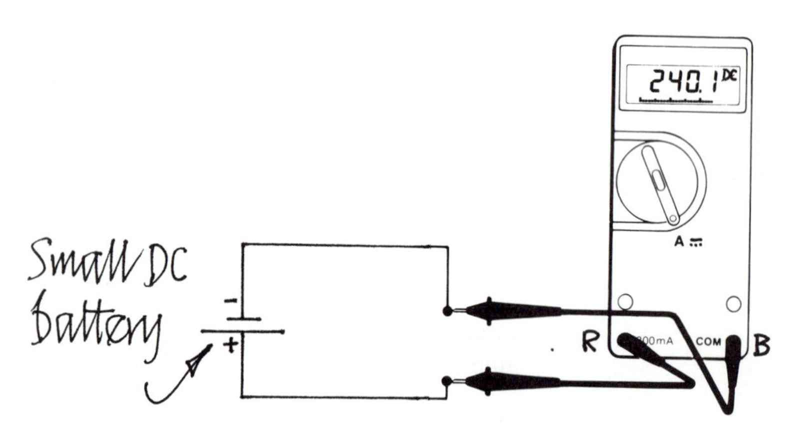

17. Use the two short leads with clips to set up a circuit containing the resistor, battery, and meter and measure the current in the circuit.

    - Now use Ohm's law to calculate the current.
    - The measured and calculated current do not agree! Why?

You should now be reasonably familiar with the use of a digital/analog multimeter in making various measurements.

Section C. A Semiconductor Silicon Temperature Sensor System

Goals: *(1) To explore some of the sensor characteristics of a temperature sensitive semiconductor device. (2) To compare it with a classical liquid-in-glass thermometer. (3) To plot appropriate calibration curves from manufacturer supplied data in order to be able to convert resistance measurements into temperature.*

Experimental Steps:

1. Obtain a silicon temperature sensor from your instructor.

   NOTE: Please be careful with the device. The semiconductor junction is packaged in glass and is somewhat fragile.

   The sensor dimensions are:

Package material: glass

The sensor must operate in the forward biased condition for proper performance.

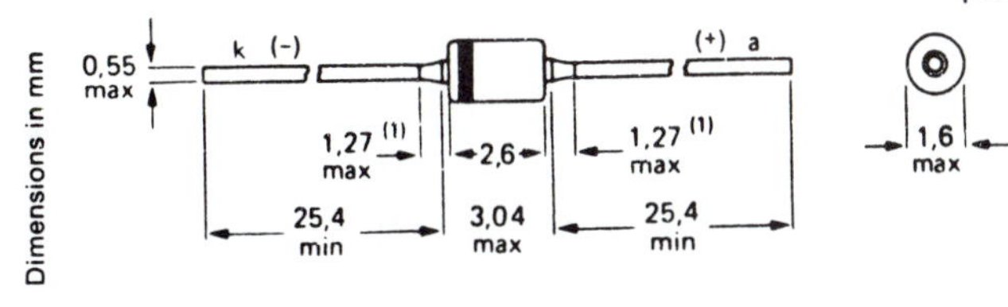

mm	inch
.55	.022
1.27	.050
1.6	.063
2.6	.102
3.04	.120
25.4	1.000

(1) Lead diameter in this zone uncontrolled.

The sensor has already been bent into a U-configuration for ease of use.

NOTE: The film-stripped pictures below illustrate steps 2 through 4.

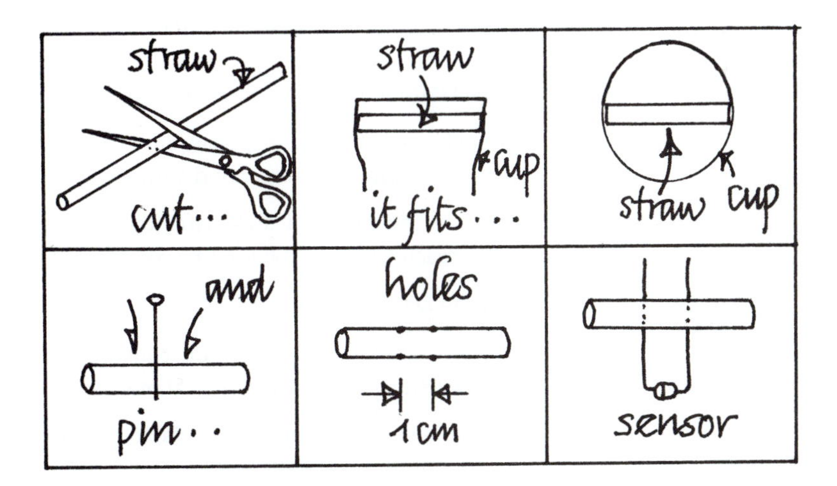

2. Make a sensor support by cutting a straw so that the straw will fit snugly by friction into a styrofoam cup.

3. Place the straw piece on your notebook and pierce holes about 1 cm apart through the straw.

4. Now *gently* push the sensor through the straw holes. The sensor should stay in the support by friction.

5. Place the straw into one of the wells of a tray.

6. Connect the device in the forward-biased mode. To do this clip a lead to the sensor cathode wire (the wire on the black ring side of sensor — the k, or negative, side) (see the sensor diagram in step 1). Connect the lead to the *black* needle probe of DMM. Use the other lead to connect the other sensor wire to the red needle probe.

   The manufacturer (Amperex — a North American Philips Co.) has supplied the following calibration information for this device.

T_A °C	resistance Ω	T_A °C	resistance Ω
-55	495	50	1206
-50	520	60	1295
-40	572	70	1387
-30	627	80	1483
-20	687	90	1583
-10	750	100	1687
0	817	110	1794
10	887	120	1905
20	961	125	1962
25	1000	130	2020
30	1039	140	2138
40	1121	150	2260

Temperature versus resistance values (KTY83)

7. Turn the DMM to resistance measurement.

8. Try to get a feel for the response time of the sensor by breathing on it or holding it between the thumb and finger and letting go.

   NOTE: In the electronics industry the thermal time constant is the time the sensor needs to reach 63.2% of the total temperature difference.

9. Switch the DMM off.

   - Make a brief list of what you consider to be the basic differences between this temperature sensor and a typical mercury-in-glass thermometer.
   - If you have a thermometer available you could certainly design simple experiments to highlight these differences.
   - Plot a full-scale graph of resistance versus temperature for the range 10 °C to 60 °C and draw a smooth curve through the points.

     NOTE: The curve is *not* a straight line!

- Make yourself a table of resistance versus temperature in 1 °C intervals from 18 °C to 35 °C by taking readings from the graph.
- Write this table on the graph paper so that you have easy access to it.

Section D. The Construction and Calibration of a Gas Thermometer

Goal: *To construct and calibrate a simple gas thermometer based on the expansion or contraction of humid air.*

Discussion: Calibration is accomplished with the sensor-DMM system developed in the last experiment. Although the gas thermometer is basically a simple instrument, it is experimentally tricky to calibrate and use. You will need to be careful, thoughtful, and patient in order to get good results.

Experimental Steps:

1. Obtain 2 styrofoam cups.
2. Fill one cup with water at about 20 °C and the other cup with water at about 30 °C.

   NOTE: Use cold tap water and small volumes of hot water to obtain water of the correct temperature.
3. Measure the temperature of the 20 °C water and the 30 °C water by placing the sensor (which is attached to the support) into the water in each cup.
4. Obtain a new, clean, thin-stem pipet. With your scissors, cut off 0.5 cm of the stem, measuring from the tip so that the thin-stem part remains exactly 11.0 cm long.
5. Place the thin-stem pipet into the 30 °C water, with the bulb down and completely immersed in the water.
6. Hold the pipet in the water for about a minute so that the air inside the bulb will expand and reach thermal equilibrium.
7. Quickly remove the pipet and immediately turn it over and place the thin-stem end into the 30 °C water.
8. Allow a small plug (about 0.5 cm) of water to be sucked up into the thin stem. When the plug has been sucked up, remove the thin-stem pipet from the water immediately.
9. Turn the thin-stem pipet over and completely immerse the bulb in the 20 °C water.

   NOTE: The indicator plug in the thin stem is now self-calibrated to be positioned right at the top of the stem at the upper working temperature.
10. Put the thin-stem pipet aside for a moment.
11. Construct a straw clamp and stand system as instructed in Section B of Chapter 4, "The Use and Abuse of Aluminum."
12. Add more hot water to the warm water cup in order to raise the temperature and bring it back to 30 °C. Stir the water in the cup with a straw.

For illustrations of the apparatus and procedures discussed in Steps 13, 14, and 15; refer to the pictures below.

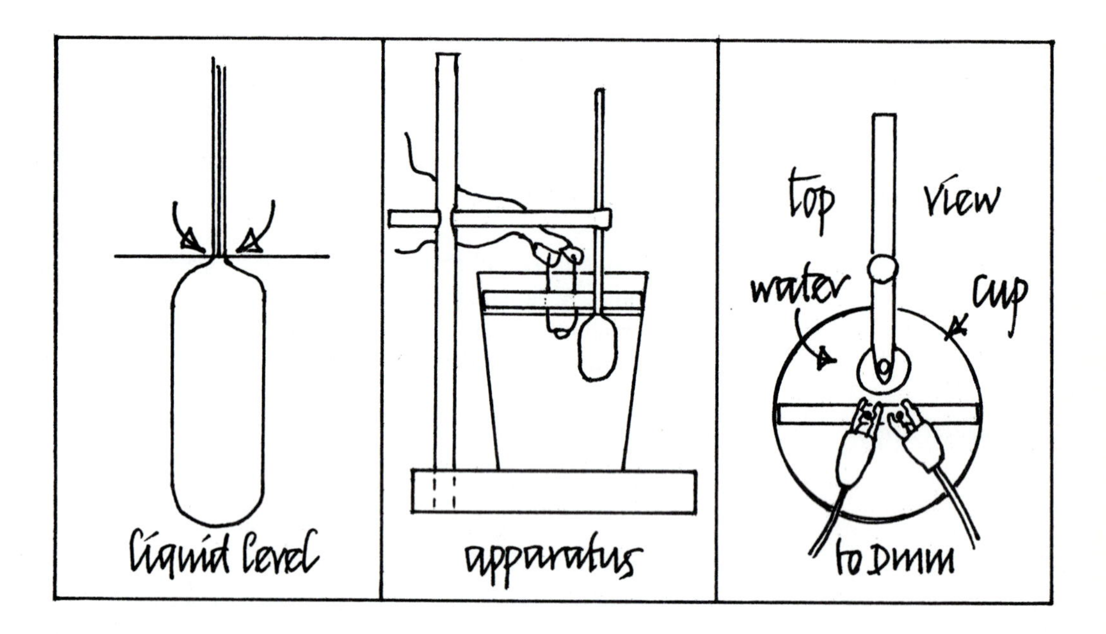

13. Lift the thin-stem pipet (that had been set aside in Step 10) by the *stem* and carefully place the thin-stem part into the straw clamp.
14. Slide the clamp down so that the bulb of the pipet is just completely immersed (as shown in the first picture below) in the 30 °C water.
15. Adjust the sensor support so that the sensor is in the water alongside the bulb of the pipet.
16. Wait until the plug stabilizes at some position in the stem.
    - Read the resistance of the sensor by means of the DMM and measure the distance of the plug (in cm, to an estimated 0.5 mm) from the tip of the thin stem.
17. Cool the water in the cup by about a degree by adding cold water to the cup and by removing some of the 30 °C water. Use another pipet to accomplish the cooling step. Stir.
18. Wait until the plug stabilizes again.
    - Take readings of resistance and distance of the plug from the tip of the thin stem.
19. Repeat Steps 17 and 18 until resistance and plug distance readings at 4 more temperatures (about 1 °C apart) have been obtained.
    - Subtract the plug distances from 11.0 cm in order to obtain the air volume decrease at each point of decreasing temperature.

- Plot subtracted distance versus absolute temperature (K). You can obtain the temperature (in °C) from your accurate resistance versus temperature graph that you prepared earlier.
- Compare your distance versus absolute temperature graph with Figure 5.4 in the Background Chemistry section.
- Which of the gas laws have you verified in this experiment?

The gas thermometer that you have just constructed and calibrated is a very different temperature measuring instrument from the semiconductor temperature sensor.

- Compare and contrast the characteristics of these two instruments. Your comparison should include a brief discussion of cost, possible applications, range, flexibility, ease of use, response time, etc.
- Write the most concise user manual you can that explains how to use the gas thermometer to measure the temperature of a water sample.

Section E. An Extended Range Gas Thermometer

Goal: *To build and calibrate an extended range gas thermometer.*

Discussion: The principle of the change in volume of a gas with changes in temperature can become the basis for developing an extended range (> 30 °C) gas thermometer. The only design change that is necessary from the previous instrument is to reduce considerably the initial volume of gas contained in the instrument. The goal of this experiment is to construct and calibrate an extended range instrument of this type.

Experimental Steps:

1. Use the pipet from the previous experiment. Suck up room temperature water so that the bulb is about 3/4 full.
2. This time in order to get a 0.5 cm water plug into the stem, you must hold the bulb and bend the stem in an arc so that the tip is pointing downwards. Squeeze the top part of the bulb to expel any water in the stem.
3. While maintaining the arc, dip the tip in water, gently expel 3 bubbles of air, and suck up a 0.5 cm plug.

   NOTE: It is best to have the plug near the bulb end of the stem at room temperature.
4. Fill (to within 1 cm of the top) a styrofoam cup with water at around 50–60 °C.
5. Set up the sensor, support, DMM system, and the straw clamp for the pipet.
6. Hold the pipet bulb in the hot water to ensure that the plug does not come out of the top.

   NOTE: If the plug looks as though it is going to come out, then simply add cold water to reduce the temperature.

7. Calibrate your instrument as in the last experiment.
   - Draw a calibration graph for the extended range instrument.

Section F. A Novel Pressure Gauge Balance

Goal:

To construct a novel and inexpensive pressure gauge balance by utlilizing Boyle's law, which states that volume is inversely proportional to pressure. NOTE: This experiment is optional.

Experimental Steps:

1. Use the technique described in Section D to place a 0.5 cm plug of water about 3/4 down the length of the thin stem of an empty pipet. Don't worry about temperature effects.
2. Hold the thin stem vertically and heat the tip until it just melts.
3. Blow out the match and then squeeze the tip firmly between thumb and index finger. This will flatten and seal the tip.
4. Place the pipet horizontally on the bench and lean on the bulb.
5. How could you calibrate this pressure gauge?

Chapter
6

Thermochemistry and Solar Energy Storage

An evaluation of water, granite rocks, and sodium thiosulfate pentahydrate as solar energy storage materials.

Introduction

Solar energy represents the only totally nonpolluting inexhaustible energy resource that can be utilized economically to supply the world's energy needs forever (or for at least two billion years). The technology necessary for the collection of solar radiation is here and is steadily being improved. Currently, the three most important collection methods are flat plate collection, lenses and mirrors, and photovoltaics. Of these, *photovoltaics* — i.e., the direct conversion of solar radiation into electricity — is perhaps the best method. Unfortunately, in spite of massive research expenditures, photovoltaics is still not economically feasible for such large-scale applications as home heating, although photovoltaic devices are used extensively in such special applications as calculators, spacecraft, and ocean buoys. Lenses and mirrors are capable of producing high temperatures for research purposes, but again are not cost effective for large-scale use. This leaves the inexpensive, low-technology method of flat plate collection as the method of choice for domestic and industrial heating and cooling.

The intermittent nature of solar radiation requires that the collected energy be stored in some suitable system. A number of potentially useful systems are available, but the practical requirements of low cost and the storage of large amounts of heat energy in a low volume are very difficult to meet. The most widely used methods of storing collected solar energy are the following:

- Solar energy may be stored by raising the temperature of inert substances, such as air, water, or rocks, with subsequent recovery of the stored energy by heat transfer.
- Solar energy may be stored in reversible chemical processes, such as the dehydration of salt hydrates or the melting of low-melting salts.
- Solar energy may be directly converted into electricity by photovoltaic cells and then stored in fuel cells or batteries. Many of the commercially available solar house heating systems now in production use either water or rock storage. Heat energy is transferred to a transport fluid at the back of the flat plate collector. The heated fluid is then circulated by a pump throughout the house or through to the water or rock storage system contained in a large insulated box in the basement. The main problem with these storage methods is the very large volume of water or rocks required.

Storage of heat in chemical systems is very attractive because much larger amounts of heat energy may be stored at much lower temperatures than with other methods. Chemical storage systems must meet the following requirements:

- The chemical must be inexpensive, nontoxic if possible, and easily obtained and transported.
- The chemical process must be reversible so that it can be used over and over again.
- The chemical process must happen with sufficient speed, or the method will not be practical.
- The chemical process must occur over the normal temperature range encountered in solar collection.

The simplest and most well researched chemical systems are those of the salt hydrates — i.e., salts with water of crystallization. A good example is sodium sulfate decahydrate, ($Na_2SO_4 \cdot 10H_2O$), which has a melting point of 32.3 °C and can store 84.5 kilocalories per liter. The chemical process you will be studying involves the melting (fusion) and subsequent crystallization of a salt hydrate called sodium thiosulfate pentahydrate, ($Na_2S_2O_3 \cdot 5H_2O$). Solar energy storage in this system is very feasible because the chemical is cheap (it is used extensively in photography as a fixer), is stable, relatively nontoxic, and melts at 48 °C. The thermal characteristics of this salt hydrate will then be compared with water and granite rocks in order to evaluate the optimum storage system.

Background Chemistry

The key questions to be answered in any evaluation of solar energy storage materials and processes are:

- How fast can energy be put into and removed from the storage system?
- How reversible is the system? Will it cycle for the lifetime of the collection system?
- How much energy can be stored in unit mass or volume of the material?

It is this last question that you can answer quantitatively in this series of laboratory experiments, although some qualitative approaches to the other questions are also possible. The materials to be investigated include liquid water, granite rocks, and the hydrated salt, sodium thiosulfate pentahydrate, ($Na_2S_2O_3 \cdot 5H_2O$). Liquid water is interesting from a thermal storage point of view. When water is heated, the energy goes into increasing the amount of molecular motion of the water molecules at the expense of hydrogen bonding (which is relatively strong intermolecular bonding). It is the considerable energy required to break hydrogen bonds that gives liquid water its very high specific heat (4.18 J g^{-1} °C^{-1}).

Sodium thiosulfate pentahydrate has a low melting point (48.3 °C), and at this point the applied energy goes into breaking the ionic bonds in the crystal structure and also into releasing fixed (bonded) water molecules into the disordered liquid state:

$$\text{Heat energy} + Na_2S_2O_3 \bullet 5H_2O_{(s)} \underset{\text{heat of crystallization}}{\overset{\text{heat of fusion}}{\rightleftharpoons}} Na_2S_2O_{3(aq)} + 5H_2O_{(l)}$$

White "glassy" crystals of $Na_2S_2O_3 \cdot 5H_2O_{(s)}$, when heated above 48.3 °C, melt to give a colorless, slightly cloudy liquid melt. The heat energy is stored in the liquid melt. If the melt is allowed to cool, some interesting things begin to happen. The melt becomes rather viscous, and the crystallization process is slowed so much that it often will not occur at all! The persistence of a melt at temperatures well below the melting point of the substance is a good example of the phenomenon of *supercooling* (a state which has been found to last for 50 years in some instances!). Thermal shock or the addition of a seed crystal will inevitably catalyze the crystallization process and produce a solid crystalline mass in a short time.

The system described above is used to store solar energy by allowing the heat energy from a solar collector to melt the hydrated salt to a liquid. The liquid is stored in an insulated box. When energy is needed, the liquid is allowed to crystallize, giving up the stored energy as heat of crystallization in the process. The study of these heat energy changes in various processes is called *thermochemistry* and is carried out experimentally by "trapping" the energy in a liquid in an insulated container called a *calorimeter*. The subsequent temperature changes can be determined by suitable graphing techniques.

CALORIMETRIC TECHNIQUES

One of the most important techniques of measuring the amount of heat energy evolved or absorbed in some process is that of *adiabatic calorimetry*. In this technique, an attempt is made to conserve all the evolved heat by carrying out the process in a calorimeter. The calorimeter usually contains a suitable liquid that makes good thermal contact with the process under investigation. If the process evolves heat energy, then the energy is transferred to the liquid and to the material of which the calorimeter is made, causing the temperature of the liquid to rise. The *temperature change* (ΔT) can be measured by any of the very accurate, sensitive temperature sensors that are now available. The value of ΔT is the difference between the final and initial temperatures of the liquid in the calorimeter. In thermochemistry, temperature should really be expressed in an absolute temperature unit — the Kelvin (K). However, because the Celsius degree and the Kelvin are the same size, temperature differences (ΔT) have the same value in °C as in K.

Unfortunately, there are a few practical problems in carrying out experimental adiabatic calorimetry. First, it is impossible to have perfect insulation. Some of the heat energy is lost to the surroundings — e.g., to the material from which the calorimeter is constructed. The problem of heat losses is usually solved by calibrating the calorimeter before using it to make measurements on an unknown system. A known amount of heat energy from a known process is released into the calorimeter system, and the temperature change is measured. A simple calculation can then be done to determine the amount of heat energy loss, called the heat capacity of the calorimeter. The second practical problem is that heat energy exchanges do not occur instantaneously — i.e., it takes time for energy to move from a hot object to a cold one. An excellent solution to this problem is to obtain a cooling curve for the heat energy exchange in question and then extrapolate the data back to the exact time that the exchange began.

The experimental calorimetric measurements described above provide the basis for a quantitative definition of the thermal properties of materials. The magnitude of the heat energy flow that accompanies an increase in temperature depends on the mass and the identity of the substance involved and is called the *heat capacity* of the substance. The heat capacity is therefore defined as the amount of heat energy required to raise the temperature of a given amount of substance or system by 1 °C. Heat capacities of substances are reported in two ways: as a molar heat capacity or as the *specific heat*. *Specific heat* is defined as the amount of heat energy required to raise the temperature of one gram of the substance by 1 °C and is given in the unit joules per gram per degree Celsius (J g^{-1} °C^{-1}). The specific heat is the direct link between the amount of energy, the mass of the substance, and changes in temperature. For any process involving a temperature change ΔT,

$$\text{Amount of heat energy absorbed or released} = \text{mass} \times \text{specific heat} \times \text{temperature change}$$

For phase change processes, e.g., crystallizations, melting, condensation, vaporization, etc., that involve no temperature change, the relationship is,

$$\text{Amount of heat energy absorbed or released} = \text{mass} \times \text{enthalpy change per unit mass}$$

where the enthalpy change (ΔH) is the change in enthalpy for the phase change under consideration. One of the experiments in this laboratory module involves a crystallization and the ΔH is called the heat of crystallization (and has units of $J\ g^{-1}$). Normally, it is important in thermochemistry to distinguish between the system and the surroundings to ensure that the sign of heat energy flow, q, is *positive* when heat energy flows into the system from the surroundings, and *negative* when heat energy flows out of the system to the surroundings. In this series of experiments, the direction of heat energy flow is from the hot water, crystallizing salt, and hot rocks to the calorimeter and its contents. Therefore, based on this knowledge, you do not have to worry about the sign of q. Remember, also, that the calorimeter system used throughout these experiments is made out of styrofoam, and is at atmospheric pressure. The calorimeter system may be used to measure specific heats at constant pressure — it is not a bomb calorimeter!

Pre-Laboratory Quiz

1. What is the name of the most common type of domestic solar energy collection system?

2. Give 2 requirements that a chemical process must meet in order to be useful as a solar energy storage system.

3. What 3 substances are you going to compare as solar energy storage systems in this laboratory?

4. Give a value for the specific heat of liquid water.

5. Give a definition of the heat capacity of a substance.

6. What is meant by the term *adiabatic calorimetry*?

7. How many joules of heat energy are required to raise the temperature of 10 grams of liquid water from 25 °C to 35 °C?

8. Name a practical problem encountered in experimental calorimetry and briefly discuss how it is overcome.

9. Why does liquid water have such a high specific heat?

10. What is the difference between °C and K? Please don't say 273.15!

Laboratory Experiments

Flowchart of the Experiments

Section A.	Calibration of a Styrofoam Microcalorimeter
Section B.	Determination of the Specific Heat of Granite Rock Used for Solar Energy Storage
Section C.	Determination of the Heat of Crystallization of Sodium Thiosulfate Pentahydrate, $Na_2S_2O_3 \cdot 5H_2O$
Section D.	An Assessment of the Microcalorimetric Methodology and the Storage Information Obtained

Requires one three-hour class period to complete

Section A. Calibration of a Styrofoam Microcalorimeter

Goal: *To measure the heat capacity of a styrofoam microcalorimeter.*

Discussion: The heat capacity measurement is accomplished by using a cooling curve analysis of an experiment involving the addition of a known volume of hot water to a known volume of cold water in the microcalorimeter. The temperature changes can be monitored by a semiconductor temperature sensor and digital multimeter combination.

Experimental Steps:

1. At your place you will find a styrofoam microcalorimeter.
2. Construct a straw support system for the temperature sensor: Cut a straw so that it just fits inside the top part of the microcalorimeter. Use a pin to make 2 holes (about 1 cm apart) that go completely through both walls of the straw. *Gently* push the sensor leads through the straw. Place the straw and sensor into the microcalorimeter. It should fit snugly.
3. Attach 2 clip leads to the sensor and clip the sensor cathode to the black DMM lead. Clip the other lead to the red DMM lead.
4. Switch on the DMM and turn the rotary switch to measure resistance.

   NOTE: When the DMM is not in use, switch it off to conserve the battery. To extend the battery life, the instrument automatically shows a blank display if left on after 1 hour of not being used. When you wish to resume operation, turn the rotary switch on.
5. Transfer a known volume of distilled water to the calorimeter. A convenient way to do this is to use a thin-stem pipet. Suck water up and completely fill bulb and stem with water. This volume is 4.0 mL. Add 4 bulb volumes (i.e., a total of 16.0 mL) of water to the microcalorimeter. Make sure that the sensor is in the water that is in the microcalorimeter.
   - Monitor and record the sensor resistance reading *in ohms* every minute for 3 or 4 minutes.
6. During this time completely fill the thin-stem bulb and stem with water again and place it in the communal hot water bath at the end of the bench.
   - Record the temperature of the hot water (T_3).
7. At a chosen minute (the 5th or 6th), add the total hot water contents of the bulb and stem to the microcalorimeter. Quickly suck up some water from the calorimeter to wash out the pipet and expel back into the microcalorimeter.
   - Read and record the sensor resistance in ohms every minute for about 10 minutes.
   - Plot a mixing-cooling curve by graphing resistance (vertical axis) versus time (horizontal axis).

8. The curve should look something like the one below.

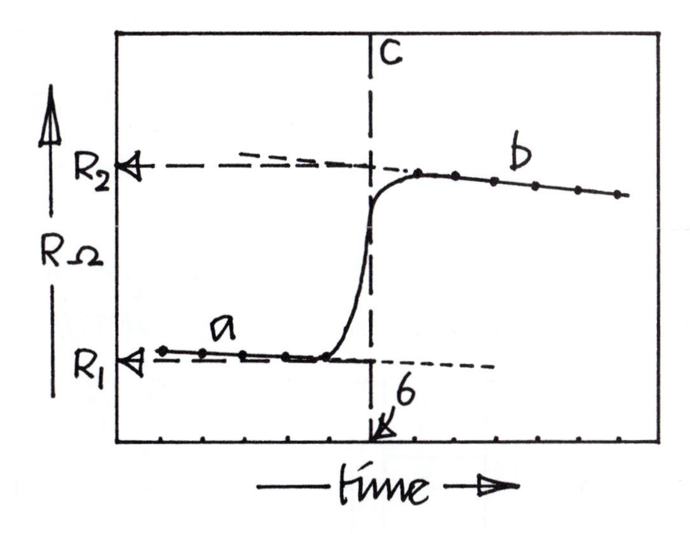

- We need to find from the graph the temperature change that occurred upon the instantaneous mixing of the hot water with the cold water in the microcalorimeter.
- With a ruler draw lines a and b through the experimental points and extrapolate the lines past a vertical line c drawn at the exact time of mixing (the 5th or 6th minute, depending on when you did it).
- Read off and record the resistances R_1 and R_2 from the graph.
- Make a graph of resistance versus temperature (from 10 °C to 40 °C) from the sensor resistance versus temperature data listed in the table below.

Ambient temperatures and corresponding resistance values of sensor

T_A °C	resistance Ω
-55	495
-50	520
-40	572
-30	627
-20	687
-10	750
0	817
10	887
20	961
25	1000
30	1039
40	1121

T_A °C	resistance Ω
50	1206
60	1295
70	1387
80	1483
90	1583
100	1687
110	1794
120	1905
125	1962
130	2020
140	2138
150	2260

- Use this graph to convert resistances R_1 and R_2 to temperatures T_1 and T_2 (with an accuracy of 0.2 °C).

NOTE: Now you have all the information to calculate the heat capacity of the calorimeter. All volumes of water can be converted to mass by using the density of water (1.0 g mL^{-1}). For example, 4.0 mL of hot water has a mass of

$$4.0\text{ mL} \times \frac{1\text{g}}{\text{mL}} = 4.0\text{ g}$$

To summarize the heat exchange,

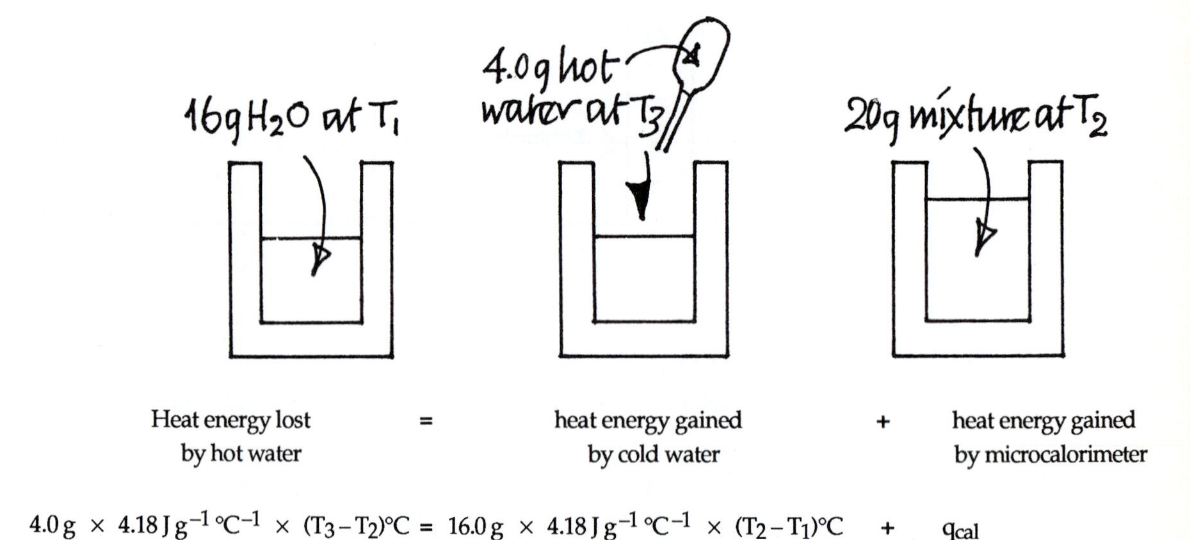

$$4.0\text{ g} \times 4.18\text{ J g}^{-1}\,°\text{C}^{-1} \times (T_3 - T_2)°\text{C} = 16.0\text{ g} \times 4.18\text{ J g}^{-1}\,°\text{C}^{-1} \times (T_2 - T_1)°\text{C} + q_{cal}$$

where the specific heat of water is 4.18 J g^{-1} $°C^{-1}$, where q is the heat energy gained by the microcalorimeter in going from a temperature of T_1 to T_2. The assumption can be made that the microcalorimeter is at the same temperature as the water contained in it.

- Calculate q_{cal}.

The heat capacity of the microcalorimeter (C_{cal}) is defined as the amount of heat energy required to raise the temperature of the microcalorimeter by 1 °C and is given by

$$C_{cal} = \frac{q_{cal}}{(T_2 - T_1)}$$

9. Empty the microcalorimeter and wash with 2 pipets of cold water. Empty the water and dab dry with a paper towel.
10. Refill the microcalorimeter with 4 × 4.0 mL of water as you did earlier in Step 5. Replace the sensor support, etc. Make sure the sensor is in the water. Allow the microcalorimeter, water, and sensor to come to equilibrium by continuing with Section B.

Section B. Determination of the Specific Heat of Granite Rock Used for Solar Energy Storage

Goal:

To determine the specific heat of granite rock by placing hot rocks into the microcalorimeter and by observing the subsequent temperature changes.

Experimental Steps:

- On graph paper set up the axes for a cooling curve graph so that you can make the plot as you carry out the experiment.

1. Make sure that the sensor is in the water in the calorimeter you had set up in Step 10 of Section A.
2. Monitor the sensor resistance at minute intervals for 3 to 5 minutes. Plot the points on the graph paper.
   - Go to the oven and record the temperature shown on the thermometer.
4. Quickly remove 6 rocks from the oven (assume that the rocks are also at the same temperature that you recorded for the oven — call it T_6) and wrap them in a paper towel. Take them to your place and slide them quickly into the calorimeter at a minute mark.
5. Gently stir the water in the calorimeter with a straw to ensure a uniform temperature around the sensor.
6. Monitor the resistance for 8 to 10 minutes or until you obtain a cooling curve. Stir occasionally.
   - Plot the points on the graph.
7. When you have a good cooling curve, remove the sensor and switch the DMM off.
8. Carefully pour off the water and remove the rocks. Dry them with a paper towel.
   - When the rocks are dry, weigh them and record the weight.
9. Analyze the cooling curve shown below in the same manner as in Section A. Use the sensor resistance versus temperature graph that you made earlier to determine the 2 temperatures, T_4 and T_5, associated with the resistances R_4 and R_5.

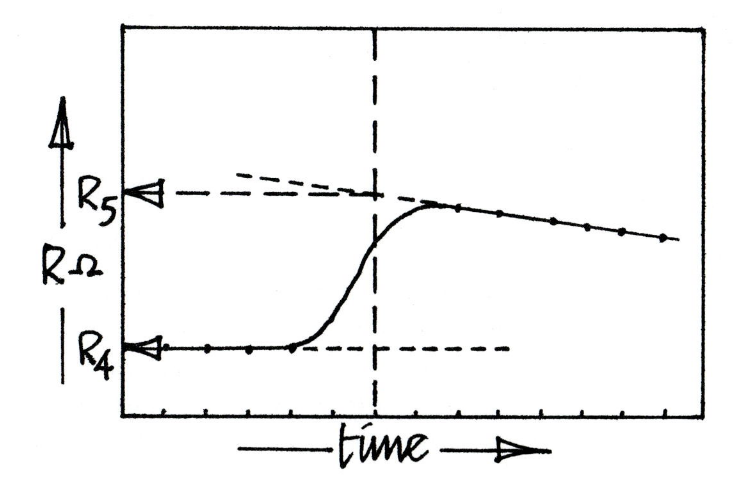

Referring to the equation below, let s be the specific heat of the granite rock and m_R be the mass of the rocks. A summary of the heat energy exchange is:

Heat energy given out by the 6 granite rocks	=	heat energy taken in by cold water	+	heat energy taken in by microcalorimeter

$$m_R\,g \times s\,J\,g^{-1}\,°C^{-1} \times (T_6 - T_5)°C = 16.0\,g \times 4.18\,J\,g^{-1}\,°C^{-1} \times (T_5 - T_4)°C + C_{cal}\,J\,°C^{-1} \times (T_5 - T_4)°C$$

- Calculate the specific heat (in $J\,g^{-1}\,°C^{-1}$) of the granite rock.

10. Dab the calorimeter dry with a paper towel and refill with 16.0 mL of cold water, as in Step 5 of Section A, and set aside to allow it to equilibrate until called for in Section C.

Section C. Determination of the Heat of Crystallization of Sodium Thiosulfate Pentahydrate, $Na_2S_2O_3 \cdot 5H_2O$

Goal: *To determine the heat of crystallization of sodium thiosulfate pentahydrate.*

Discussion: This ΔH determination can be carried out by melting a known weight of the salt in a capped straw and by then allowing the salt to crystallize (at its melting point) in the microcalorimeter system.

Experimental Steps:

- Set up the 2 axes, resistance and time, on graph paper so that you can add the experimental points to the graph as you take readings in the experiment.

  It's fun to be able to see the cooling curve develop as you go along.

1. Cut 5 cm off the end of a straw and discard the 5 cm piece.

   You now have a piece about 15 cm long.

2. Insert a plastic cap *tightly* into one end of the 15 cm piece.

   - Weigh the capped straw and record the weight.

3. Make a mark on the straw 6 cm from the cap.

4. Using a thin-stem pipet, transfer liquid salt from the bath of molten sodium thiosulfate pentahydrate into the straw until it reaches the 6 cm mark. Set aside for a few minutes in order to allow the salt to cool to the crystallizing temperature (~ 48 °C).

5. Place the sensor into the water (as instructed in Section A and B earlier) and read the resistance at 1-minute intervals for 3 minutes.

   - Record the resistance readings.

6. Check on the salt. It should still be liquid.

7. At a minute mark, quickly add a seed crystal and make sure that the salt begins to crystallize.

8. (Refer to the following filmstripped illustration while completing this Step.) Bend the straw at the center of the liquid salt and place doubled up into the water in the calorimeter.

NOTE: The liquid salt level in the straw *must* be covered by the water in the calorimeter!

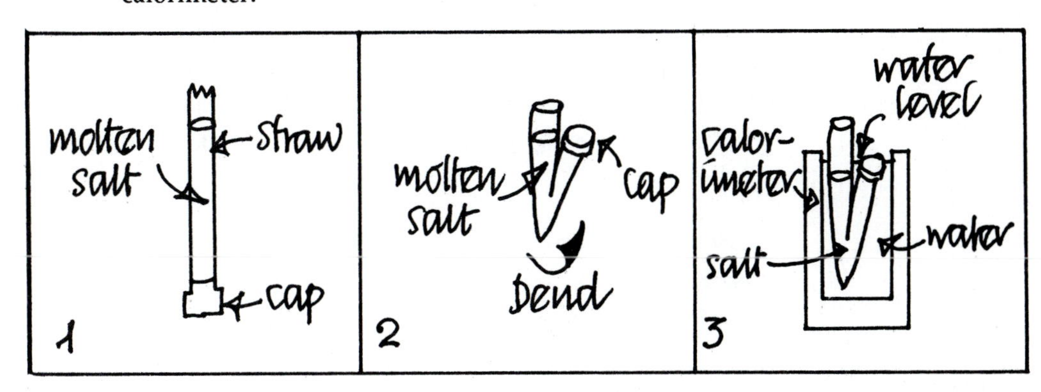

9. Read the sensor resistance at 1-minute intervals. The sensor must be in the water in the calorimeter *and* you must stir the water by *gently* moving the bent straw up and down a little.

   - Record and plot the resistance readings.

   The crystallization of the salt occurs rather slowly and therefore the transfer of heat energy is rather slow.

10. Continue stirring, measuring resistance, and plotting until you have a cooling curve.

11. Remove the sensor from the calorimeter. Remove the bent straw and dry it with a paper towel.

    - Weigh the straw with the crystallized salt and determine the weight of the salt.

12. Analyze your cooling curve in the same manner as before. Use the sensor resistance versus temperature graph to determine the two temperatures, T_7 and T_8, necessary for the calculation of the heat of crystallization.

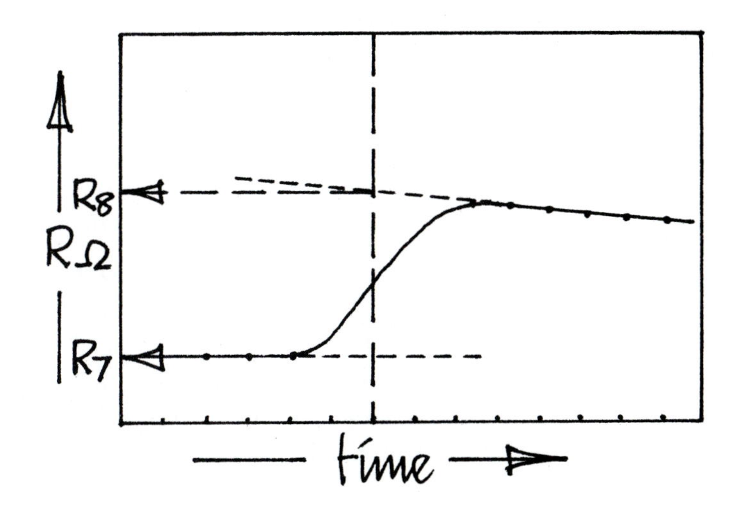

NOTE: Don't forget that the crystallization of the salt is a phase change and is occurring at the melting point. The temperature of the salt remains at its melting point during this process!

In reference to the equation below, assume that the specific heat of liquid water is 4.18 J g^{-1} °C^{-1}, and that the density of water is 1.0 g mL^{-1}. Let the weight of the salt in the straw be m_{salt}, and the heat of crystallization of the salt be ΔH. The heat energy exchange is

Heat given out by $Na_2S_2O_3 \cdot 5H_2O$ crystallizing at 48 °C	=	heat energy taken in by cold water	+	heat energy taken in by microcalorimeter

$$m_{salt}\,g \times \Delta H\,J\,g^{-1} = 16.0\,g \times 4.18\,J\,g^{-1}\,°C^{-1} \times (T_8 - T_7)°C + C_{cal}\,J\,°C^{-1} \times (T_8 - T_7)°C$$

- Calculate the heat of crystallization per gram of $Na_2S_2O_3 \cdot 5H_2O$.
- Calculate the molar heat of crystallization of $Na_2S_2O_3 \cdot 5H_2O$.
- Is the crystallization endothermic or exothermic?
- What is the sign of ΔH?

13. Empty the calorimeter and dab it dry. Your instructor will dispose of the straws containing the salt.
14. Don't forget to switch off the DMM.

Section D. An Assessment of the Microcalorimetric Methodology and the Storage Information Obtained

- Do you think that a classical mercury thermometer could have been used to carry out the microcalorimeter experiments? Give reasons for your answer.
- Do you think that a heat-insulating top for the microcalorimeter would have improved the results?
- Why is styrofoam such a good heat-insulating material for adiabatic microcalorimetry?
- Straws happen to be excellent containers for the determination of the heat of crystallization of molten salts. Why?
- Calculate and compare the amount of heat energy stored by 1 pound of water, 1 pound of granite, and 1 pound of sodium thiosulfate pentahydrate. Assume a ΔT of 40 °C for the temperature change of the water and the rock.
- If you could not afford to pay elves (or students) to add seed crystals at the appropriate time to the solar energy storage tank to make the molten salt crystallize, how could you insure crystallization?

- Give some advantages and disadvantages for each of the storage materials that you investigated.
- Just for kicks! What is the specific heat of styrofoam?

Chapter 7

Solutions and Reactions

Introduction

Did you ever stand and shiver
Just because
You were lookin' at a river.

Ramblin' Jack Elliott
(told to the author on a plane from Detriot to Denver)

Water is by far the most common and the most important liquid on our planet. It is vital to all living organisms, and it is indispensable to civilization. Water covers most of the earth and through its global cycle — the hydrologic cycle — permeates the atmosphere and the rocks. It shapes the face of the planet as it slowly dissolves and transports the land into the sea. Much of the importance of water lies in its ability to dissolve and suspend so many substances and then transport them to different places. It is indeed the universal solvent for gases, liquids, and solids.

It is probably safe to say that all biochemical reactions in living systems require water as a solvent medium and often use water as a reactant or as a catalyst. For example, the human cardiovascular system is a marvelous multipurpose medium for the exchange and transport of dissolved gases, sugars, suspended organelles, and so on. The oceans, a global example, contain about 3.5% of dissolved inorganic solids, mostly in the form of hydrated cations and anions. Millions of species have evolved to make this salty solution the perfect medium for life. The billions of tiny bubbles continuously trapped in the crashing waves burst and fling microscopic droplets of sea solution into the atmosphere. The droplets evaporate, leaving solid particles that are so small that they remain in the air for long periods of time. These marine aerosols play a large part in controlling the weather because they act as cloud condensation nuclei, catalyzing the formation of the natural precipitates: rain, snow, and hail. The vast global ocean currents transport everything ranging from microscopic plankton to carbon dioxide and air pollutants, such as fluorocarbons, from one hemisphere to the other. At every scale, from the microscopic to the macroscopic, the solvent and fluid properties of water are essential to the removal of the toxic wastes generated by cells, organisms, people, factories, and cities.

Of equal importance are the substances that are not very soluble in water. In many biological systems, the fluid bag of cells is supported by a skeletal structure whose composition can vary from silicates in Radiolaria (tiny sea creatures) to calcium hydroxyphosphate in humans. The transport of hydrated calcium and phosphate ions in the complex system of bone and teeth formation is a marvel of solution chemistry. The survival of birds depends upon a fragile shell composed of tiny calcite crystals laid down on a light network of protein collagen. When things go awry and substances precipitate in the wrong places, serious health problems can occur. Cholesterol plaques in the blood stream can end the flow of life. Kidney, gall, and bladder stones and calcified deposits in the joints can cause severe pain. Microscopic silica and asbestos and smoke particles of all kinds injure lung cells and can cause cancer and death. Much is now known about the solution chemistry of these complex biological processes, but there is still much to be learned. We are living in the modern era of microelectronics and photonics* in which the control of the flow of electrons and photons in water-insoluble solids is the basis for our communication and manufacturing machines. It is intriguing to speculate that the study of the chemistry and flow of ions in aqueous solution may lead to a revolution in the understanding of communication and evolution in living systems. Microionics may rival microelectronics in importance in the scientific future. In this series of experiments, you have the opportunity to study the fascinating chemistry of dissolution, reaction, and precipitation in aqueous solution.

---

* The use of laser light in communication systems, sometimes termed *optronics*.

Background Chemistry

A homogeneous mixture of two or more substances is called a *solution*. Usually, the substance that is present in a smaller amount is called the *solute*, and the substance present in the larger amount is called the *solvent*. Solutions may be gaseous (e.g., air), liquid (e.g., blood), and solid (e.g., steel).

Solutions are the places where chemical and biological reactions happen. Solutions are also the means by which substances are transported from one place to another. Liquid solutions are particularly important in geological, environmental, biological, and chemical systems; the universal solvent for these solutions is water. The solution is the medium that brings reactants close together long enough to allow new associations and bonds to form. Products appear as bond formation moves toward completion. The solvent almost always plays an active role in all of these processes, especially when water is the solvent. The solvent must first dissolve the substances that are eventually going to react. This deceptively simple process, called *dissolution*, also involves the breaking of bonds and the formation of new associations. The solubility of a substance reflects how easily the solvent can make these changes occur. The *solubility* of a substance is, in fact, defined as the concentration of solute in a saturated solution at a specified temperature. The reverse of dissolution is the "coming out" of solution in which the product(s) of a reaction exceed the solubility and form a new phase that could be a gas, liquid, or solid. Again, the solvent often plays an active part in this process.

At the molecular level these dynamic aqueous solution processes can be pictured as shown in Figure 7.1.

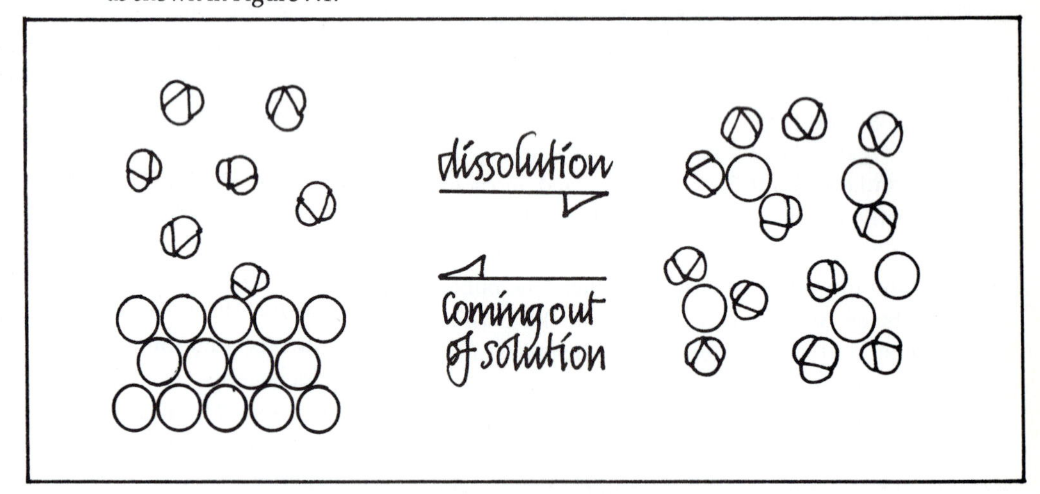

Figure 7.1 Dynamic Aqueous Solution Processes

where ○ are solute and ⊘ are water molecules. The extent to which each of these processes happens can be described in terms of three major interactions:

- Solute-solute interactions ○ $\leftrightarrow$ ○

- Solute-water interactions
- Water-water interactions

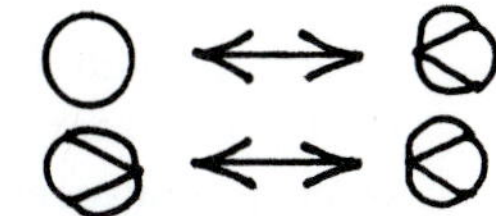

If the solute-solute interaction is dominant, then solute will come out of solution and two phases will form. Which phase goes up (perhaps a gas) and which phase goes down (perhaps a solid) depends on external forces, such as gravity. If the water-water interaction is dominant, then the solute will be "squeezed out" of the solution and form another phase. In between these extremes, water-solute interactions can assure a stable, homogeneous solution.

Water is an excellent solvent for many ionic and polar covalent compounds because the water molecule is a polar covalent molecule. Each H_2O molecule has a small negative charge on the oxygen atom and a small positive charge on the hydrogen atoms. (Refer to Figure 7.2 for illustration.)

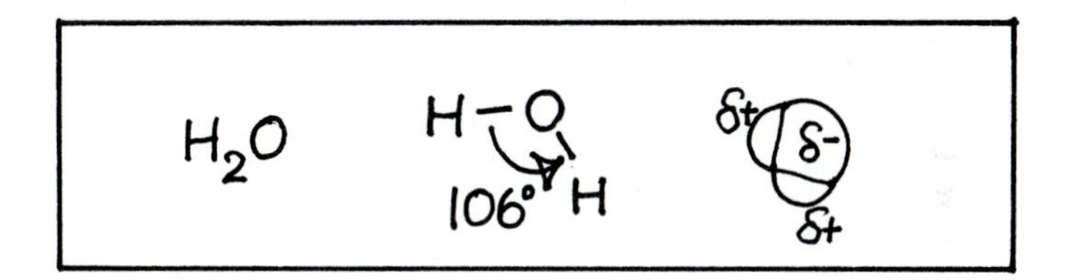

Figure 7.2 Three Representations of the Water Molecule

This permanent dipole (two poles, a positive one and a negative one) is due to the electronegativity of the oxygen atom. The electrons in the covalent bonds between O and H are pulled slightly towards the O atom, creating the dipole. In liquid water (and ice), the water molecules are fairly close together and attract each other, forming weak bonds called hydrogen bonds, which are pictured in Figure 7.3.

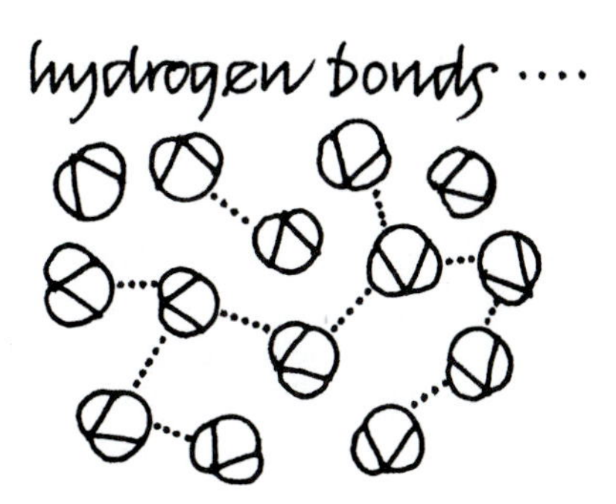

Figure 7.3 Hydrogen Bonding in Liquid Water

Although these hydrogen bonds are weak, they are responsible for many of the interesting and unusual properties of water, including some of its versatile solvent properties. Solutes that interact with the charges on the water molecule are generally soluble provided that the solute-solute interactions are not very strong. Ammonia (NH_3), for example, is very soluble in water, forming hydrogen bonds in the process, whereas dinitrogen (N_2), a nonpolar molecule, is not (see Figure 7.4).

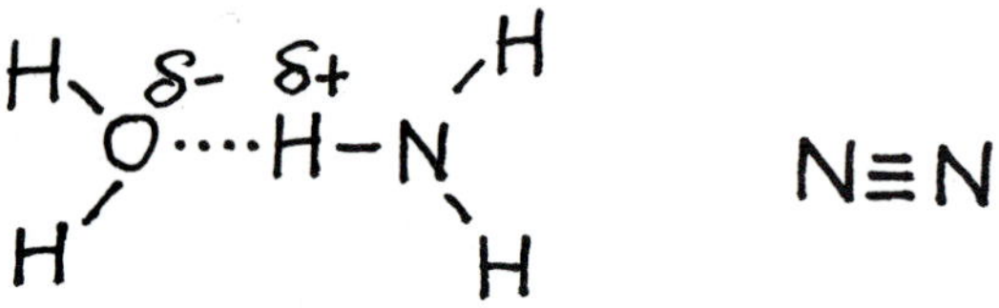

$$H_2\overset{\delta-}{O}\cdots\overset{\delta+}{H}-NH_2 \qquad N\equiv N$$

Figure 7.4 Hydrogen Bonding to NH_3 and not to N_2

Large nonpolar substances, such as gasoline, are not very soluble in water because the long nonpolar hydrocarbon chains disrupt the hydrogen bonding between water molecules. Solvent-solvent interactions dominate, and the gasoline molecules are "squeezed out" of solution and form a separate liquid floating on top of the water, as shown in Figure 7.5.

Figure 7.5 Gasoline and Water Form Two Layers

The solubility of solids in water is also governed by the three major interactions discussed earlier. A simple example is the dissolution of sodium nitrate ($NaNO_3$) in water. Solid sodium nitrate has a crystal structure that consists of sodium ions (Na^+) ionically bonded to nitrate ions (NO_3^-). Water is a good solvent for salts such as sodium nitrate because water molecules are able to move in between the cation and anion and screen the charges from each other. When this happens the ionic bond weakens, and the water molecules can then orient around the dissociated (separated) ions with the negative end of the water dipole pointed towards the cation and the positive end pointed towards the anion. The water molecules are bonded quite strongly to the ions by the attraction of unlike charges. All cations and anions dissolved in aqueous solution are surrounded by bound water molecules, and they are said to be *hydrated*. Figure 7.6 illustrates the process of dissolution and hydration.

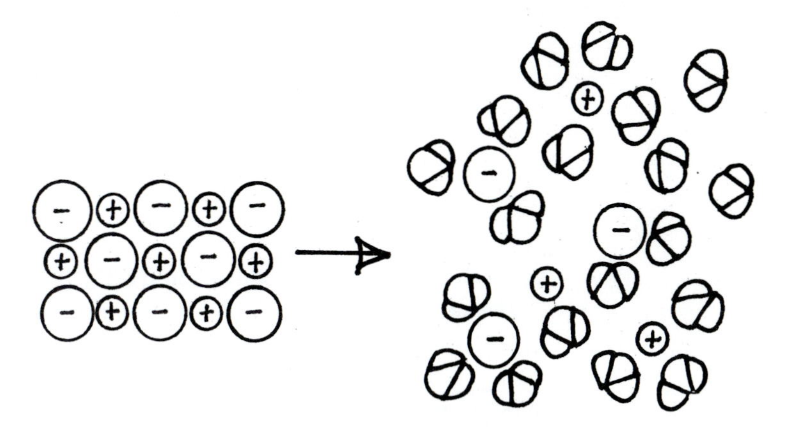

Figure 7.6 Dissolution and Hydration of an Ionic Crystal

NOTE: Most chemistry texts, including this one assume that the reader understands that cations and anions in aqueous solution exist as hydrated ions; therefore the formulas for ions are usually written *without* the bonded solvent molecules. The extent to which the dissolution process occurs — i.e., how soluble the salt is — depends on the magnitude of the attraction between ions and water molecules (*hydration energy*) compared with the attraction between ions in the solid salt (*lattice energy*). The battle of forces within the crystal versus those between ions and water molecules is discussed in a later section on precipitation reactions.

The strength of the attraction between ions and water molecule dipoles depends on the charge and the radius of the ion. Small ions with high charge and electronegativity have the greatest attraction for water molecules; conversely, large ions with a single charge have the smallest attraction. If the attraction of the metal ion for the negative end of the water molecule dipole is strong enough, the water molecule may be ripped apart, releasing a hydrogen ion (H^+) that is then hydrated. A good example is the aluminum ion (radius 67 pm and a +3 charge), which can react with water to produce the hydroxy cation $[Al(H_2O)_5OH]^{2+}$ and the hydronium ion H_3O^+:

$$[Al(H_2O)_6]^{3+} + H_2O \rightleftharpoons [Al(H_2O)_5OH]^{2+} + H_3O^+$$

Compounds that react with water to produce hydronium ions are called *acids*. The aluminum ion is therefore called an *acidic cation*, whereas the sodium ion discussed earlier is called a *nonacidic cation*. This type of reaction can continue with more water molecules to produce hydrated cations containing several hydroxy groups. Eventually, these hydroxy cations can generate insoluble hydroxides — e.g., $Al(OH)_{3(s)}$ — that precipitate from solution.

Anions in aqueous solution also interact with water, except of course in this instance, the positive end of the water dipole is attracted to the anion. If the anion has a small size and high charge, then water molecules are pulled apart and the hydrogen atom bonds to the anion, releasing a hydroxide ion (OH^-). A simple example is the reaction of a carbonate ion (CO_3^{2-}) with water,

$$CO_3^{2-} + H_2O \rightleftharpoons HCO_3^- + OH^-$$

which produces a bicarbonate ion (HCO_3^-) and a basic (the opposite of acidic) solution. The carbonate ion is called a *basic anion*, whereas the nitrate ion (NO_3^-) discussed earlier is called a *nonbasic anion*.

The chemistry of ionic solutes in aqueous solution becomes considerably more interesting and complicated when two solutions, each containing a soluble ionic compound, are mixed. Generally, in this situation the resulting solution, at the instant of mixing, contains two different hydrated cations and two different hydrated anions. If strong solute-solute interaction occurs between any two of the ions, then a chemical reaction results and a new product is formed. Often the two remaining ions do not interact and remain in solution in the same hydrated state as before the reaction. These unreacted ions are often called *spectator ions*. Chemical reactions in aqueous solution may be divided into four types: precipitation, acid-base, complexation, and redox reactions. A brief discussion of the main characteristics of each of these reaction types follows.

PRECIPITATION REACTIONS

A *precipitation reaction* is defined as a reaction that produces a new compound that is not soluble in an aqueous solution. Precipitation can be regarded as the reverse of dissolution. The factors mentioned earlier — i.e., hydration energy and lattice energy — play a major role in determining whether an ionic compound will precipitate from solution. Ionic compounds (salts) in which the cation and anion have approximately the same size and the same charge tend to form especially stable crystal structures and precipitate from aqueous solution. Acidic cations — e.g., Al^{3+} — and nonbasic anions — e.g., NO_3^- — give rise to soluble salts because these ions are quite different in size and have much smaller lattice energy than hydration energy. Although these generalizations are useful, it is often difficult to make exact predictions, and it is worthwhile to remember a few simple aqueous solubility rules:

- Most alkali metal compounds are soluble.
- Most nitrates are soluble.
- All common inorganic acids are soluble.
- All ammonium salts are soluble.

You will certainly come across other solubility generalizations as you carry out the chemical reactions in this series of experiments.

ACID-BASE REACTIONS

In aqueous solution an acid reacts with a base to give a salt and water. An *acid-base reaction,* sometimes called *neutralization,* is characterized by the formation of covalent, neutral water molecules from hydrated hydrogen ions (H_3O^+) and hydroxide ions (OH^-).

COMPLEXATION REACTIONS

Complexation reactions are closely related to acid-base reactions. Earlier, it was pointed out that some metal cations that have small size and high charge can act as acids — in fact they are called acidic cations. These acidic cations can react with electron-rich species called *ligands* to form *complexes.* A *complex* may be defined as a chemical compound in which there is one or more coordinate-covalent bonds. A *coordinate-covalent bond* is a covalent bond in which the shared pair of electrons is provided by the ligand. A complexation reaction is therefore defined as a reaction in which one or more coordinate covalent bonds are produced during the formation of product.

REDOX REACTIONS

Redox (an abbreviation of reduction-oxidation) *reactions* are defined as chemical reactions in which electrons are *transferred* from one species to another. *Oxidation* is defined as loss of electrons; *reduction* is defined as gain of electrons. A transfer of electrons means that the oxidation number of some of the atoms involved in the redox reaction must have changed. The oxidation number is defined by a simple set of rules based on arbitrarily counting electrons:

- The sum of the oxidation numbers of all the atoms in a molecule is zero.
- The sum in an ion equals the charge on the ion.
- Oxidation numbers are on a per-atom basis.
- Oxygen usually has an oxidation number of 2 (except in peroxides), and hydrogen has an oxidation number of 1.
- Monatomic ions (e.g., Na^+) have an oxidation number equal to the charge on the ion.

WRITING AND INTERPRETING CHEMICAL EQUATIONS

Chemical reactions in aqueous solution are described very efficiently by writing a chemical equation in which the component ions of dissolved ionic compounds are written as separate ions — e.g., K^+ and I^- rather than KI. These chemical equations are called *ionic equations.* Reactions that involve spectator ions can be written in a form in which the spectator ions are deleted (do not appear). These equations are called *net ionic equations.* The following simple rules should enable you to write most net ionic equations for a reaction:

- Soluble ionic salts are written as separate ions.
- Insoluble compounds are written as complete formula units with the subscript (s).
- Covalent compounds (e.g., CO_2, H_2O) are written as molecules.

The example that follows develops the net ionic equation for a precipitation reaction between silver nitrate solution and hydrochloric acid solution.

Since the two solutes $AgNO_3$ and HCl are obviously soluble and therefore dissolved, we can write the reactants as

$$Ag^+ + NO_3^- + H^+ + Cl^-$$

understanding that all the ions are hydrated. In this reaction an off-white precipitate comes out of solution. The precipitate must be either $AgCl_{(s)}$ or $HNO_{3(s)}$. In this instance it is easy to make the decision because we know that all common inorganic acids such as HNO_3 are soluble in water. The precipitate must be $AgCl_{(s)}$, and the spectator ions are H^+ and NO_3^-. The ionic equation is

$$Ag^+ + NO_3^- + H^+ + Cl^- \rightleftharpoons AgCl_{(s)} + H^+ + NO_3^-$$

The net ionic equation is

$$Ag^+ + Cl^- \rightleftharpoons AgCl_{(s)}$$

The long form of this equation would be

$$AgNO_3 + HCl \rightleftharpoons AgCl_{(s)} + HNO_3$$

Not only is the net ionic equation more concise, but it also suggests that any soluble Ag^+ salt and any soluble Cl^- salt would give the same reaction. Hence, many chemical reactions are summarized in this single net ionic equation.

Pre-Laboratory Quiz

1. Who asked the question "Did you ever stand and shiver Just because You were lookin' at a river?"

2. What is the chemical name for bone?

3. Give a definition of a solution.

4. In an aqueous solution the solubility of the solute is dominated by 3 interactions. What are they?

5. What makes water have such strong hydrogen bonding?

6. Give a chemical equation that describes the reaction of an aluminum ion with water.

7. Give an example of a nonacidic cation.

8. What is the formula for a carbonate ion?

9. Write the net ionic equation for the reaction of dissolved $AgNO_3$ with dissolved NaCl in aqueous solution.

10. What is the definiton of oxidation?

Laboratory Experiments

Flowchart of the Experiments

Section A.	Naming and Making Solutions
Section B.	Solubility and Solutions
Section C.	Solutions and Reactions
Section D.	Four Major Types of Chemical Reaction in Aqueous Solution
Section E.	A Chemical Reaction Survey: The Ion Reaction Chart
Section F.	Identification of Three Unknowns
Section G.	Five Unknowns: Solo
Section H.	Finding General Solubility Rules

Requires two three-hour class periods to complete

In this two-laboratory-period sequence of experiments, you can investigate a) the chemical factors that control solubility; b) the properties of solutions; c) chemical reactions in aqueous solutions; d) the correlation of chemical phenomenology, nomenclature, and reaction writing; and e) the relationships between chemical reactivity and the periodic chart. You will have ample opportunity to test your understanding of the chemical principles by analyzing the data from several experiments that involve identifying unknown solutions.

CAUTION: Assume that all the chemicals and solutions used in these laboratories are either toxic or corrosive, including those chemicals that are the products of chemical reactions. Even though you are using very small amounts and volumes, exercise caution and follow the appropriate methodologies for cleanup and disposal (discussed in each section). If you get chemicals on any part of your body, wash with cold water and check with your instructor.

Section A. Naming and Making Solutions

Goal: *To write the chemical formula for fourteen solutes.*

Discussion: Most of the equipment and the solutions which are required for this series of experiments is set out at your table.

Equipment List

- Straws
- Microstirrer
- Wash bottle filled with distilled water
- Microtowels
- Waste disposal cup
- Hand lens
- Scissors
- Tweezers
- Plastic reaction surface

Other chemicals, etc., will be passed around the class or are available at Reagent Central.

Experimental Steps:

1. Have your hand lens, scissors, and tweezers with you.
2. Remove the reaction matrix chart labelled Figure 7.7, located after Section H, at the end of this chapter. In order to keep the sheet flat, cut off the serrated edges with your scissors. Insert the chart between the outer sheets of the plastic reaction surface.
3. Obtain a 24-well tray containing 14 empty and 2 filled microburets.

Each empty microburet is labelled with the formula for one of the 14 solutions that you will be using in this chapter's experiments. The 2 filled microburets contain test solutions — universal acid-base indicator (labelled UI) in one, and starch/KI solution (labelled starch) in the other.

4. Use good transfer technique to fill each empty microburet with the appropriate solution.

CAUTION: Make sure you fill each microburet with the correct solution indicated by the label. A mistake in filling the microburets will cause all the chemical reactions to be incorrect.

5. Now, let's see how many chemical names you know. Complete the table below, matching the formula on the label of the microburet to a name in the table.

NOTE: The test solutions are not listed in the table.

Name	Chemical Formula of Solute	Concentration
Sulfuric acid		1 M
Sodium phosphate		0.2 M
Sodium hydroxide		1 M
Sodium carbonate		0.2 M
Silver nitrate		0.1 M
Potassium iodide		0.2 M
Nitric acid		1 M
Lead nitrate		0.2 M
Hydrochloric acid		1 M
Iron (III) chloride		0.1 M
Copper (II) sulfate		0.2 M
Barium chloride		0.1 M
Ammonia		1 M
Aluminum chloride		0.3 M

6. How well did you do? Check with your instructor for the answers.

These solutions, which were made by the laboratory staff, are *aqueous* solutions. The correct amount of solute was dissolved in distilled water (the solvent), and the resulting solution was diluted with water to a known volume to produce a solution of exactly known *concentration* (which the table identifies). Before you start exploring chemical reactions, it is necessary to know some of the chemical properties of the *individual* solutions. In order to make a solution in the manner described earlier, you must know something about the *solubility* of the solute in water. Some general principles about the factors controlling solubility and about the nature of solutions were presented in the Background Chemistry section of this chapter.

Section B. Solubility and Solutions

Goal:

To examine some of the general principles governing the solubility of various solute in water and other solvents.

Discussion:

Unless special techniques are used (they weren't), solutions which are made in the presence of gases (e.g., the atmosphere) will contain dissolved gases.

- Which of the following two gases, ammonia or carbon dioxide, do you think would have the greatest solubility in distilled water? Explain your answer with a simple molecular picture.

(a) $H-\ddot{N}(-H)-H$

ammonia
(polar covalent molecule)

(b) $O=C=O$

carbon dioxide
(nonpolar covalent molecule)

The solubility of liquids in water follows the same chemical principles. Here is some interesting data for the solubility of different alcohols in water.

Name	Formula	Solubility (moles L^{-1} at 25 °C)
Ethanol	C_2H_5OH	Infinite
1-pentanol	$C_5H_{11}OH$	0.25
1-decanol	$C_{10}H_{21}OH$	0.00024

- Explain the data on the basis of a simple molecular picture.
- Speaking of the solubility of liquids, how can you remove ball point pen ink from the best Grateful Dead T-shirt?

Now, more to the point of the solutions in front of you. Of these solutions, 10 were made by dissolving ionically bonded solids in water and 4 by diluting pure liquid reagents (HCl, HNO_3, H_2SO_4, and NH_3) with water. The only solute that is not appreciably dissociated upon dissolution in water is ammonia.

- Give a simple explanation (with a picture) of the dissolution of solid potassium iodide in water to give a solution.
- Three of the solutions are going to be *acidic*. They are made by dissolving acids in water. Which are they?

Experimental Steps:

1. Use the plastic surface with the inserted reaction matrix chart, Figure 7.7, to carry out the following tests: Drop 1 drop of each of the acids (about 1 cm apart) onto a white background in the *test* area (indicated on the reaction matrix chart) of the plastic sheet.

NOTE: Use the test area to mix solutions until indicated otherwise, in Section E.

2. To each drop add 1 drop of UI (universal indicator).
   - Record color changes. What happens with distilled water and UI?
3. Now test 1 drop of the ammonia solution with UI.
   - Note changes.
4. Try this one; put 1 drop UI *next* to a NH_3 drop.
   - What do you conclude?

   NOTE: Recall that ammonia is a base.
5. Try testing sodium hydroxide solution.

   To clean up the waste, add water from a wash bottle to dilute and mix all the drops. Suck the waste liquid up with a large-drop microburet and eject to waste cup. Wipe with a damp microtowel. This is a general procedure for disposing of all tests.

   NOTE: If the waste cup is full, transfer the solution to the appropriately labelled container at the side of the room.

   The UI test solution gives you a practical way to distinguish between acids and bases. A simple way of writing the dissociation of an acid (e.g., HCl) in aqueous solution is

$$HCl \rightarrow H^+ + Cl^-$$

NOTE: The test solution is responding to the H^+ ions.

Similarly, the dissociation of a base, e.g. NaOH, is written

$$NaOH \rightarrow Na^+ + OH^-$$

and the test solution responds to OH^- ions.

The table below is a list of ions and molecules that you will encounter during these experiments.

Cations		Anions		Molecules	
Formula	Name	Formula	Name	Formula	Name
H^+	hydrogen ion	Cl^-	chloride ion	H_2O	water
NH_4^+	ammonium ion	I^-	iodide ion	NH_3	ammonia
Na^+	sodium ion	OH^-	hydroxide ion	CO_2	carbon dioxide
Ag^+	silver ion	NO_3^-	nitrate ion	I_2	iodine
K^+	potassium ion	HCO_3^-	bicarbonate ion		
Ba^{2+}	barium ion	CO_3^{2-}	carbonate ion		
Pb^{2+}	lead ion	SO_4^{2-}	sulfate ion		
Cu^{2+}	copper (II) ion	PO_4^{3-}	phosphate ion		
Fe^{3+}	iron (III) ion				
Al^{3+}	aluminum ion				

Notice that one of the cations (NH_4^+) and several anions (e.g., NO_3^-) are polyatomic ions — i.e., 2 or more atoms are covalently bonded together in the ion. Water does not break the covalent bonds between the nonmetal atoms of a polyatomic ion, and these ions will generally stay intact in chemical reactions.

6. The remaining 9 untested solutions are solutions of salts as solutes. Test 1 drop of each solution with the UI test solution.
   - Note any color changes.
   - Write a simple dissociation and a reaction to explain your results for $AlCl_3$ and for Na_2CO_3 solutions.
   - Give a simple picture of the actual state of the ions present in a solution of KI.
7. Clean up and dispose of the waste as before.
   - Here is a strange question for you, but one you might be able to answer if you think about the general principles involved in this last series of experiments. Skywriting from airplanes is done by spraying a fine spray of $TiCl_4$ into the atmosphere. The radius of the Ti^{4+} ion is 74 pm. Explain the chemistry of skywriting.

Section C. Solutions and Reactions

Goal: *To observe the results of dissolution, diffusion, and reaction of KI and $Pb(NO_3)_2$.*

Experimental Steps:

1. Plastic containers of solid, crystalline KI and $Pb(NO_3)_2$ will be circulated around the room. Use a *clean,* dry, straw spatula to transfer 2 or 3 small crystals of each compound to the plastic surface.

   CAUTION: Cut off the end of the straw or clean it and dry it before going from one solid to the other!

2. Push 1 crystal of KI and 2 or 3 crystals of $Pb(NO_3)_2$ together with a straw. Observe with your hand lens. Add 1 drop of water. Stir with a plastic stirrer.
   - What happens?
3. Make a pool of water about 1.5 cm in diameter on the plastic by expelling water from a microburet.
4. Carefully push 1 crystal KI near one side of the pool, but *not* in it. Push a few crystals of $Pb(NO_3)_2$ near the other side. The position of the crystals is illustrated below.

$Pb(NO_3)_{2(s)}$ H_2O $KI_{(s)}$

5. Gently push the $Pb(NO_3)_2$ crystals onto the edge of the pool. Wait a few seconds. Now do the same with the KI. Use your hand lens to examine the reaction.
   - Explain the chemistry of what you see.
   - Predict exactly what will happen if 1 drop of KI solution is added to 1 drop of $Pb(NO_3)_2$ solution.
   - What would happen if they were added the other way around?
   - What is the most important reason for using solutions to carry out chemical reactions of this type?
6. Clean up as usual, remembering that lead compounds are toxic!

Section D. Four Major Types of Chemical Reaction in Aqueous Solution

Goal: *To investigate the four major types of chemical reactions which occur in aqueous solutions.*

Discussion: In the last section you carried out a chemical reaction between KI and $Pb(NO_3)_2$ in an aqueous medium. The reaction between these two reactants produced a solid product. In general, products that are not soluble in water and precipitate from solution are called *precipitates*, and the reactions are called *precipitation reactions*. Here we are, back at the principle of solubility again! In order to write the *net ionic reaction*, you first need to identify the precipitate. There are two possible choices.

- What are they?
- Which one of these is not soluble in water?
- Perhaps it's easier to ask which one *is* soluble in water.
- Now write the net ionic reaction.

Experimental Steps:

1. Use the UI test solution to investigate the reaction between HCl and NaOH solutions.

   This type of reaction is called an *acid-base reaction.*
   - Write the net ionic reaction, which by the way is the *same* for all acid-base reactions in aqueous solution.
   - Why is a test solution necessary in order to follow this reaction?
2. Now explore the reaction between $CuSO_4$ and NH_3 solutions by dropping 3 drops of NH_3 onto the plastic and adding 1 drop of $CuSO_4$. Stir.

   A change has obviously occurred. The net ionic reaction is

$$Cu^{2+} + 4NH_3 \rightarrow Cu(NH_3)_4^{2+}$$

   This is an example of a *complexation reaction.* The deep-blue soluble product $Cu(NH_3)_4^{2+}$ (illustration follows) is a *complex* called tetraamminecopper(II) ion.

Complexes are formed between cations and molecules or ions that have lone pairs of electrons (e.g., NH_3 in the above reaction).

$$Cu^{2+} + 4\,:NH_3 \rightleftharpoons [Cu(:NH_3)_4]^{2+} \quad \text{complex ion}$$

The bond between each NH_3 and Cu^{2+} arises from the lone pair on the N of NH_3 being *shared* with Cu^{2+}. Sometimes these bonds are called *coordinate-covalent bonds.*

3. Now add 2 more drops of $CuSO_4$ to the complex ion. Stir.

   The net reaction is

$$Cu^{2+} + 2OH^- \rightarrow Cu(OH)_{2(s)}$$

   - Now why did that happen? *Hint*: We swamped it with $CuSO_4$!

4. One more type of reaction. React 1 drop $CuSO_4$ and 1 drop KI. Stir. Ugly.

5. Now drop 1 drop of starch test solution very close to (but *not* in) the reaction mixture. Wait and watch.

   Starch/KI is an excellent probe for molecular iodine I_2. The reaction must have produced I_2 as a product. Since the KI solution contains I^-, then during the reaction

   (i) $2I^- \rightarrow I_2 + 2\,\text{electrons}$

   electrons are produced as I^- goes to I_2. Something must react with the electrons and it is Cu^{2+}:

   (ii) $2\,Cu^{2+} + 2\,\text{electrons} \rightarrow 2Cu^+$

   If we add these two reactions, (i) and (ii), we will get the net ionic reaction

$$2\,Cu^{2+} + 2I^- \rightarrow 2Cu^+ + I_2$$

   This type of reaction is called a *reduction-oxidation* reaction (redox) because it is a combination of reduction [gain of electrons by Cu^{2+}, reaction (ii)] and oxidation [loss of electrons by I^-, reaction (i)]. *Redox* reactions are reactions in which electrons are transferred from one chemical species to another. In this example, from I^- to Cu^{2+}.

6. Clean up as usual.

Section E. A Chemical Reaction Survey: The Ion Reaction Chart

Goals:

(1) To binary-mix the solutions. (2) To describe and record chemical changes. (3) To write net ionic reactions and identify the type. (4) To name the products of chemical reactions. (5) To correlate reactivity with trends in the periodic table.

NOTE: *Before* beginning with Step 1, read Section E completely.

Experimental Steps:

1. Binary-mix the 14 solutions provided in Section A (mix one solution with one other) in a dropwise manner on the rectangles (not the test area) of the plastic reaction surface (with the inserted reaction matrix chart).

   NOTE: Each rectangle corresponds to one mixing. The background is white and black so that you can easily see color changes and/or precipitates.

   - Figure 7.8, located at the end of the chapter, is a reaction description chart (actually a reaction matrix chart without heavy lines), which you should use to record what you see (the phenomenology) in the binary mixing.
   - Devise an accurate code that can be used to describe color changes, precipitate formation or disappearance, textures of precipitates, gas evolution, etc.
   - Write the net ionic reactions in your laboratory record as you go. You may need to review the background chemistry section to do this. Identify the type (e.g., precipitation) and give the name of the product.

   NOTE: The best way to mix is to drop 1 drop of a solution onto the rectangle. Then add 1 drop of the second solution, stir, add a second, stir.

   If you are in doubt about a reaction, feel free to play in the test area. Make sure you understand the arrangement of the reaction chart *before* you start!

   Use cotton swabs and microtowels to keep things neat. Don't write on the plastic sheet — it will destroy your flat reaction surface and drops will roll all over the place.

   CAUTION: *Drop* the drops — *do not* touch the solution on the plastic with the tip of the buret or you will contaminate everything!

2. When you have finished, leave the mixed solutions on the reaction chart.

Section F. Identification of Three Unknowns

Goal:

To identify three unknowns by comparison with the reactions of the 14 known solutions.

Experimental Steps

1. Obtain your 3 unknowns and place them into the 24 well tray.

   NOTE: One unknown is one of the 14 solutions. Do this one first.

   - Write the unknown number in your lab record and on the line provided in Figure 7.8, the reaction description chart, in the first rectangle of the third row from the bottom.

2. Binary-mix the first unknown with the other 14 solutions in the same manner as you mixed the known solutions.

   - Compare the results of the reactions of the 14 solutions listed on the reaction matrix chart with the reaction of your unknown. Remember that some time has elapsed and some changes may have occurred — e.g., bubbles may have flown away.
   - Identify the unknown solution.

3. Binary-mix the second unknown solution with the 14 known solutions.

   NOTE: The second unknown contains a *cation* that is one of the known solution cations and an anion that it is not associated with in one of the known solutions.

   - Identify the cation in the second unknown.

4. Binary-mix the third unknown solution with the 14 known solutions.

   NOTE: The third unknown contains an *anion* that is one of the known solution anions and a cation that it is not associated with in the known solutions.

   - Identify the anion in the third unknown.
   - Record the numbers and chemical identities of your unknowns in your laboratory record.

5. Clean the entire reaction surface.

Section G. Five Unknowns: Solo

Goal: *To identify five unknowns by simply binary-mixing them with each other.*

Experimental Steps:

1. Obtain a set of 5 unknowns (your instructor will assign them).

   The 5 unknowns are selected from the known solutions.

2. Binary-mix the 5 unknowns *only with each other* in the chart space labelled 5 Unknowns.

3. Identify the unknown solutions.

   - Record the results in your laboratory record.

Section H. Finding General Solubility Rules

Goal: *To look at all of your data in order to determine a minimum set of simple solubility rules that would be operative for all the combinations of ions that you have investigated.*

NOTE: This is not as difficult as it looks — e.g., you might begin by counting how many insoluble nitrates you have seen, and so on.

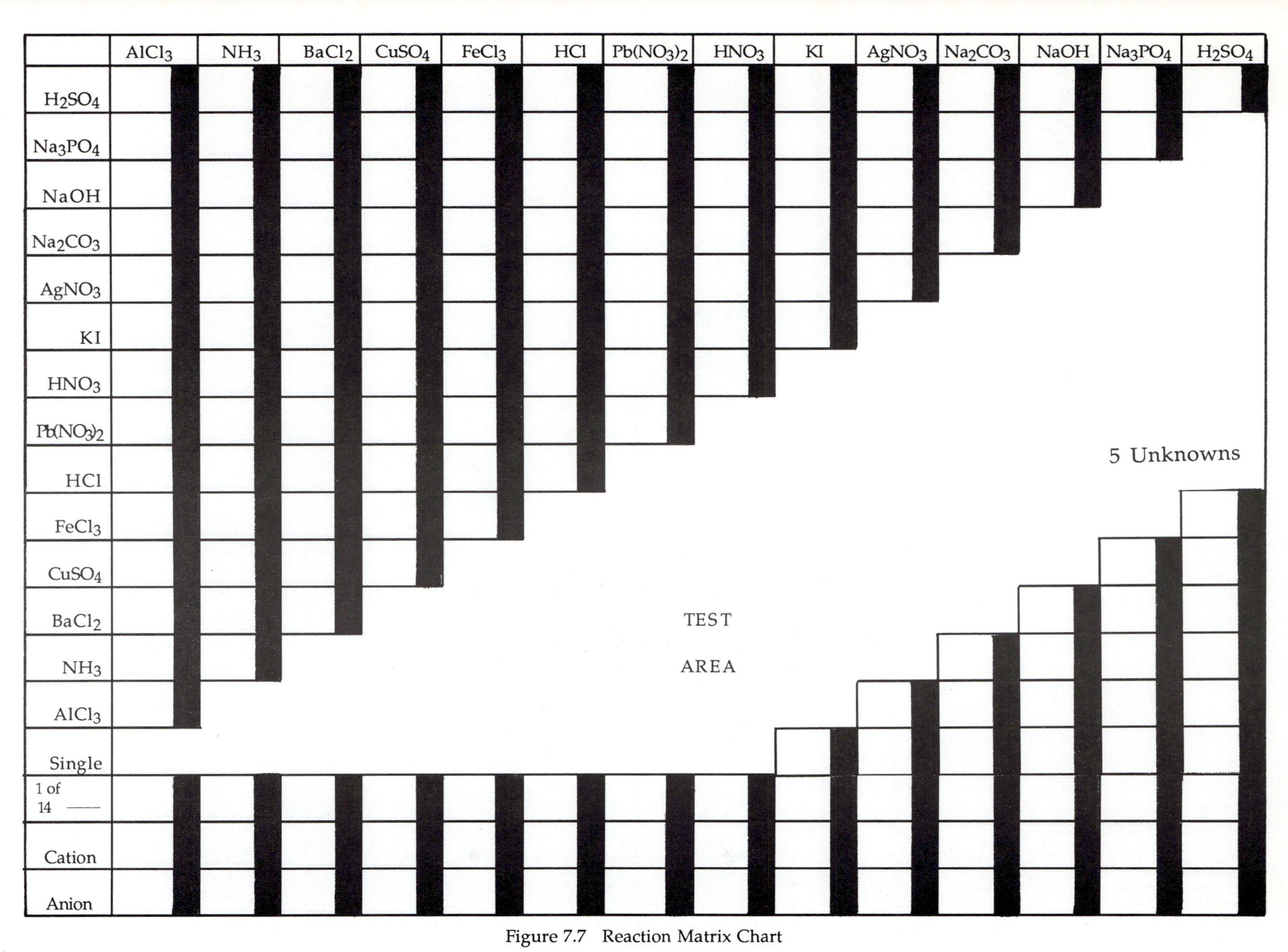

Figure 7.7 Reaction Matrix Chart

	$AlCl_3$	NH_3	$BaCl_2$	$CuSO_4$	$FeCl_3$	HCl	$Pb(NO_3)_2$	HNO_3	KI	$AgNO_3$	Na_2CO_3	NaOH	Na_3PO_4	H_2SO_4
H_2SO_4														
Na_3PO_4														
NaOH														
Na_2CO_3														
$AgNO_3$														
KI														
HNO_3														
$Pb(NO_3)_2$														
HCl														
$FeCl_3$														
$CuSO_4$														
$BaCl_2$														
NH_3														
$AlCl_3$														
Single														
1 of 14 ___														
Cation														
Anion														

5 Unknowns

TEST AREA

Figure 7.8 Reaction Description Chart

Chapter
8

An Introduction to Acids and Bases

Introduction

The subject of acids and bases is so enormous that it virtually comprises the whole of chemistry. The definition of an acid or base has evolved continuously throughout the history of chemistry. Acids have been characterized as substances that produce a sour taste on the tongue, whereas bases have a bitter taste. The development of nonsensory (and safer!) definitions began in the 1880s when Swedish chemist Svante Arrhenius developed a theory of ionization in aqueous solutions. He defined an acid as a substance that produced H^+ ions in solution, whereas a base produced OH^- ions. This chemical description stimulated later attempts to generalize the theory so as to include more substances. The Lowry-Brønsted theory (1932), which is still used extensively, defines an acid as a proton donor and a base as a proton acceptor. The Lewis theory emphasizes the electronic characteristics of substances, an acid being defined as an electron acceptor and a base as an electron donor. The most recent theory — the hard and soft acid-base theory (HSAB) of Pearson — is a general empirical system based on thousands of experimental observations. Whatever the definitions, the concept of acids and bases is a very useful classification that is widely applied in almost all areas of science and technology.

The widespread use and enormous industrial production of acids and bases is reflected in the fact that every year, six of the top ten chemicals produced in the United States are acids or bases. Billions of pounds of these compounds are used in an extraordinary variety of applications. Table 8.1 shows some of the major industrial uses of acids and bases. Sulfuric acid is so widely used in industry that its increased or decreased consumption has proved to be a dependable barometer of general business conditions.

The supermarket is full of acids and bases in all kinds of disguises. The detergent aisle is stacked with large cardboard packages of bases — e.g., sodium polyphosphate salts and anionic surfactants, which are the major ingredients in fabric and dishwashing detergents. Drain cleaners and oven cleaners often contain high concentrations of sodium hydroxide (lye). Window, floor, and brass cleaners almost always contain the base ammonia. And then of course there is the antacid aisle! Typical antacid products for the neutralization of gastric hydrochloric acid contain one or more bases — e.g., aluminum hydroxide, magnesium hyroxide, calcium carbonate, etc. Toilet bowl cleansers and lime deposit removers are highly acidic products.

Acids and bases are present in and often added to many food and beverage products. The "real" taste of cola beverages has more to do with the pleasantly sour taste of phosphoric acid than the other secret ingredients. Jams, jellies, preserves, sundae and yogurt fruit toppings often contain added malic or citric acid. Gelatin desserts, fruit popsicles, and some types of hard candy all contain acidulants. Acidulants are used for a variety of purposes in the manufacture of products for baking and of baked goods. They are employed, for instance, in the production of baking powders, refrigerated biscuit dough, fruit fillings for pies and cake, angel food cake, rye bread, and frozen pies. Strangely enough, the base sodium hydroxide is used in the production of maraschino cherries and grits!

And now it's raining...guess what?!

	Formula	Industrial Use
ACID	Sulfuric acid H_2SO_4	fertilizers, chemical and petroleum production, steel treatment
ACID	Phosphoric acid H_3PO_4	fertilizers, detergents, acid catalyst, acidulant for beverages
ACID	Nitric acid HNO_3	fertilizers, explosives, organic chemicals, dyes
BASE	Ammonia NH_3	manufacture of HNO_3, synthetic fibers, fertilizers, chemicals
BASE	Lime CaO, $Ca(OH)_2$	agriculture, water and waste treatment
BASE	Sodium hydroxide $NaOH$	chemicals, alumina, food production
BASE	Sodium carbonate Na_2CO_3	chemicals, soaps, glass, textiles, water softening

Table 8.1 The Major Industrial Uses of Common Acids and Bases

Background Chemistry

One of the earliest methods of differentiating acids, bases, and salts was by observing their reaction with naturally occurring plant pigments called anthocyanins. Robert Boyle, in his book entitled *Experiments and Considerations Touching Colours* (1664), described an extensive survey of the color changes produced by the action of acids, bases, and salts on flower and vegetable extracts. In the pages of the book reproduced in Figure 8.1, Boyle describes not only the acid-base reactions, but also an ingenious magic writing trick (which is worth trying when you get to the laboratory!).

To decipher Boyle's experiments, read *f*, which is printed in the middle or beginning of words, as *s*. The modern names for some of the compounds used are:

- Spirit of vitriol - sulfuric acid, H_2SO_4
- Juice of limmons - lemon juice (citric acid)
- Spirit of salt - hydrochloric acid, HCl
- Solution of potashes - potassium carbonate, K_2CO_3
- Spirit of vinegar - acetic acid, CH_3COOH
- Spirit of harts-horn - ammonia, NH_3
- Spirit of urine - ammonia, NH_3

(246)

who have produc'd the like, by Spirit of Vitriol, or juice of Limmons, but have Groundlefsly afcrib'd the Effect to fome Peculiar Quality of thofe two Liquors, whereas, (as we have already intimated) almoft any Acid Salt will turn Syrrup of Violets Red. But to improve the Experiment, let me add what has not (that I know of) been hitherto obferv'd, and has, when we firft fhew'd it them, appear'd fomething ftrange, even to thofe that have been inquifitive into the Nature of Colours; namely, that if inftead of Spirit of Salt, or that of Vinegar, you drop upon the Syrrup of Violets a little Oyl of Tartar *per Deliquium*, or the like quantity of Solution of Potafhes, and rubb them together with your finger, you fhall find the Blew Colour of the Syrrup turn'd in a moment into a perfect Green, and the like may be perform'd by divers other Liquors, as we may have occafion elfewhere to Inform you.

Annotation upon the twentieth Experiment.

The ufe of what we lately deliver'd concerning the way of turning Syrrup of Violets, Red or Green, may be this; That, though it be a far more common and procurable

(247)

curable Liquor than the Infufion of *Lignum Nephriticum*, it may yet be eafily fubftituted in its Room, when we have a mind to examine, whether or no the Salt predominant in a Liquor or other Body, wherein 'tis Loofe and Abundant, belong to the Tribe of *Acid* Salts or not. For if fuch a Body turn the Syrrup of a Red or Reddifh Purple Colour, it does for the moft part argue the Body (efpecially if it be a diftill'd Liquor) to abound with Acid Salt. But if the Syrrup be made Green, that argues the Predominant Salt to be of a Nature repugnant to that of the Tribe of Acids. For, as I find that either Spirit of Salt, or Oyl of Vitriol, or *Aqua-fortis*, or Spirit of Vinegar, or Juice of Lemmons, or any of the Acid Liquors I have yet had occafion to try, will turn Syrrup of Violets, of a *Red*, (or at leaft of a *Reddifh* Colour, fo I have found, that not only the Volatile Salts of all Animal Subftances I have us'd, as Spirit of Harts-horn, of Urine, of Sal-Armoniack, of Blood, &c. but alfo all the Alcalizate Salts I have imploy'd, as the Solution of Salt of Tartar, of Pot-afhes, of common Wood-afhes, Lime-water, &c. will immediately change the Blew Syrrup, into a perfect Green. And by the fame way (to hint that upon

R 4 the

Figure 8.1 Robert Boyle's Description of the Reactions of Acids and Bases with Some Plant Pigments

(248)

the by) I elsewhere show you, both the changes that Nature and Time produce, in the more Saline parts of some Bodies, may be discover'd, and also how ev'n such Chymically prepar'd Bodies, as belong not either to the Animal Kingdome, or to the Tribe of *Alcali's*, may have their new and supern...'d Nature successfully Examin'd. In this place I shall only add, that not alone the Changing the Colour of the Syrrup, requires, that the Changing Body be more strong, of the Acid, or other sort of Salt that is Predominant in it, than is requisite for the working upon the Tincture of *Lignum Nephriticum*; but that in this also, the Operation of the formerly mention'd Salts upon our Syrrup, differs from their Operation upon our Tinctures, that in this Liquor, if the Cæruleous Colour be *Destroy'd* by an Acid Salt, it may be *Restor'd* by one that is either Volatile, or Lixiviate; whereas in Syrrup of Violets, though one of these contrary Salts will *destroy* the Action of the other, yet neither of them will *restore* the Syrrup to its native Blew; but each of them will Change it into the Colour which it self doth (if I may so speak) affect, as we shall have Occasion to show in the Notes on the twenty fifth Experiment.

EXPE-

(249)

EXPERIMENT XXI.

There is a Weed, more known to Plowmen than belov'd by them, whose Flowers from their Colour are commonly call'd *Blew-bottles*, and *Corn-weed* from their Growing among Corn. These Flowers some Ladies do, upon the account of their Lovely Colour, think worth the being Candied, which when they are, they will long retain so fair a Colour, as makes them a very fine Sallad in the Winter. But I have try'd, that when they are freshly gather'd, they will afford a Juice, which when newly expres'd, (for in some cases 'twill soon enough degenerate) affords a very deep and pleasant Blew. Now, (to draw this to our present Scope) by dropping on this fresh Juice, a little Spirit of Salt, (that being the Acid Spirit I had then at hand) it immediately turn'd (as I predicted) into a Red. And if instead of the Sowr Spirit I mingled with it a little strong Solution of an Alcalizate Salt, it did presently disclose a lovely Green; the same Changes being by those differing sorts of Saline Liquors, producible in this *Natural juice*, that we lately mention'd to have

Herbarists are wont to call this Plant Cyanus vulgaris minor.

(250)

have happen'd to that *factitious Mixture*, the Syrrup of Violets. And I remember, that finding this Blew Liquor, when freshly made, to be capable of serving in a Pen for an Ink of that Colour, I attempted by moistning one part of a piece of White Paper with the Spirit of Salt I have been mentioning, and another with some Alcalizate or Volatile Liquor, to draw a Line on the leisurely dry'd Paper, that should, ev'n before the Ink was dry, appear partly Blew, partly Red, and partly Green: But though the latter part of the Experiment succeeded not well, (whether because Volatile Salts are too Fugitive to be retain'd in the Paper, and Alcalizate ones are too Unctuous, or so apt to draw Moisture from the Air, that they keep the Paper from drying well) yet the former Part succeeded well enough; the Blew and Red being Conspicuous enough to afford a surprizing Spectacle to those, I acquaint not with (what I willingly allow you to call) the *Trick*.

Annotation upon the one and twentieth Experiment.

But lest you should be tempted to think (*Pyrophilus*) that Volatile or Alcalizate Salts

(251)

Salts change Blews into Green, rather upon the score of the easie Transition of the former Colour into the latter, than upon the account of the Texture, wherein most Vegetables, that afford a Blew, seem, though otherwise differing, to be Allied, I will add, that when I purposely dissolv'd Blew Vitriol in fair Water, and thereby imbu'd sufficiently that Liquor with that Colour, a Lixiviate Liquor, and a Urinous Salt being Copiously pour'd upon distinct Parcels of it, did each of them, though perhaps with some Difference, turn the Liquor not Green, but of a deep Yellowish Colour, almost like that of Yellow Oker, which Colour the Precipitated Corpuscles retain'd, when they had Leisurely subsided to the Bottom. What this Precipitated Substance is, it is not needfull now to Enquire in this place, and in another, I have shown you, that notwithstanding its Colour, and its being Obtainable from an Acid *Menstruum* by the help of Salt of Tartar, it is yet far enough from being the true Sulphur of Vitriol.

EXPERIMENT XXII.

Our next Experiment (*Pyrophilus*) will perhaps seem to be of a contrary Nature to

Figure 8.1 (continued) Robert Boyle's Description of the Reactions of Acids and Bases with Some Plant Pigments

The use of red cabbage extract as an acid-base indicator was first described by James Watt in 1784 (by the way, he also invented the steam engine). These natural extracts are still used as indicators, but the chemistry is very complicated and was not understood until the work of Dubois in 1978 and Thompson in 1984. Synthetic dyes are now used extensively as acid-base indicators in the field of acid-base chemistry, including acid-base titrations, pH measurement, etc.

Indicators may be used to follow the progress of any acid-base reaction. Consider the reaction that occurs when a solution of the base sodium hydroxide (NaOH) is added to a solution of the acid hydrochloric acid (HCl). The reaction proceeds according to

$$NaOH + HCl \rightarrow NaCl + H_2O$$

which may be summarized by the net ionic equation

$$OH^- + H^+ \rightarrow H_2O$$

This type of reaction is called a *neutralization*. The above chemical equation says that 1 mole of NaOH will react with 1 mole of HCl to produce 1 mole of NaCl (sodium chloride, salt) and 1 mole of H_2O. The ratio 1:1:1:1 is called the *stoichiometry* of the reaction. If the concentration (moles per liter) of one of the reactant solutions is known accurately, then an analysis for an unknown concentration of the other can be carried out by volumetric analysis. *Volumetric analysis* is the process whereby the amount of a substance is determined by measuring the volume of a solution of known concentration (called a standard solution) that reacts with it. The standard solution is usually added by means of a buret to the second solution until all the reactant in the second solution has been consumed. This operation is called a *titration*. The point in the titration when the exact stoichiometric amount of reagent has been added to react with the other reactant in solution is called the *equivalence point*. It is necessary to find some way of determining when this point has occurred. Acid-base color indicators may be used to monitor the progress of a titration. A change in color of the indicator occurs at the end point of the titration, which is usually close to the equivalence point.

In the specific titration introduced at the beginning of the last paragraph, the solution starts off being acidic because HCl is present, and an indicator would show its acid color. At the equivalence point the only substance in solution would be sodium chloride (salt), and the indicator would be in the middle of a change from its acid color to its base color. Just after the equivalence point, the solution is basic due to a slight excess of NaOH, and the indicator exhibits its basic color. A sharp color change should occur at the end point.

The traditional apparatus that is used to carry out volumetric analysis consists of calibrated pyrex glass pipets and burets. Most of these classical volumetric containers are analog devices in the sense that the measurement of volume is made by estimating the position of a liquid meniscus on a graduated scale. In most of the experiments in this book, the quantitative delivery of incremental volumes of reagent solutions is carried out by digital pipets and microburets. These nontraditional devices can easily be made from commercially available, nonwetted polyethylene pipets. The volume increment is the drop, and total volume is measured by counting and calibration of the microburet. Calibration can be achieved in a number of ways, depending on several factors: (a) how accurate you wish to be, (b) how much time you are willing to spend on the calibration, (c) what type of balance equipment you have access to, and (d) how much money you wish to spend. It is important to point out that the successful use of plastic digital microburets in volumetric analysis requires a fundamental knowledge of the principles of liquid transfer. It is also worth noting that the use of microprocessor-controlled digital microburets is quite the norm in industrial, clinical, and pharmaceutical research laboratories.

One of the major experiments in this laboratory is the determination of the calcium carbonate content of eggshells. Most birds produce eggs that have a shell consisting largely of tiny crystals of calcium carbonate laid down on a collagen-protein network. Good shell formation is critical to the survival of the chick embryo and, therefore, the species. The shell not only provides a strong protective covering, but also acts both as a source of calcium and as a respiratory membrane of the embryo. Any environmental factor leading to a decrease in the amount of calcium carbonate in the shell could result in shells too thin and fragile to survive. Unfortunately, there is now firm evidence that halogenated hydrocarbons can upset the delicate mechanism of shell formation.

Birds have evolved extraordinary biochemical processes that enable them to rapidly produce the massive quantities of calcium carbonate required for shell formation. The breeding cycle of birds starts when sex hormones trigger changes in the reproductive organs and promote the storage of a supply of calcium for the eggs. The calcium is stored in the form of new bone growth in the marrow cavities of most of the hen's bones. After ovulation the yolk travels slowly down the oviduct where layers of albumen (egg white) are laid down (four hours) around the yolk, followed by two membranes. Fibrous growths, consisting of a protein-mucopolysaccharide material appear on the outside of the membrane. This material binds and orients calcium ions so that the ions act as seeds for the later growth of calcium carbonate crystals that form the shell. Over the next five hours, water and salts move through the membranes and into the egg in a process called *plumping*. The main process of shell formation occupies the next 15 – 16 hours. This process takes place in the shell gland, which has a rich supply of blood containing large amounts of calcium ions bound to a protein called phosvitin and some free calcium ions. The calcium ions react with carbonate ions produced in the shell gland to form calcium carbonate crystals.

The average weight of the shell of a chicken egg is about 5 g. About 2 g of this is calcium. Since shell formation takes 15 hours, the calcium is deposited at the rate of 2000 mg per 15 hours or 133 mg per hour. The *total* calcium content of the blood of a chicken is about 25 mg. This means that every 11 minutes the total amount of calcium circulating in the blood disappears to form calcium carbonate crystals in the shell.

The calcium must come from somewhere or the shell would not form. Food is one source, but it has been shown that it is not the complete answer because the digestive system cannot supply the calcium fast enough. The deficit is made up by the breakdown of the marrow bone growths. The released phosphate is excreted in the urine. These coordinated biochemical processes have been shown to be disrupted by chemicals present in the environment. Chlorinated hydrocarbon pesticides (e.g., DDT and Dieldrin) and polychlorinated and polybrominated biphenyls (PCBs and PBBs which are used as hydraulic fluids, electrical transformer fluids, and plasticizers), have all been found to accumulate in birds. Research has shown that these substances lead to abnormally late breeding in birds, a reduction in the number of eggs, and a dramatic reduction in calcium carbonate content. The latter causes thinning and breakage of the shell — a sure way to kill the fetus. The problem has been found to be general, and a variety of species have been decimated. Peregrine and prairie falcons and other raptor species have almost become extinct on the American continent.

The environmental problems associated with halogenated hydrocarbons were recognized as a result of an incredible amount of research in many disciplines, from agriculture to oology (the study of eggs). Extremely useful information was obtained from chemical research on eggshell thickness and from data on the calcium carbonate content of eggshells. The culmination of this research led to the banning of DDT in the United States (1972) and to reductions in the production of PCBs and PBBs by Monsanto, the sole supplier.

The analytical determination of calcium carbonate in eggshell cannot be carried out simply by titrating the carbonate with acid to the end point because the reaction is too slow. A good way around this problem is to first add an excess of hydrochloric acid to completely dissolve the calcium carbonate and then to titrate the unreacted acid with a standard solution of a soluble base, e.g., sodium hydroxide. This indirect method is sometimes called a *back titration*. The dissolution reaction is

$$CaCO_{3(g)} + 2HCl \rightarrow CaCl_2 + CO_{2(g)} + H_2O$$

and the titration reaction is

$$NaOH + HCl \rightarrow NaCl + H_2O$$

Acid-base volumetric analysis can therefore provide a simple, rapid, and inexpensive method of determining the calcium carbonate content of eggshells. This type of analysis is equally useful for determining the calcium carbonate content of seashells.

Pre-Laboratory Quiz

1. Give one definition of an acid and a base.

2. What is one major use of sulfuric acid?

3. Name a supermarket product that contains a base.

4. Cola drinks contain an acid. What is the acid's formula?

5. Which chemist first differentiated acids, bases, and salts by using color reactions of anthocyanins?

6. What is meant by the stoichiometry of an acid-base reaction? Give an example.

7. Define the equivalence point of a titration.

8. How many grams of sodium hydroxide does 1.0 mL of 1.0 M NaOH solution contain?

9. What types of chemical compounds have been found to interfere with the biological production of calcium carbonate in bird eggshell formation?

10. Give an example of a sparingly soluble base.

Laboratory Experiments

Flowchart of the Experiments

Section A.	Common Laboratory Acids and Bases: Necessary Facts and Some Questions

Section B.	Indicator Color Probes for Acids and Bases

Section C.	Microburet Construction

Section D.	Acid-Base Titrations; Stoichiometry and Molarity

Section E.	Egg- and Seashell Analysis

Section F.	Acid Concentration in Fruits: Pucker Order

Requires one three-hour class period to complete

CAUTION: **You will be investigating the chemical reactions of several acids and bases in this sequence of experiments. Regard all of these solutions as being potentially corrosive and dangerous. Even though you will be using very small volumes of these dilute solutions, you should exercise due care when transferring, storing, and reacting solutions. If you spill any of these chemicals, wash with cold water and check with your instructor.**

Section A. Common Laboratory Acids and Bases: Necessary Facts and Some Questions

Goal:

To give you some important facts about the common laboratory acids and bases so that you can become familiar with safe procedures for working with these substances.

Discussion:

The common laboratory acids (see the table below) are all very corrosive, highly dangerous liquids when in concentrated form.

	Name	Formula	Formula Weight	Color	Density g mL^{-1}	Percent Weight	Molarity Mol L^{-1}
COMMON LABORATORY ACIDS (concentrated)	Sulfuric acid	H_2SO_4	98.08	colorless	1.841	95–98	18.0
	Hydrochloric acid	HCl	36.46	colorless liquid	1.18	36.5–38	12.0
	Nitric acid	HNO_3	63.01	colorless liquid	1.503	70.3	16.8
	Acetic acid	CH_3COOH	60.05	colorless liquid	1.049	99–100	17.5
	Phosphoric acid	H_3PO_4	98.00	colorless liquid	1.834	85	14.7
COMMON LABORATORY BASES	Sodium hydroxide	$NaOH_{(s)}$	40.00	white pellets	2.130	--------	--------
	Potassium hydroxide	$KOH_{(s)}$	56.11	white pellets	2.044	--------	--------
	Ammonia (ammonia solution)	$NH_{3(g)}$ $NH_{3(aq)}$	17.03 17.03	colorless gas colorless liquid	0.000771	28–30 NH_3	14.8
	Sodium carbonate	$Na_2CO_{3(s)}$	105.99	white granular powder	2.532	--------	--------

These chemicals are generally delivered to laboratories in small (5 pint) glass bottles. A typical label from a bottle of acid is shown below.

Cat. # **2876-9x6** Size **9 LB.** (2.3L) **AR®** Analytical Reagent

LOT 3553 KBLX

It is hereby certified that the following is a true copy of the actual analysis of the lot indicated.

C. E. Elmore
Quality Control

Meets ACS Specifications	H_2SO_4
Certificate of Lot Analysis	F.W. 98.08
Assay (H_2SO_4)	96.0%
Ammonium (NH_4)	0.0001 %
Arsenic (As)	0.0000003%
Chloride (Cl)	<0.000005 %
Color	APHA 5
Heavy Metals (as Pb)	0.00002 %
Iron (Fe)	0.000001 %
Mercury	<5 ppb
Nitrate (NO_3)	<0.00001 %
Residue after Ignition	<0.0003 %
Specific Gravity	1.84
Substances Reducing $KMnO_4$(as SO_2) Test	Passes

Sulfuric Acid H_2SO_4

Poison

Corrosive

Storage Code
White

LABGUARD™
Health Hazard: Severe
Flammability: None
Reactivity: Moderate
Contact Hazard: Extreme

Protective Equipment:
Protective Eyewear
Hand Protection
Safety Clothing
Laboratory Hood

A/2 Printed in U.S.A. 487157-1 M

DANGER! Corrosive. Liquid and mist cause severe burns to all body tissue. May be fatal if swallowed. Harmful if inhaled. Inhalation may cause lung damage.

Do not get liquid in eyes, on skin, on clothing. Wash thoroughly after handling. Avoid breathing vapor. Keep container closed. Use with adequate ventilation. Do not add water to contents while in container because of violent reaction. Store in tightly closed container. KEEP OUT OF REACH OF CHILDREN.

Distributed by
AMERICAN SCIENTIFIC PRODUCTS
McGaw Park, Illinois 60085-6787
Packaged by
MALLINCKRODT INC.
Paris, Kentucky 40361
CAS 7664-93-9
Proper Shipping Name-
SULFURIC ACID, UN1830

FIRST AID: If swallowed, do NOT induce vomiting. Give large quantities of water or milk if available. **CALL A PHYSICIAN.** Never give anything by mouth to an unconscious person. **In case of contact,** immediately flush eyes or skin with plenty of water for at least 15 minutes while removing contaminated clothing or shoes. **CALL A PHYSICIAN. If inhaled,** remove to fresh air. If not breathing, give artificial respiration. If breathing is difficult, give oxygen. **CALL A PHYSICIAN.**

EPA: HWDC - Corrosive
*See LabGuard Guide and Material Safety Data Sheet for additional information.

NOTE: The warnings and instructions on the label should always be followed. Concentrated acids such as these often require dilution with water before use. Dilution should always be carried out by pouring the acid into the water with stirring. It is particularly important that concentrated sulfuric acid be diluted in this way. The dilution process generates an enormous amount of heat, which necessitates that the dilution be carried out in a pyrex container cooled by an ice bath.

- What chemical reaction generates all that heat?
- Given the following information, explain with the aid of a simple diagram what would happen if sulfuric acid were diluted incorrectly (with water):

Density of pure H_2SO_4	=	$1.86\ g\ mL^{-1}$
Density of H_2O	=	$1.00\ g\ mL^{-1}$
Boiling point of H_2SO_4	=	290 °C
Boiling point of H_2O	=	100 °C

The common laboratory bases are also highly corrosive and dangerous in the pure form. Sodium and potassium hydroxides are manufactured in the form of 0.5 cm pellets and are extremely *hygroscopic* — i.e., the solid attracts water from the atmosphere. This happens to such an extent that in a humid atmosphere, a pellet of the solid turns into a pool of solution in a short time.

- Explain why these bases are so hygroscopic.

Most pure bases are packaged in plastic containers because they dissolve glass. Ammonia is sold as a 30% solution of the gas dissolved in water. Great care must be exercised when opening and diluting concentrated ammonia solutions. The solution has a very high vapor pressure of NH_3 gas, which has a pungent Jovian odor and which quickly attacks the eyes and lungs. For these reasons ammonia solutions should be diluted in a hood.

- Weighing solid sodium hydroxide accurately on a balance is impossible. Why?
- Give a reason (and a reaction) why the dissolution of solid potassium hydroxide generates an enormous amount of heat.

Accurately known concentration solutions of the above acids and bases cannot be made by the usual procedure of weighing out the substance, dissolving it in solvent, and then diluting to a known volume. First, a solution of a very approximate concentration is made (with appropriate precautions taken); then further accurate dilution, along with standardization, is required.

- Describe exactly how you would make a 1.000 M solution of sulfuric acid starting from 100% pure liquid H_2SO_4.

Section B. Indicator Color Probes for Acids and Bases

Goal: *To investigate the use of a variety of color indicators as probes for acids and bases.*

Discussion: Some of the reagents for these and subsequent experiments are in microburets in a 24-well tray assigned to you. Others can be found at Reagent Central, and a few will be put in plastic bottles that may be passed around the class.

Experimental Steps:

1. Use a plastic surface with a white background for viewing color changes.
2. First let's investigate one of the oldest color probes known. Drop 1 drop of red cabbage extract (or if you don't like cabbage, red rose extract) into each of several wells of a clean 1 × 12 well strip.
   - Note the color of the extract in your laboratory record. Use your hand lens!
3. Drop 1 drop 0.1 M HCl into a well and stir with a microstirrer.
   - Observe and record any color change by comparison with the extract in the next well.

NOTE: It is worth pointing out here that one of the beauties of small-scale experiments is that comparisons of this type are remarkably easy to do, and subtle differences can often be discerned.

- Does the color change if more acid is added?

4. Now add 1 drop 0.1 M NaOH to another well and stir.
   - Observe and record the color change.

   NOTE: You now have a simple method of distinguishing between acids and bases.

5. Make a permanent record of these changes by putting 2 or 3 drops of extract onto filter paper, blotting off the excess, and hanging the paper up to dry (with a straw hanger perhaps!).

   While it's drying...
   - What color would the following solutions give with the extract?

     a) A sodium carbonate solution
     b) An ammonia solution
     c) Automobile battery fluid

   - Devise an experiment to show that the color probe is reversible. Report your results.

6. Drop 5 or 6 drops of 0.1 M NaOH into a well containing 1 drop of extract and stir.

7. Now carefully and gently try to layer a few drops of 0.1 M HCl over the NaOH so that the acid does not mix with the base. Tilt the strip.
   - Record what you see.

8. Stir again.
   - What happens?
   - Write down an equation for the chemical reaction you just probed (Steps 6, 7, and 8).
   - Do you think that the reaction between the acid and the base was fast or not? Explain.
   - Do you think that the probe response is fast?

9. Repeat the last experiment using a synthetic indicator probe, such as bromothymol blue (BTB).
   - Do you notice any differences compared to the extract?

Section C. Microburet Construction

Goal: *To construct and calibrate two microburets for use in a series of volumetric acid-base titrations.*

Experimental Steps:

1. Use a microburner to make a small-drop microburet and a pair of scissors to make a large-drop microburet.

   NOTE: You should be familiar with the required techniques for microburet construction. The detailed instructions are given in the earlier laboratory on small-scale techniques, colorimetry, and spectrophotometry.

2. Calibrate the microburets by the appropriate methods.

Section D. Acid-Base Titrations; Stoichiometry and Molarity

Goals:

(1) To carry out a series of acid-base titrations using microburets and synthetic indicator color probes. (2) To determine the stoichiometry of the acid-base reactions by suitable calculations. (3) To determine the molarity of unknown concentration solutions of acids and bases.

Experimental Steps:

1. Clean a 1 x 12 well strip thoroughly with water and shake out any liquid by slapping the strip against a paper towel held in your hand.
2. Clean a small-drop microburet by sucking distilled water into it and expelling it several times.

   NOTE: In all quantitative volumetric experiments it is necessary to use the *same microburet* for all dilutions and for delivering standard and unknown solutions.

   - Why?

   NOTE: It is also important to use good transfer and storage technique in order that the concentration of solutions not be changed inadvertently. All these techniques are used in the next steps in this section.

3. Fill a clean, dry (dried with paper towel) well of a 24-well tray about 3/4 full with 0.1 M HCl. The HCl solution is in a plastic bottle that will be passed around.
4. With a clean, empty small-drop microburet, suck up a little HCl and shake the bulb so that the liquid rinses all of the inside of the bulb. Expel to the waste cup.
5. Repeat with a little more HCl solution.
6. Now suck up HCl solution so that the bulb is about 1/2 full.
7. Use good delivery technique to deliver 5 drops of the 0.1 M HCl to each of 10 wells of the 1 x 12 well strip.
8. Add 1 drop BTB indicator to each well.
9. Now use good wash, storage, and transfer techniques to fill the same microburet about 1/2 full of 0.1 M NaOH solution.
10. Carry out a *serial* titration of the HCl in the 1 x 12 well strip: Add 1 drop 0.1 M NaOH to the first well and stir; wipe the stirrer clean; add 2 drops NaOH to the second well, stir; and so on.

    - Note the color changes and the number of drops at which the *end point* occurs.
    - Write down the overall equation and the net ionic equation for the acid-base reaction in this titration.
    - What experimental fact confirms the stoichiometry that you show in your reaction?

11. Obtain a sample of the unknown concentration solution and determine its molarity. Use 1 drop of the unknown solution for your titration.
    - Show how you calculate the molarity.
12. Carry out a duplicate analysis of the unknown concentration solution by titrating in a single-well titration rather than in a serial titration.

Section E. Eggshell and Seashell Analysis

Goal:

To determine the percentage of calcium carbonate in eggshells and seashells by dissolving a weighed shell sample in excess hydrochloric acid, and then backtitrating unreacted hydrochloric acid with sodium hydroxide.

Experimental Steps:

1. Obtain a powdered sample of dry eggshell or seashell from Reagent Central.
2. Use tweezers to place a clean, dry weighing scoop (made from a thin-stem pipet) onto the pan of an analytical balance.

   NOTE: Your instructor will show you how to use the analytical balance.
3. Weigh the weighing scoop.
    - Record the weight (preferably to the nearest 0.0001 g).
4. Use a slim straw spatula to transfer about 0.050 g of the shell powder to the scoop.
5. Weigh the shell powder plus scoop.
    - Record the weight.
6. Carefully transfer the shell powder to a well in the middle of a clean, 24-well tray by gently tapping the scoop. Don't worry if some particles of shell stick to the scoop.
7. Take the scoop back to the balance and reweigh the scoop (plus any particles that may be stuck to it).
    - Calculate the weight of the shell powder delivered to the well.
8. Add a drop of alcohol to the shell (it acts as a wetting agent).
9. Use good transfer technique to fill a calibrated, large-drop microburet with 2.00 M HCl and deliver exactly 20 drops to the shell powder in the well.

   NOTE: Use standard delivery technique to ensure that the drops are all of the same volume!
10. Use a thin-stem pipet to transfer some hot tap water to the spaces in the tray that are around the well containing the reaction mixture.
11. Wait about 10 minutes for the reaction to go to completion. Stir occasionally with a microstirrer and leave the microstirrer in the reaction mixture.

12. While you are waiting for the reaction to finish you should carry out an acid-base titration to standardize an approximately 1.0 M NaOH solution with 2.00 M HCl, using bromthymol blue as indicator.
    - Calculate the molarity of the sodium hydroxide solution.
13. Add one or two drops of bromthymol blue to the mixture in the well and stir.
14. Wash the large-drop microburet used to deliver the HCl in Step 9.
15. Use good transfer technique to fill the washed large-drop microburet with standardized 1.0 M NaOH.
16. Titrate the unreacted HCl in the well with the 1.0 M NaOH until an endpoint is obtained.
    - Note the number of drops of 1.0 M NaOH required.
17. If you have time, you should carry out a duplicate analysis on the same shell material.

CALCULATION OF CALCIUM CARBONATE CONTENT OF SHELL

- First calculate the number of moles of HCl added to the shell powder in Step 9.

  To calculate the number of moles of HCl, multiply the number of drops of HCl (20 drops) by the volume of 1 drop (in L, obtained from the calibration of the microburet), and by the molarity of the HCl (2.00 M).

  $$\text{moles HCl} = 20 \text{ drops HCL} \times \frac{\chi \text{ L}}{1 \text{ drop HCl}} \times 2.00 \frac{\text{mol}}{\text{L}}$$

  where $\frac{\chi \text{ L}}{1 \text{ drop HCl}}$ is the calibration factor.

- Now calculate the number of moles of NaOH required to titrate the unreacted HCl that was left over from the reaction with the $CaCO_3$ in the shell.

  The number of moles of NaOH is calculated by multiplying the number of drops of NaOH required in the titration by the volume of 1 drop (in L) and then by the molarity of the NaOH (obtained from the standardization).

- Calculate the number of moles of unreacted HCl left over from the reaction with the $CaCO_3$ in the shell.

  The number of moles of unreacted HCl is easy to calculate. It is simply equal to the number of moles of NaOH required in the titration, because the stoichiometry of the reaction

  $$HCl + NaOH \rightarrow NaCl + H_2O$$

  is 1:1:1:1.

- Calculate the number of moles of HCl that reacted with the $CaCO_3$ in the shell by subtracting the moles of the unreacted HCl from the moles of HCl originally added to the shell in Step 9.

- Calculate the number of moles of $CaCO_3$ that reacted with the HCl.

  The number of moles of $CaCO_3$ that reacted with the HCl is one-half of the number of moles of HCl that reacted with the $CaCO_3$ because of the stoichiometry of the reaction

$$CaCO_{3(s)} + 2HCl \rightarrow CaCl_2 + CO_{2(g)} + H_2O$$

- Calculate the number of grams of $CaCO_3$ by multiplying the number of moles of $CaCO_3$ by the molar mass of $CaCO_3$ (100 g mol^{-1}).
- Calculate the percentage $CaCO_3$ in the shell by dividing the number of grams of $CaCO_3$ by the number of grams of shell weighed out and multiplying by 100.
- Report your duplicate values of $CaCO_3$ percentage in your notebook.

Section F. Acid Concentration in Fruits: Pucker Order

Goal: *To design and execute an experiment to test the hypothesis that pucker power is proportional to the concentration of organic acids present in fruit.*

Discussion: There is a hypothesis that the pucker-promoting potential of fruits and fruit juices is directly proportional to the concentration of organic acids present in the fruit.

Experimental Steps: Design and execute an experiment to test this hypothesis. Your analysis should include acid-base chemistry and taste-panel sensory-evaluation methodologies. Check with your instructor before you start this experiment.

Chapter
9

Halogens and Their Compounds

Introduction

The chemistry of the halogens and their compounds is extraordinarily rich and complex. Although the free elements are far too reactive to be naturally occurring, all the halogens are manufactured industrially and have a wide variety of uses. Chlorine (Cl_2) ranks in the top ten of industrial chemicals with an annual production of about 20 billion pounds, most of which is generated by the electrolysis of brine (concentrated sodium chloride) by the chloralkali industry. More than 50% of this production of free chlorine is utilized in the manufacture of other chemicals — e.g, dry-cleaning fluids, pharmaceuticals, refrigerants, herbicides, pesticides, and plastics (principally, polyvinylchloride, or PVC). Other major uses are as a bleach in the paper and textile industries and as a disinfectant in municipal water supplies and sewage treatment. The use of chlorine for disinfection of water is largely responsible for the almost complete eradication of water-borne diseases — e.g., cholera and hepatitis — that plagued societies for centuries. The active chemical that kills bacteria in disinfection is not chlorine, but hypochlorous acid (HClO), which is formed from the reaction of chlorine gas with water:

$$Cl_{2(g)} + H_2O \rightarrow HCl + HClO$$

Uncharged hypochlorous acid can then diffuse easily through bacterial cell walls and destroy the enzyme system of the pathogen.

Fluorine (F_2) is industrially produced by electrolysis in a molten salt reaction because the process must be carried out in the absence of water. Fluorine is extremely reactive and very dangerous; the technology to produce, store, transport, and use it originated primarily from the production of fissionable uranium for atomic bombs. Fluorine is used to produce many products that contain carbon-fluorine covalent bonds. The *chlorofluorocarbons*, known by the trade name Freons, are nontoxic, inert, and nonflammable and are extensively used as heat transfer fluids in refrigerators and air conditioners. Unfortunately, there is definite evidence that Freons are not inert in the stratosphere, and it has been shown that photochemical breakdown of these compounds produces chlorine radicals that catalytically destroy ozone. Recently, there has been unprecedented international agreement to phase out the production of chlorofluorocarbons and to carry out research to find substitutes for these fluids. The stability of carbon-fluorine bonds is also utilized in a variety of nonstick coatings, such as Silverstone®. These coatings are fluorocarbon polymers, such as Teflon (polytetrafluoroethylene), and are extremely inert and resistant to high temperatures.

The *halides*, such as sodium chloride, iodide, and calcium fluoride, are by far the most common halogen compounds found in nature. The oceans contain a remarkably constant concentration of common salt (sodium chloride), with very much smaller concentrations of various bromides and fluorides. It is interesting to note that the salt composition of the ocean is the result of a gigantic, global, acid-base titration between acids that have leaked out of the interior of the earth (e.g., HCl) and bases that have been set free by the weathering of primary rock. The production of common table salt for human consumption was undoubtedly one of the first chemical industries. It has been argued that the earliest roads were made for the transportation of salt and that the earliest cities were established as centers of the salt trade. Sources of salt have even been the object of military campaigns! Sodium chloride is certainly one of the most important electrolytes in body fluids, particularly in blood plasma and interstitial fluid. The addition of small amounts of other halides (e.g., iodide and fluoride) in the human diet is also now very common. Iodide ion is necessary for the production of thyroxine in the thyroid gland, and an insufficiency of iodide ion in the diet leads to a condition known as goiter. To assure the presence of enough iodide ion in the diet, sodium or potassium iodide is often added to table salt. The fluoridation of water is now a relatively common practice. At the very low concentration of one part per million, fluoride ion helps prevent tooth decay without causing discoloration.

One of the most controversial aspects of halogen chemistry is the toxicity problem associated with many of the organic compounds containing carbon-halogen bonds. Halogenated hydrocarbons, such as chloroform, the Freons mentioned earlier, ethylene dibromide in gasoline, DDT, PCB, and PBBs have all been shown to be toxic in animal systems. Although the complete biochemistry is not yet clear, it seems that the damaging effects are caused as the carbon-halogen bond is broken down in the liver, forming toxic substances. The knowledge of the potential health problems of these halogenated products has led to limits and, in some instances, a complete ban in production of these substances.

It is hoped that you have gained a perspective on the incredible diversity and importance of the chemistry of the halogens and their compounds from this brief introduction. The experiments that follow can only touch the surface of this area of science, but they do represent some of the major principles of halogen chemistry.

Background Chemistry

The elements in the halogen group are all nonmetals that exist at normal temperatures and pressures as diatomic molecules. The trends in properties expected for a family of elements in the periodic table are very evident in the halogens, as can be seen in Table 9.1.

Property	Fluorine (F_2)	Chlorine (Cl_2)	Bromine (Br_2)	Iodine (I_2)
Color	pale yellow	greenish yellow	reddish brown	black
State	gas	gas	liquid	solid
Melting point (°C)	–220	–101	–73	113
Boiling point (°C)	–188	–34	59	184
Atomic radius (pm)	71	99	114	133
Ionic radius (x^-, pm)	136	181	196	220

Table 9.1 Some Properties of the Halogens*

*Astatine, the other member of the halogen group, does not appear in Table 9.1 because it is extremely radioactive and has never been obtained in the pure form.

The outstanding characteristic of the halogens is the large number of compounds they form with other elements. In chemical reactions, the halogens readily accept one electron per halogen atom to form singly charged anions (e.g., $F^-$and Cl^-) or readily share their single, unpaired electron to form covalent bonds. As a result of this strong tendency to attract electrons, the halogens are all strong oxidizing agents. Fluorine (F_2), the first member of the family, is somewhat different from the other halogens primarily because of the small size and very high electronegativity of the fluorine atom. Fluorine is the most reactive of all the nonmetals and is one of the strongest oxidizing agents known. Fluorine cannot be prepared in aqueous solutions because it reacts rapidly with water:

$$2F_{2(g)} + 2H_2O \rightarrow 4HF + O_{2(g)}$$

Chlorine, bromine, and iodine all undergo redox reactions with water. For example, chlorine, with an oxidation number of 0, reacts as follows:

$$Cl_2 + H_2O \rightarrow HCl + HClO$$

In this reaction, it can be seen that in one product, HCl, the oxidation number is –1, and in the other, HClO, the chlorine has an oxidation number of +1. As a result, a saturated solution of chlorine in water (chlorine water) is about 30% hypochlorous acid (HClO). The disinfectant properties of chlorine when used in water and sewage treatment are due to the presence of the hypochlorous acid, which is also a powerful oxidizing agent. Commercial liquid bleach is made by bubbling chlorine into sodium hydroxide solution:

$$Cl_2 + 2NaOH \rightarrow NaCl + NaOCl + H_2O$$

In the resulting reactions, the active ingredient is sodium hypochlorite (NaOCl). Iodine has the unique property of reacting with excess iodide in aqueous solution to give the triiodide ion

$$I_2 + I^- \rightleftharpoons I_3^-$$

This ion has the very unusual property of reacting with soluble starch to give a blue-black-colored complex in which I_3^- and additional I_2 molecules form I_5^- chains:

$$I_3^- + I_2 \rightleftharpoons \cdots[\,I-I-I-I-I\,]^-\cdots$$

These I_5^- chains just happen to be the right size to fit down the middle of the amylose sugar helix (pictured in Figure 9.1).

Schematic structure of the starch–iodine complex. The amylose sugar chain forms a helix around the nearly linear iodine chain.

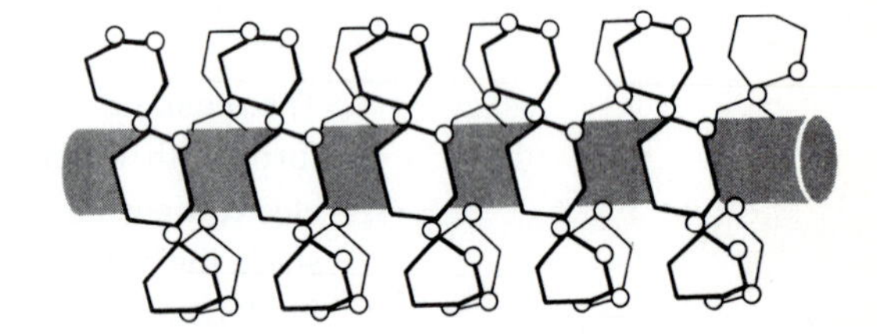

Figure 9.1 The Starch-Iodine Complex

The pronounced color of this complex has led to its use as an excellent detection reaction for iodine and for oxidizing agents that can oxidize I^- to I_2.

One of the most common types of reactions of the halogens is the formation of *halides* in which each halogen atom gains one electron to become F^-, Cl^-, Br^-, or I^-. Almost all metals, metalloids, and nonmetals will react with halogens to produce halides. The properties of halides vary tremendously depending on which element enters into combination with the halogen. Alkali metal halides (e.g., sodium chloride) are ionically bonded, white, crystalline solids that are very soluble in water and dissociate to give hydrated ions. Nonmetal halides (e.g., phosphorus trichloride) are generally covalent compounds with polar covalent bonds because of the high electronegativity of the halogen atom. The combination of halogens with carbon produces covalent organic compounds, many of them having properties that make them useful as solvents, refrigerants, anesthetics, pesticides, and plastics.

There are many compounds in which the halogen atoms have a positive oxidation number, the great majority of them in the form of oxyanions and oxyacids. Table 9.2 shows most of the known compounds of this type, together with the nomenclature and some examples of common stable salts.

Oxyacid	Name	Can it Be Isolated?	Oxidation No. of Halogen	Example of Salts
$HClO$	hypochlorous acid	no	+1	hypochlorites ($NaClO$)
$HBrO$	hypobromous acid	no	+1	hypobromites (not stable)
HIO	hypoiodous acid	no	+1	hypoiodites (not stable)
$HClO_2$	chlorous acid	no	+3	chlorites (not stable)
$HClO_3$	chloric acid	no	+5	chlorates ($KClO_3$)
$HBrO_3$	bromic acid	no	+5	bromates ($NaBrO_3$)
HIO_3	iodic Acid	yes	+5	iodates (KIO_3)
$HClO_4$	perchloric acid	yes	+7	perchlorates ($LiClO_4$)
$HBrO_4$	perbromic Acid	no	+7	perbromates ($KBrO_4$)
HIO_4	periodic acid	yes	+7	periodates (KIO_4)

Table 9.2 Halogen Oxyacids and Their Salts

Laboratory Experiments

Flowchart of the Experiments

Section A. From Fluorine to Astatine: A Basic Introduction to the Halogens

Section B. The Synthesis and Reactions of Chlorine

Section C. A Small-Scale Pilot Plant for the Manufacture of Chlorine by the Industrial Process

Section D. Electrochemical Writing with a Halogen

Section E. Precipitation Reactions and Titration of a Halide

Section F. Redox Analysis of Commercial Bleach

Requires one three-hour class period to complete

CAUTION: All of the halogens are very toxic and should be handled with great care — even iodine! Even though you will be working with extraordinarily small amounts of all the materials in this series of experiments, please be careful. Sulfuric acid is corrosive, and silver compounds will turn the skin black. If you get any of these chemicals on your skin, wash well with cold water and check with your instructor.

Section A. From Fluorine to Astatine: A Basic Introduction to the Halogens

Goal:

To become familiar with some of the structures, properties, and reactions of the halogens (listed in the table below).

Group VIIA

Symbol	Name	Atomic No.
F	fluorine	9
Cl	chlorine	17
Br	bromine	35
I	iodine	53
At	astatine	85

Discussion:

This family of elements, collectively called the halogens, represents one of the most typical groups of nonmetallic elements in the periodic chart. The electron configurations of the halogens reveal that each element in this group has 7 valence electrons, one short of an octet.

- Write the electron configuration for an atom of chlorine.
- Which are the valence electrons?

At normal temperatures and pressures, all the halogens form diatomic molecules in which they complete their octet by sharing electrons.

- Write the chemical formula for a molecule of iodine and draw its Lewis structure.

At room temperature and pressure, fluorine and chlorine are gases, bromine is a red-brown liquid, and iodine is a shiny black solid that sublimes readily to give a violet vapor. Astatine is so radioactive that it has never been isolated in the pure form. The longest-lived isotope ($^{210}_{85}At$) has a half-life of 8.3 hours.

- Why should a group of elements exhibit a progression from the gaseous to the solid state?

CAUTION: All the halogens have pungent and irritating odors, and all can cause serious burns to the skin (particularly bromine). They are also highly reactive and form a large number of compounds with other elements. Fluorine is the most reactive and iodine the least.

- Explain why the halogens are so reactive.

Although the halogens form a family (Group VIIA in the Periodic Chart) with periodic properties, fluorine (F_2), the first member, exhibits some differences from the other elements. These differences are largely due to the small size and high electronegativity of the fluorine atom. F_2 is extraordinarily reactive because of the very strong tendency to grab electrons and become fluoride ion F^-.

If a metal such as sodium is exposed to gaseous fluorine, there is an instantaneous violent reaction that produces a white saltlike product. The reaction is

$$2Na_{(s)} + F_{2(g)} \rightarrow 2Na^+ + 2F^-$$

This is an example of a redox reaction in which the Na atom has lost an electron in becoming a Na^+ ion:

$$2Na \rightarrow 2Na^+ + 2e^-$$

This part of the reaction is a *loss* of electrons (by Na) and is called *oxidation*. The fluorine atoms in F_2 each gained an electron to become fluoride ions F^-:

$$F_2 + 2e^- \rightarrow 2F^-$$

This gain of electrons by F_2 is called *reduction*. If the two parts, reduction and oxidation, are added up, we obtain the redox reaction

$2Na \rightarrow 2Na^+ + 2e^-$	oxidation
$F_2 + 2e^- \rightarrow 2F^-$	reduction
$2Na + F_2 \rightarrow 2Na^+ + 2F^-$	redox

NOTE: The product is the ionically bonded salt sodium fluoride (NaF).

One more piece of jargon — but a useful one. In the above redox reaction, the chemical species that grabs the electrons (F_2) is called the *oxidizing agent*, and the species that gives the electrons (Na) is called the *reducing agent*. F_2 is one of the most powerful oxidizing agents known.

Fluorine is far too nasty to make in this laboratory, but you can at least write some of its reactions.

- Write a balanced net ionic equation for the reaction between F_2 and Cl^-. Divide the reaction into 2 parts and decide which is reduction and which is oxidation.
- What species is the reducing agent in this reaction?

When aluminum metal is heated and dropped into fluorine gas, there is an explosion.

- Write the equation for this reaction.
- What color do you think the product is?

Section B. The Synthesis and Reactions of Chlorine

Goals:

(1) To carry out a microscale preparation of chlorine gas ($Cl_{2(g)}$). (2) To investigate some of its redox reactions.

Discussion:

The solutions you will need for this experiment and the following experiments can be found in the 24-well tray or at Reagent Central. At your place is a plastic reaction surface that provides a white and a black background for viewing the chemistry. The synthesis of gases will be done in a plastic petri dish, which acts as a miniature environmental chamber.

The chemical reaction for the synthesis of chlorine is the action of dilute acid (hydrochloric acid) on commercial bleach solution (e.g., regular Clorox®). Bleach (according to the label) is 5.25% active ingredient, which is sodium hypochlorite (NaOCl), and 94.75% inactive ingredient, which is presumably water. The reaction equation is

$$NaOCl + 2HCl \rightarrow Cl_{2(g)} + NaCl + H_2O$$

It is worth noting that the misuse of bleach solutions in household situations is common enough to warrant a hazardous chemical warning on the label. The label reads: "Strong oxidizer. Flush drains before and after use. *Do not use or mix with other chemicals,* such as toilet bowl cleaners, rust removers, acid- or ammonia-containing products. To do so will release hazardous gases."

Experimental Steps:

1. Place the petri dish onto the plastic surface against a white background.
2. Drop 1 drop of diluted bleach solution (already diluted to 50%) into the center of the dish.
3. In a circle around the drop of bleach, drop separately 1 drop of 0.1M KI (potassium iodide), 1 drop of dye (bromocresol green), and 1 drop of starch/0.1M KI solution.
4. Also place in the dish, close to the center, a small circle of filter paper that has been wet with 1 drop of 0.1M KI and that has had the excess liquid removed by dabbing it with a piece of folded microtowel.

   NOTE: The paper circles are in a box at Reagent Central.
5. Get ready to put the top on. Drop 1 drop of HCl onto the bleach solution. Quickly put the top on.
   - Watch and record what happens.
   - Make a picture of which solution is where — it's easy to get them mixed up.
   - From your observations, what can you deduce about the solubility of chlorine in an acidic solution?

- Write a balanced equation for the redox reaction of chlorine with potassium iodide solution and with potassium bromide solution.
- What happened to the starch/KI solution? Do you remember this reaction from an earlier lab?
- Describe carefully the appearance of the filter paper circle. What must be happening here?

  HINT: Chlorine gas is still present in excess in the dish!
- Describe the color change of the dye.

  NOTE: This reaction is typical of the oxidizing power of chlorine towards organic dyes in general.

6. Terminate the reaction by adding 1 drop of 2M NH_3 to the dish.

   CAUTION: Do *not* add the NH_3 to any of the other drops.
7. Place the petri dish onto a black background.
   - Describe what happens.

   NOTE: You have formed an aerosol of ammonium chloride (NH_4Cl) that consists of tiny white particles of the salt settling on the plastic surface.
8. Flood the dish by adding distilled water from your wash bottle. Swirl to dilute the remaining drops.
9. Pour the liquid into the waste cup.
10. Wash thoroughly with tap water at the sink.
11. Rinse once with distilled water, and dab dry with a microtowel.

Section C. A Small-Scale Pilot Plant for the Manufacture of Chlorine by the Industrial Process

Goal: *To construct a small-scale pilot plant for the manufacture of chlorine.*

Discussion: Chlorine is manufactured on a very large scale by the electrolysis of sodium chloride solution (brine). *Electrolysis* is a process in which a redox reaction is made to occur by means of an outside source of electrical energy in the form of direct current. The design of the pilot plant is critical because it is necessary to separate the two products of the electrolysis — chlorine gas and sodium hydroxide — before they can react and disappear.

Experimental Steps: The pilot plant is pictured in the following diagram. Refer to the diagram as you complete each step to ensure that your pilot plant is built properly.

NOTE: The design features of the pilot plant have been carefully chosen to duplicate those in the industrial process.

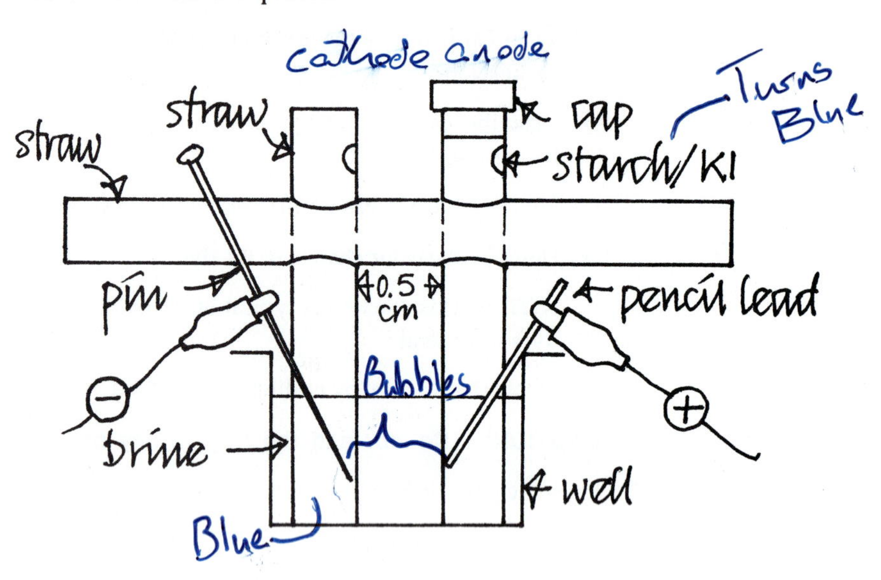

1. Obtain pins and pencil lead from Reagent Central.
2. To contruct the plant, cut 3 straw pieces, 1 piece about 8 cm long, and 2 pieces about 6 cm long.
3. Use your 1/4" punch to make 2 holes about 0.5–0.7 cm apart in the 8 cm straw.
4. Push the short straws through the holes in the long straw.
5. Make a hole in one of the 6 cm straws with the pin so that you can insert the pencil lead without breaking it.
6. Now push the pin into first the 8 cm straw and on through the other 6 cm straw.

   NOTE: The pencil lead and pin in the straws is the electrode system. The pin is the cathode, defined as the place where reduction occurs. The pencil lead is the anode, defined as the place where oxidation occurs.
7. Fill a clean well of a 24-well tray about 3/4 full with brine. Add 3 drops of dye (bromocresol green) to the brine and stir.
8. Dip the electrode system into the brine.
9. Clip the alligator clips to the electrodes but do *not* attach the clips to the battery yet.
10. Place a small drop of starch/KI solution on the wall inside each straw (about 1 cm down).
11. Place a cap on the anode straw.
12. Now attach the clips to the battery.
    - Report your observations.

13. Remove one clip from the battery terminal after 3 or 4 seconds (or you will generate too much chlorine gas).

    - Now write the part of the redox reaction that is occurring at the *anode.*
    - What evidence do you have for the reaction?

    The part of the redox reaction at the *cathode* is the reduction of water:

$$2H_2O + 2e^- \rightarrow H_{2(g)} + 2OH^-$$

    - Explain why the indicator dye in the brine changed color.
    - Explain carefully how the products of the electrolysis reaction were kept from reacting with each other.
    - Write the equation for the redox reaction accomplished during this electrolysis.

14. Terminate by unclipping all clips.
15. Lay the wires to one side.

    NOTE: Try not to get brine on the clips or they will corrode!

16. Carefully lift off the cap, add 1 drop of NH_3 to it, and put it back on the same straw.

    - What will be formed in the straw?

16. Leave the straw for a few minutes and then clean up. Remove the pencil lead, wash it, wipe it with a towel, and retain for the next experiment. Put the straw pieces into the waste straw box.

Section D. Electrochemical Writing with a Halogen

Goal: *To carry out an electrolysis to generate iodine by using a pencil lead as a stylus.*

Discussion: This experiment is an interesting variation on the electrolytic production of a halogen that you just completed.

Experimental Steps:

1. Obtain a piece of aluminum foil and a piece of filter paper from Reagent Central.
2. Lay the foil down on the plastic surface and place the paper on top of it.
3. Drop 2 or 3 drops of starch/KI solution onto the paper. Let it spread out into the pores of the paper.
4. Clip 1 electrical wire to the edge of the foil. Clip the pencil lead to the other wire.
5. Decide how you are going to clip the wires to the battery and which electrode will be the stylus, then do so.
6. Make contact between the stylus and the paper and write your name.

7. Unclip the battery.
8. With your tweezers, turn the paper over and add 1 drop of dye solution (bromocresol green) to the paper.
   - Write the redox equation and the two parts that are the reduction and the oxidation.
   - Is the aluminum foil the cathode or the anode in this experiment?

Section E. Precipitation Reactions and Titration of a Halide

Goal: *To investigate, both qualitatively and quantitatively, the reaction of halides with silver nitrate ($AgNO_3$) to produce sparingly soluble silver halides as precipitates.*

Discussion: The precipitation reactions are the basis for a simple quantitative analysis of halides by means of volumetric titrations.

Experimental Steps:

1. Use the plastic surface with a black background to carry out a dropwise study of the reaction of silver nitrate solution with HCl, NaCl, KBr, and KI solutions.
   - Describe colors and textures and write the net ionic equations wherever you see a chemical reaction.

   In this next experiment, you are going to be doing a quantitative volumetric experiment as accurately as possible. This means that all the drops used in the reaction have to be exactly the same volume.

   NOTE: The only way to insure this is to use the same microburet for all solutions. Good wash and transfer technique is critical to obtaining accurate results.
2. Clean and dry a well in the 24-well tray.
3. Transfer about 1 mL of 0.1M NaCl solution to the well.
4. Wash a microburet several times with distilled water and expel all water. Suck up a little NaCl solution, roll it around the bulb to wash the bulb completely, then expel it. Now suck up the rest of the NaCl solution.
5. Carefully clean a 1×12 well strip.
6. Now drop 1 drop of the 0.1M NaCl solution from the microburet into each of the 12 wells, making sure that the drop actually *drops* into the well.
7. Add 1 drop of 1% potassium chromate indicator solution to each well.

   NOTE: The volume of this solution is not critical, so just go ahead and use the dropper that the solution is in.
8. Expel the NaCl solution from the microburet, wash it several times with distilled water, and then use good transfer technique to rinse and fill it with some 0.02M $AgNO_3$ solution.

9. Now you're ready to carry out a serial titration. This involves adding 1 drop of $AgNO_3$ to the first well, 2 drops to the second, and so on. Stir the solution in each well with a microstirrer.

   - Describe what you see and report the number of drops of 0.02M $AgNO_3$ required to make the indicator change color.

   The overall chemical reaction for this titration is

$$NaCl + AgNO_3 \rightarrow AgCl_{(s)} + NaNO_3$$

   with a simple 1:1:1:1 stoichiometry.

10. The reddish-brown color that occurs at the *endpoint* is due to the precipitation of silver chromate,

$$K_2CrO_4 + 2AgNO_3 \rightarrow Ag_2CrO_{4(s)} + 2KNO_3$$

   which will only occur when all the chloride ion has reacted with silver ion, and there is an *excess* of $AgNO_3$ in the solution.

   - What is the limiting reagent in the titration before the endpoint?

   This precipitation titration (sometimes called a Mohr titration) is a quantitative analysis for the halide (NaCl), provided the exact concentration of the titrant ($AgNO_3$) is known.

   - Calculate the molarity of the NaCl solution.

11. Wash the 1 × 12 well strip thoroughly with water. If any solid remains, add 1 drop of NH_3 and remove with a cotton swab. Place swab in waste solution to dilute the ammonia, then wash again with cold water.

Section F. Redox Analysis of Commercial Bleach

Goal: *To carry out an analysis of a commercial chlorine bleach.*

Discussion: The sodium hypochlorite (NaOCl) in bleach is allowed to oxidize I^- to I_2. The I_2 concentration formed in this reaction can be determined by titration with sodium thiosulfate ($Na_2S_2O_3$) solution using starch as an indicator.

Experimental Steps: NOTE: Again, this experiment is a quantitative volumetric analysis and you must be careful to use good wash, rinse and transfer techniques and the same microburet for the dilutions and titration. Commercial bleach (e.g., Clorox®) is available in a plastic bottle at Reagent Central.

1. Fill a clean, dry well in a 1 × 12 well strip with bleach.
2. Clean a microburet and suck some of the bleach up into the microburet.

3. Carry out a 10-fold dilution of the bleach by dropping 2 drops onto the plastic surface.
4. Wash the microburet several times with water and fill it 1/2 full with distilled water.
5. Add 18 drops of water to the 2 drops of bleach.
6. Expel the water from the microburet and suck up the diluted bleach solution from the surface. If you like, you can easily assure that it is mixed by expelling and sucking up again.
7. Drop 1 drop of the diluted bleach solution in 2 places on the plastic.

   NOTE: You can do duplicate titrations very easily by this method!
8. Expel the diluted bleach solution from the microburet into a waste cup.
9. Wash the microburet thoroughly with water and use good wash and rinse technique to fill it 1/2 full with 0.01M $Na_2S_2O_3$ solution.
10. Now carry out a "pool" titration on the plastic by adding 1 drop of acetic acid and 2 drops of starch/KI solution to one of the drops of diluted bleach solution on the plastic. Stir and leave the stirrer lying in the pool.
11. Titrate the blue-black pool with the 0.01M $Na_2S_2O_3$ with constant stirring.

    NOTE: Don't forget to count drops of titrant as you go!
12. Do the same titration on the duplicate sample.

    The chemistry is a little complicated, but it's not too bad. Here are the reactions that occur. First, the bleach oxidizes the I^- to I_2:

$$ClO^- + 2I^- + 2H^+ \rightleftharpoons Cl^- + I_2 + H_2O$$

In the presence of I^- (which was deliberately added in excess), triiodide ion I_3^- is formed:

$$I_2 + I^- \rightleftharpoons I_3^-$$

The I_3^- then reacts with the starch to form the blue-black complex. The I_3^- blue-black complex is then titrated with sodium thiosulfate, which reduces the I_3^- back to colorless I^-, forming the tetrathionate ion $S_4O_6^{2-}$ as a product:

$$I_3^- + 2S_2O_3^{2-} \rightleftharpoons S_4O_6^{2-} + 3I^-$$

CALCULATION OF SODIUM HYPOCHLORITE CONCENTRATION IN BLEACH SAMPLE

- First, calculate the molarity of the I_3^- solution.

Let's say that it took 12 drops of 0.01 M $Na_2S_2O_3$ to titrate the blue-black I_3^- to a colorless end point. Then the I_3^- molarity is

$$0.01\ \text{M}\ Na_2S_2O_3 \times \frac{12\ \text{drops}}{1\ \text{drop}} \times \frac{1\ \text{mol}\ I_3^-}{2\ \text{mol}\ Na_2S_2O_3} = 0.06\ \text{M}$$

The reason for the last factor is that each I_3^- reacts with two $S_2O_3^{2-}$ ions in the titration reaction.

The calculated molarity is the molarity of the diluted bleach solution. You diluted the original bleach by 10 times. The molarity of the original commercial bleach is $10 \times 0.06\text{M} = 0.6$ M. The original bleach contains NaClO, which has a formula weight of 74.5 g mol^{-1}. Now all you have to do is calculate the number of grams of NaClO per liter, divide by 10, and obtain the number of grams per 100 mL of bleach. Phew! Check your bleach sample container to see if you can find the percentage of sodium hypochlorite in the commercial sample.

Chapter 10

The Chemistry of Natural Waters

Introduction

The availability of high-quality water is essential to the survival of all living species. On a global scale, the hydrologic cycle ensures the continuous cycle of evaporation, condensation, and precipitation that brings water to the land and moves the winds over the earth. The constant supply of incoming solar energy drives the water cycle at a relatively constant rate although lately it seems that we humans can influence global kinetics, even to the point of changing the climate. The annual amount of relatively pure water precipitating onto the land is therefore quite constant, and it is important to note that almost all of the good-quality water from this precipitation is now being used. About 80 liters of water per person per day is required to sustain a reasonable quality of life. However, the average consumption ranges from 5.4 liters a day in Madagascar (it takes 5 liters simply to survive!) to more than 500 liters a day in the United States. As the world's population doubles over the next 15 years, meeting the challenge of providing quality usable water will be critical to the survival of billions of people.

The quality of water and its suitability for particular uses are almost entirely evaluated on the basis of its chemical and biological composition. The chemical composition of natural water derives from many different sources of solutes, including gases and aerosols from the atmosphere, weathering and erosion of rocks and soils, and solution or precipitation reactions occurring below the land surface and from anthropogenic wastes. Rainwater generally contains dissolved gases and aerosols, man-made pollutants, and particles from sea spray. Seawater, which is about 98% of the total water on earth, has a remarkably constant composition. The average concentrations of the 10 major dissolved elements or ions in seawater are shown in Table 10.1. Saltiness (NaCl) of seawater precludes its use as drinking water. Other natural waters, such as streams, rivers, lakes, and groundwaters, contain dissolved substances that come mainly from the dissolution of minerals as the water flows over or percolates through various soil and rock strata.

Species	Concentration (mg L^{-1})
Cl^-	19,000
Na^+	10,500
SO_4^{2-}	2,700
Mg^{2+}	1,350
Ca^{2+}	410
K^+	390
HCO_3^-	142
Br^-	67
Sr^{2+}	8
SiO_2	6.4

Table 10.1 Some Ion Concentrations in Seawater

Two of the most important chemical species that come from mineral dissolution are the cations Ca^{2+} and Mg^{2+}. Calcium is the most abundant of the alkaline earth metals and is a major constituent of many common rock minerals. It is an essential element for plants and animals and is a major component in most natural waters. Most of the soluble calcium ion comes from the more soluble sedimentary rocks, such as limestone ($CaCO_3$), dolomite ($CaMg(CO_3)_2$), and gypsum ($CaSO_4 \bullet 2H_2O$), rather than from the very much less soluble igneous rocks. The dissolution of carbonates is dominated by the aqueous chemistry of dissolved carbon dioxide (CO_2). Aqueous solutions of CO_2 are slightly acidic, and calcium carbonate slowly dissolves under these conditions, forming soluble calcium bicarbonate. Magnesium is also a common element and is essential in plant and animal nutrition. Rock sources of magnesium ion include olivine, pyroxenes, the dark colored micas, and the dolomites.

The presence of the dissolved divalent cations Ca^{2+} and Mg^{2+} (and other polyvalent cations) gives natural waters chemical properties that are often evaluated as *hardness*. Water that contains a large combined concentration of Ca^{2+} and Mg^{2+} are said to be hard, and waters with a low concentration are said to be soft. The terms hard and soft are very old. For example, they are contained in a discourse on water quality by Hippocrates (460–377 B.C.). The modern concept of hardness pertains to the use of natural waters in washing processes involving soaps and detergents and in the use of waters in industrial boilers and evaporators.

Most soaps are anionic surface-active agents that work by reducing the surface tension of wash water and by weakening chemical bonds between dirt and the surface being cleaned. If the concentration of a divalent cation is high, the soap anion will react with Ca^{2+} and Mg^{2+} to produce an insoluble, greasy scum. The scum precipitates and coats the cleaned surface. Soaps can be precipitated by other ions — e.g., H^+ and all polyvalent cations — but these ions are present in insignificant amounts in waters that are used domestically. The importance of the hardness of water is reflected in the fact that manufacturers of fabric and dishwashing detergents formulate their products with ingredients specifically designed to complex Ca^{2+} and Mg^{2+}. These water-softening ingredients are generally polyphosphates or citrates. Many detergents are also formulated to work well with water, which has a natural hardness of about 6 grains per gallon. (See Section D for a discussion of units.)

The problem in the use of hard water in industrial boilers and evaporators is scale formation. When hard water is heated or evaporated, rocklike deposits consisting largely of calcite crystals ($CaCO_3$) form on the surface of pipes, boiler walls, tubes, and evaporator surfaces. Scale is one of the banes of industry. It narrows pipes, blocks jets and tubes, and is extremely expensive (and sometimes impossible) to remove. The hard layer interferes with heat transfer in boilers, leading to gross energy inefficiencies, and can often lead to metal corrosion and structural weakness. Scale formation causes problems in the utility industry, in beer brewing, in humidifying systems for computer centers, chicken hatcheries, and hog farms, in steel and auto plants – in fact everywhere where large volumes of natural waters are used. Of course, the harder the water, the bigger the problem. A number of industrial processes are in use for removing most of the divalent cations or for slowing down the rate of scale formation. A variety of chemical treatments, such as the addition of lime ($Ca(OH)_2$) or washing soda (Na_2CO_3), are effective in precipitating Ca^{2+} and Mg^{2+} ions as insoluble salts:

$$Ca(HCO_3)_2 + Ca(OH)_2 \rightarrow 2CaCO_{3(s)} + 2H_2O$$

$$Mg(HCO_3)_2 + 2Ca(OH)_2 \rightarrow 2CaCO_{3(s)} + Mg(OH)_{2(s)} + 2H_2O$$

$$CaSO_4 + Na_2CO_3 \rightarrow CaCO_{3(s)} + Na_2SO_4$$

The precipitated compounds are removed by settling and/or filtration, and the softened water is fed to the boiler or evaporator. Ion exchange is also a useful, if somewhat expensive, softening technique, particularly for small-scale domestic applications.

The importance and magnitude of the problems associated with the presence of dissolved minerals in natural waters have led to research into rapid and accurate methods of analysis for these ions. Prior to the 1940s, the most commonly used method was a soap titration. This method built many strong arms (as you will see later), but was rather time-consuming and inaccurate. The introduction of ethylenediaminetetra-acetic acid (EDTA) as a complexing agent for Ca^{2+} and Mg^{2+}, together with suitable indicators such as eriochrome black T (EBT), revolutionized the analysis of divalent ions in natural waters. It is this method that will allow you to explore the world of the chemistry of natural waters in this series of experiments.

Background Chemistry

The dissolution of many different minerals in water results in natural waters that contain Ca^{2+} and Mg^{2+} ions, with the counter ions being predominantly bicarbonate (HCO_3^-) and sulfate (SO_4^{2-}) ions. The solubility of most minerals is increased by the acidity of aqueous solutions of carbon dioxide (CO_2). In many groundwaters, the acidity may be quite high because of the large amount of CO_2 produced by biological activity:

$$\text{Biological activity} \longrightarrow CO_2 + H_2O \rightleftharpoons H_2CO_3$$

$$H_2CO_3 \rightleftharpoons H^+ + HCO_3^-$$

An example of the dissolution of an igneous rock is the decomposition of anorthite:

$$CaAl_2Si_2O_{8(s)} + H_2O + 2H^+ \rightleftharpoons Al_2Si_2O_5(OH)_{4(s)} + Ca^{2+}$$

The dissolution of a magnesium-containing rock is exemplified by the decomposition of fosterite into serpentite:

$$5Mg_2SiO_{4(s)} + 8H^+ + 2H_2O \rightleftharpoons Mg_6(OH)_8Si_4O_{10(s)} + 4Mg^{2+} + H_4SiO_4$$

Sedimentary rocks containing calcium and magnesium are most often found in the form of carbonates, which are much more soluble than igneous rocks, particularly in acidic solutions. For example,

$$CaCO_{3(s)} + H^+ \rightleftharpoons Ca^{2+} + HCO_3^-$$

$$CaMg(CO_3)_{2(s)} + 2H^+ \rightleftharpoons Ca^{2+} + Mg^{2+} + 2HCO_3^-$$

Many other cations (e.g., Na^+, K^+, Fe^{3+}, and Al^{3+}) are often present in natural water samples, but their concentrations are usually insignificant compared with Ca^{2+} and Mg^{2+}.

One of the simplest methods of finding out what solutes are present in a water sample is to evaporate the solvent (water) and examine the residue of nonvolatile solids that remains. The amount of residue left after the evaporation of a known volume of water is called the *total dissolved solids* (TDS) of the water sample. This type of analysis also allows an assessment to be made of the potential scaling problem of a water sample. Quantitative analysis of a water sample (or the residue from a TDS experiment) for total divalent cation can be conveniently carried out by means of a complexation titration. The chelating agent, ethylenediaminetetraacetic acid (EDTA), is most commonly used in this analysis:

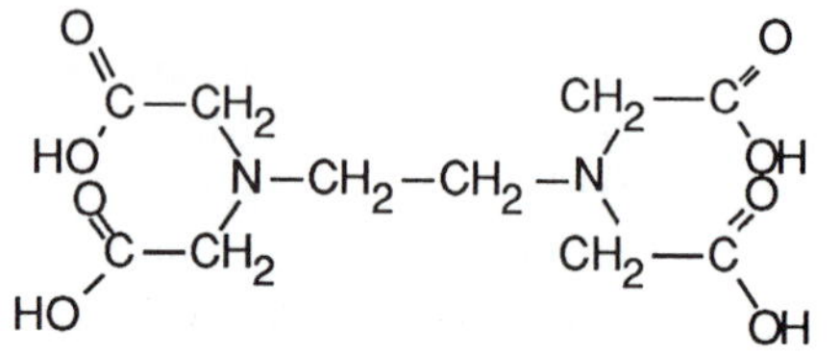

EDTA, $C_{10}H_{16}N_2O_8$

Actually, for practical reasons of solubility, the reagent used in the laboratory is the disodium dihydrate salt of EDTA ($C_{10}H_{14}N_2O_8Na_2 \cdot 2H_2O$). The indicator used in EDTA titrations is often a dye called eriochrome black T (EBT) which has the following structure:

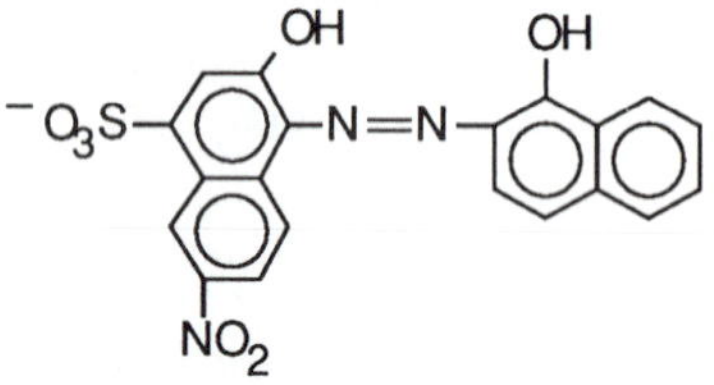

This dye is water soluble because of the negative charge on the sulfonate group ($-SO_3^-$) and is intensely colored because of delocalized electrons in the structure. EBT is also an acid-base indicator ($pK_a = 6.3$):

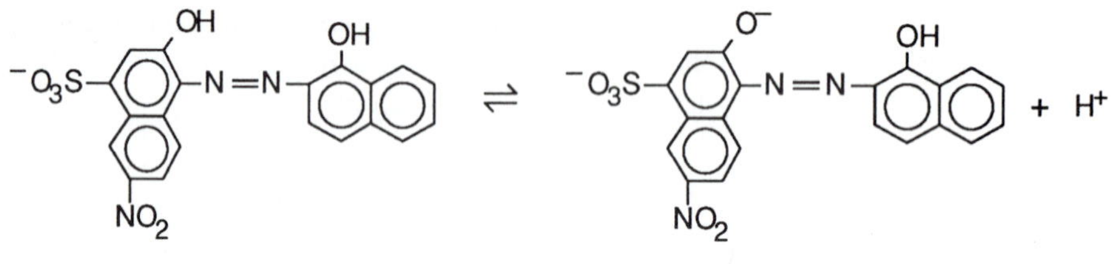

Red form (H_2D^-) ⇌ Blue form (HD^{2-}) + H^+

EBT also forms intensely colored chelates with certain metal ions (e.g., Mg^{2+}), but not with others (e.g., Ca^{2+}). It is necessary to adjust the pH of the solution in the titration so that the blue form predominates.

CHEMISTRY OF THE DETERMINATION OF DIVALENT CATION CONCENTRATION (HARDNESS) BY EDTA TITRATION: A STEPWISE DISCUSSION

1. A known volume of the natural water sample is taken, and the pH is adjusted to 10 by means of an NH_3/NH_4Cl buffer.
2. EBT indicator is added to the solution. At the high pH the indicator is in the form HD^{2-}.

3. If Mg^{2+} is present in the water sample, a reaction occurs with the indicator, i.e.,

$$HD^{2-} + Mg^{2+} \rightleftharpoons MgD^{-} + H^{+}$$

Blue Red

and a wine red chelate MgD^{-} is formed. Ca^{2+} does not react with the indicator. At the start of the titration, the solution is a wine red color.

4. EDTA solution (of known concentration) is now added to the solution from a microburet. First, the EDTA reacts with Ca^{2+}, forming a colorless chelate:

$$Ca^{2+} + EDTA^{2-} \rightleftharpoons CaEDTA$$

As soon as enough EDTA has been added to chelate all the Ca^{2+}, then the EDTA begins to react with the magnesium indicator chelate, i.e.,

$$H^{+} + MgD^{-} + EDTA^{2-} \rightleftharpoons MgEDTA + HD^{2-}$$

to produce a colorless MgEDTA chelate and the *blue* form of the dye HD^{2-}.

5. The end point in the titration, which corresponds to the complete reaction of all Ca^{2+} and Mg^{2+} with EDTA, is a definite change from a wine red color to a sky blue color.

It should be noted that the titration is difficult, if not impossible, to do if there is little or no Mg^{2+} in the natural water sample. The water sample must contain some Mg^{2+} in order for reaction to occur with the indicator to give a *wine red* color at the start. If there is no Mg^{2+} in the sample, then the color of the solution at the start and end of the titration will be blue — i.e., no end point! Even though most natural water samples contain sufficient Mg^{2+} for the color change to occur, it is important to ensure that the method work for *all* samples. The usual practice is to "spike" the buffer with a small volume of a solution that contains the MgEDTA chelate. As soon as this spiked buffer is added to the water sample, the following reaction occurs, releasing enough Mg^{2+} to produce a wine red color with the indicator:

$$MgEDTA + Ca^{2+} \rightleftharpoons CaEDTA + Mg^{2+}$$

Chelation of Ca^{2+} and Mg^{2+} by EDTA has a number of applications in commercial products. Bathroom tub and tile cleaners often contain EDTA in the form of a tetrasodium salt. Stubborn lime and scum deposits dissolve easily in the highly alkaline chelating medium, and the products that are soluble can be flushed away with water. Many salad dressings and other oil-containing products contain CaEDTA as a preservative. The calcium chelate removes any trace iron or copper ions that promote the spoilage of the oil:

$$CaEDTA + Fe^{3+} \rightleftharpoons FeEDTA + Ca^{2+} + H^+$$

CATION EXCHANGE RESIN

Another very common method of removing divalent cations from water samples is by *ion exchange*. The ion exchange materials for water softening are cation exchange resins in which monovalent cations (e.g., Na^+) are exchanged with divalent cations (e.g., Ca^{2+}). The process may be illustrated as follows

$$\text{Resin}(SO_3^-Na^+)(SO_3^-Na^+)(SO_3^-Na^+) + Ca^{2+} \rightarrow \text{Resin}(SO_3^-)(SO_3^-)Ca^{2+}(SO_3^-Na^+) + 2Na^+$$

The exchange is usually carried out by percolating the water through a column containing the ion exchange material. Note that the column will need regeneration once all the Na^+ ions have been exchanged. Regeneration is accomplished by reversing the above process. A solution containing a high concentration of salt (NaCl) is pushed through the spent column.

Pre-Laboratory Quiz

1. What is the per capita consumption of water per day in the United States?

2. What cation has the highest concentration is seawater?

3. Give an example of a sedimentary rock containing Ca^{2+}.

4. What is meant by the hardness of water?

5. Why is scale formation the bane of industry?

6. What is a common method of softening water?

7. Carbon dioxide concentrations can be very high in groundwaters. Why?

8. Give the structure of EDTA.

9. What is the name of the indicator used in EDTA titrations?

10. Why do some salad dressings contain EDTA-chelated calcium?

Laboratory Experiments

Flowchart of the Experiments

Section A.	The Evaporation of Water Samples to Give Total Dissolved Solids

Section B.	Divalent Cation Analysis by EDTA Titration

Section C.	The Dissolution of Rocks

Section D.	Important Ways of Reporting the Hardness of Water

Section E.	Determination of the Hardness of Ground-, Spring-, and Wellwater

Section F.	The Reaction of Divalent Cations with Soap; Soap Titrations

Section G.	Water Softening with Commercial Water-Conditioning Agents

Section H.	Divalent Cation Removal by Ion Exchange

Requires one three-hour class period to complete

Section A. The Evaporation of Water Samples to Give Total Dissolved Solids

Goals:

(1) To carry out a microscale evaporation of a variety of natural water samples. (2) To compare the relative amounts of nonvolatile solids contained in various natural water samples.

Experimental Steps:

NOTE: You will need a source of heat for this experiment. If you have an electric hot plate, then you can start immediately by collecting the appropriate natural water samples. If you do not, then construct a microburner system in the following way.

(1) Obtain a glass Pasteur pipet from Reagent Central.

(2) As illustrated below, light a match and hold it under the thin part of the pipet until it falls gently to a 90° angle as the glass softens.

(3) Allow the glass to cool (place it on your notebook).

(4) Carefully use a small file or scorer to cut off the end, as shown.

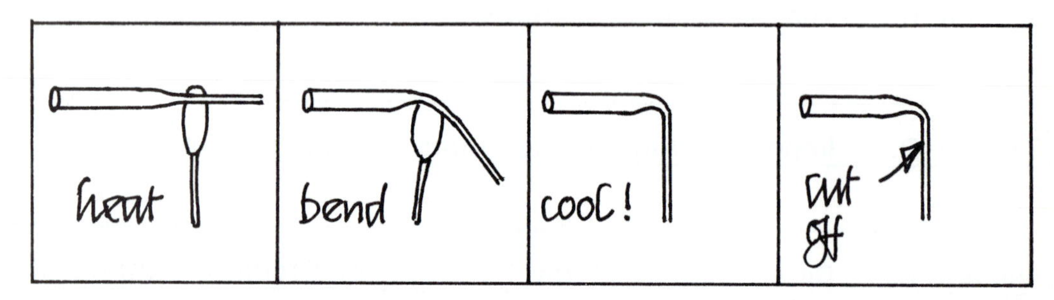

(5) Insert the wide end into a piece of latex tubing and attach a wooden clamp (clothes pin) to the pipet to keep the tip vertical.

(6) Connect the tubing to the gas tap.

(7) Turn the tap on a little, light the gas, adjusting the flame height to about 1 cm.

CAUTION: Keep your nose, hair, etc., out of the way of the flame!

(8) Place the burner underneath a cutout aluminum can, which acts as a hotplate.

CAUTION: *Remember* that metals conduct heat very well, and the can (all of it) will be *hot*.

NOTE: If you do not have a gas or electric source, then you can build a Thompson microburner (see Chapter 3, Section B), or try using a small candle.

NOTE: Plastic squeeze bottles of the natural water samples listed below will be passed around.

- Distilled water
- Tap water
- A mineral water from the Unite States
- A foreign mineral water
- A local well water

Obtain samples using clean wells in the 24-well tray and good transfer technique. Then fill clean small-drop microburets about 1/3 full with the water samples from the wells. Keep track of the location of each sample by placing the filled microburet in one of the wells of the 24-well tray.

- Record the location of the sample in the 24-well tray.

Do not deliver drops directly from the large bottles. There are many reasons why you should never use the large bottles directly. If you are tempted to do so, take a minute to think about *why* you should not.

Experimental Steps:

1. Obtain a piece of aluminum foil.
2. Place it on the table shiny side up and smooth it with your fingers.
3. Using tweezers, place the foil *shiny* side up onto the hotplate.
4. Drop 1 drop of each water sample onto the foil in a line about 0.5 cm apart.
   - Note where each water sample is located on the foil.
5. Allow the water to evaporate.

   NOTE: If it spits and splashes, the hotplate is too hot and you should turn it down.
6. After the water has evaporated, use the tweezers to remove the foil from the hotplate.
   - Describe and record what you see.

   The white solids that remain after evaporation of the water are the nonvolatile salts that were originally in solution in the water sample. The amount of nonvolatile salts in a certain volume of a water sample is called the *total dissolved solids* (TDS).

   - Compare the TDS of the various water samples that you placed on the foil.
   - Rank the various water samples in order from the lowest TDS to the highest TDS, and record the ranking in your notebook.
   - Which salts are likely to be present in the white solid residue on the foil?

6. Attach the foil to your notebook.

Section B. Divalent Cation Analysis by EDTA Titration

Goal:

To investigate the chemistry of the complexation reactions of ethylenediaminetetraacetic acid (EDTA) with divalent cations, such as Ca^{2+} and Mg^{2+}, that are present in natural water samples.

Discussion:

All of the titrations in this section should be carried out in a *serial* manner in 1 x 12 well strips. It is important to note that cleanliness is critical to achieving good results. Sources of Ca^{2+}, Mg^{2+} ions, etc., are ubiquitous. Make sure that all your apparatus is well washed with distilled water.

In this section you are investigating the factors that effect the complexation of Ca^{2+} and Mg^{2+} by EDTA, and it is therefore not necessary to use the same microburet for titrations. In later sections, where you will be doing quantitative volumetric analysis, it is necessary to use good wash, rinse, and transfer techniques and the *same* microburet for all solutions except buffer and indicator.

Experimental Steps:

1. Drop 1 drop of 10^{-3}M Ca^{2+} solution in each of 12 wells of a 1 x 12 well strip.
2. Add 1 drop of eriochrome black T indicator (EBT).
3. Titrate serially with 2 x 10^{-4}M EDTA solution — i.e., 1 drop in the first well, stir, 2 drops in the second well, stir, etc.
   - Record what you see.
4. Now add 1 drop of NH_3/NH_4Cl buffer to each well and stir.

   NOTE: *Remember* — there are *two* buffers: One is NH_3/NH_4Cl and the other is NH_3/NH_4Cl/MgEDTA.
   - Anything happen?
5. Clean the strip and carry out a serial titration of 10^{-3}M Mg^{2+} solution in the same manner.
   - What differences do you observe?
6. Now titrate a mixture of Ca^{2+} and Mg^{2+}. Clean the strip and add 1 drop each of 10^{-3} M Ca^{2+} and 10^{-3} M Mg^{2+} to each well.
7. Add 1 drop EBT and 2 drops NH_3/NH_4Cl buffer to each well and stir.
   - Comment on the color.
8. Serially titrate with EDTA.
   - How many drops are required to reach the end point? Why?
9. Clean the strip.
10. Carry out a serial EDTA titration of 10^{-3} M Ca^{2+} solution using 1 drop EBT and 1 drop of the *other buffer*, NH_3/NH_4Cl/MgEDTA.
    - What happens?
    - Calculate the molarity (approximately) of the Ca^{2+} solution.
    - Using simple chemical reactions, explain the chemistry of the last titration.
    - Summarize the method you would adopt in order to titrate an unknown water sample that might contain only Ca^{2+}, only Mg^{2+}, or a combination of the two.

Section C. The Dissolution of Rocks

Goals:

(1) To examine the processes of rock dissolution that produce dissolved divalent cations, such as Ca^{2+} and Mg^{2+}, in natural water samples. (2) To determine the concentration of divalent cations by means of EDTA titration.

Experimental Steps:

1. Thoroughly clean a petri dish and lid.

2. Select a small piece of limestone with your tweezers, wash it with distilled water, and place it into the bottom of the dish.

3. Do the same with small pieces of gypsum and granite.

4. Use a clean large-drop microburet to drop distilled water on each rock until it is well covered with a pool of water (about 5–6 drops). You will need to wait for 10 minutes before sampling the pools.

   - The dissolution process is rather slow — Why?
   - List some of the factors that you think are important in the dissolution of rocks. You might want to refer back to the discussion in the Background Chemistry section of Chapter 7, "Solutions and Reactions".
   - By the way, what is the chemical composition of limestone, gypsum, and granite?

5. Use a clean large-drop microburet to remove a small volume of the pool over the limestone.

6. Carry out an EDTA titration (it does not have to be done serially) on 2 drops of the pool water. Use 1 drop of buffer ($NH_3/NH_4/MgEDTA$ — from now on, this buffer is the one used in *all* EDTA titrations) and 1 drop of EBT indicator. Don't forget to stir and count drops.

   - Calculate the divalent cation molarity.

7. Now carry out an analysis of the granite pool.

   - Report the molarity.
   - Comment on the comparative solubility of the limestone and granite.

8. Use your hand lens and look closely at the gypsum.

   - Do you really want to try to analyze this pool?
   - What will happen if some solid gets into the water sample?

   NOTE: The following experiments in this *section* are optional and you should only proceed if you have the time (about 1/2 hour).

9. In order to investigate the effect of water temperature and acidity on rock dissolution, wash out the dish and place 3 small, washed pieces of limestone into the dish.

10. Make an ice bath.

11. Use a clean microburet to create a pool of ice-cold water on one rock. Deliver a pool of hot water to another and a pool of dilute acid (10^{-4} M) to the third. Wait 5 minutes.

    - Determine the divalent cation concentration in each pool.

    NOTE: You may need to adjust the number of drops of sample that you use!

    - Discuss your results. Write chemical reactions wherever you can.
    - Do you think that the size of the rock particles would have any effect on the dissolution? Explain.

Section D. Important Ways of Reporting the Hardness of Water

Goal:

To be able to calculate and report the hardness of water in various units.

Discussion:

The divalent cation concentration of water is often referred to as the *hardness* of water — less of a mouthful and more tactile! There are several ways of quantitatively expressing the hardness of water that are used extensively by engineers and technicians in the water industry. The two most common units are *parts per million* (ppm) and *grains per gallon* . It is not usual to carry out separate determinations for the individual $[Ca^{2+}]$ and $[Mg^{2+}]$ in an analysis of water. The combined total concentration of Ca^{2+}, Mg^{2+}, and any other polyvalent cations is what is determined by an EDTA titration.

Hardness is therefore a property of water that is imparted by several different cations, all of which are included in the single EDTA analysis. In this situation, it turns out to be very convenient to express hardness in terms of an equivalent concentration of *calcium carbonate* ($CaCO_3$). Let us make a conversion from divalent cation concentration, expressed as a molarity, into parts per million $CaCO_3$ units. You have analyzed a water sample, and let's say the hardness was found to be 10^{-3} M. Expressing the hardness as if it were due to $CaCO_3$,

$$\text{Hardness} = \frac{10^{-3}\ \text{mol CaCO}_3}{\text{liter}}$$

Now, the formula weight (molar mass) of $CaCO_3$ is very close to 100 g mol^{-1}:

$$\text{Hardness} = \frac{10^{-3}\ \text{mol CaCO}_3}{\text{liter}} \times \frac{100\ \text{g CaCO}_3}{\text{mol CaCO}_3}$$

$$= \frac{10^{-1}\ \text{g CaCO}_3}{\text{liter}}$$

$$= \frac{10^{-1}\ \text{g CaCO}_3}{\text{liter}} \times \frac{1000\ \text{mg CaCO}_3}{1\ \text{g CaCO}_3}$$

$$= \frac{100\ \text{mg CaCO}_3}{\text{liter}} \quad \text{or simply} \quad \frac{100\ \text{mg}}{\text{L}}$$

If the density of the water sample is assumed to be 1000 g L^{-1}, then we can say

$$\text{Hardness} = \frac{100\ \text{mg CaCO}_3}{1000\ \text{g water}}$$

Now,

$$\frac{1 \text{ mg } CaCO_3}{1000 \text{ g water}} = 1 \text{ ppm}$$

Therefore,

$$\frac{100 \text{ mg } CaCO_3}{1000 \text{ g water}} = 100 \text{ ppm hardness}$$

Thus, 10^{-3}M total divalent cation concentration = 100 ppm.

Conversion to grains per gallon requires a knowledge of Henry VIII units:

$$1 \text{ grain} = \text{weight of an average grain of wheat} = 64 \text{ mg}$$

Since

$$1 \text{ gallon} = 3.785 \text{ liters,}$$

Then,

$$\frac{1 \text{ grain } CaCO_3}{\text{gallon water}} = 17.1 \text{ ppm}$$

Thus, the 10^{-3} M water sample has a hardness of

$$100 \text{ ppm} \times \frac{1 \text{ grain per gallon}}{17.1 \text{ ppm}}$$

$$= 5.8 \text{ grains per gallon}$$

- Convert 2.5×10^{-3} M Ca^{2+} concentration into ppm and grains per gallon hardness.
- What is the factor that will allow you to convert molarity directly into ppm?

Section E. Determination of the Hardness of Ground-, Spring-, and Wellwater

Goals:

(1) To determine the total divalent cation concentration (hardness) of a variety of ground-, spring-, and wellwater samples by EDTA titration. (2) To compare the analyses with information obtained from a TDS determination and with label information for bottled mineral waters.

Experimental Steps:

NOTE: Your instructor will provide you with a list of the available water samples. For mineral waters, the bottles are available for your inspection; for other samples, source data are available.

1. Select 3 or 4 types of water that you wish to analyze.
2. The water samples will be passed around the class in labelled plastic containers. Use good transfer, wash, and rinse techniques.
3. Using the $NH_3/NH_4Cl/MgEDTA$ buffer, determine the hardness for your selected water samples by EDTA titration.
   - Calculate the hardness in molarity, ppm, and grains per gallon.
   - Compare the hardness values with any information you have from TDS determinations or from label information.

Section F. The Reaction of Divalent Cations with Soap; Soap Titrations

Goal: *To examine the effect of divalent cations, such as Ca^{2+} and Mg^{2+}, on typical soap solutions.*

Discussion: The reaction may actually be carried out quantitatively in the form of a dropwise titration. A standard soap solution is added dropwise to the natural water sample containing Ca^{2+} and Mg^{2+}. The solution is shaken vigorously between soap drops. The end point is suds formation!

Environmental Steps:

1. Wash 2 glass vials and caps with distilled water.
2. Add distilled water to each vial to about 1 cm in height.
3. Use a microburet to deliver 10 drops of the local well water to *one* of the vials.
4. Fill another microburet with 2% soap solution.

   NOTE: You can use a different microburet here because this titration is only semiquantitative.
5. Deliver 1 drop of soap to the vial containing only distilled water.
6. Replace the cap and shake vigorously.
   - What happens at the surface?
7. Titrate the solution in the other vial by adding soap solution dropwise until you obtain suds formation. Use the first vial for comparison.
   - Look carefully at the solutions in the two vials against a black background. What do you see?
   - What is the chemical composition of the precipitate?
   - How much water hardness is titrated by 1 drop of 2% soap solution?

Section G. Water Softening with Commercial Water-Conditioning Agents

Goal: *To investigate the softening of water — i.e., the removal of hardness by commercial water-conditioning products.*

Experimental Steps:

1. Set up this experiment exactly like the last one, through Step 6.
2. Now to the second vial (the one containing 10 drops of local well water), add about 20 mg of selected commercial conditioning product by means of a straw spatula. Swirl to dissolve.
3. Titrate with 2% soap solution until an end point is obtained.
   - Compare the results from this titration with the results from the titration where no softening agent was added.
   - Is there any information about the active ingredients on the product package?
   - Does the product contain phosphates?

Section H. Divalent Cation Removal by Ion Exchange

Goal:

To investigate the use of cation exchange resins in the removal of Ca^{2+} and Mg^{2+} from very hard natural water samples.

Experimental Steps:

1. Wash a vial and cap thoroughly with distilled water.
2. Transfer a small amount of cation exchange resin (as a slurry) using a thick-stem pipet. Your instructor will demonstrate the apparatus and technique.
3. Deliver 20 drops of local well water to the vial.
4. Replace the cap and shake gently for about a minute.
5. Set the vial down and allow the resin to settle.
6. Remove some of the supernatant liquid using a clean microburet and being careful not to suck up any resin beads.
   - Analyze the liquid for polyvalent cations. Report the hardness.
7. Test 1 drop of the liquid with wide-range pH paper.
   - What cation replaced the polyvalent cations in the solution?
   - Give a brief description of the chemistry of this exchange process.

Chapter 11

Vitamin C Analysis

Introduction

One of the most interesting and fruitful applications of chemical knowledge in the last decade has been the development of food and nutritional chemistry. The close link between diet and health and human behavior is now so firmly established that "we are what we eat" has almost become a cliché. Out of this explosion of information concerning the chemistry of nutrition have come many controversies, one of the most important of which is the problem of *recommended dietary allowance* (RDA). According to the National Academy of Sciences, RDAs are recommendations for the average daily amounts of nutrients that population groups should consume over a period of time and should not be confused with requirements for a specific individual. The key question is, Can scientists really say what levels of nutrients are necessary for a healthy human diet? The controversy has been very apparent (and at times bitter) with respect to the water soluble vitamins, especially vitamin C. In the United States, the RDA for vitamin C is 60 mg for adults of both sexes. For children up to the age of 11 years, an allowance of 45 mg per day of vitamin C is recommended, and for older children 60 mg per day. For pregnant women an additional 20 mg per day (over the regular 60 mg) is recommended, particularly for the second and third trimester of pregnancy. For lactating women, an additional 40 mg per day (over the regular 60 mg) is recommended to assure a satisfactory level of vitamin C in breast milk. It is interesting to note that the RDAs seem to vary widely in different countries. The most famous proponent of the use of a much larger daily intake of vitamin C than the RDA is the two-time Nobel Prize winner Dr. Linus Pauling. In his book *Vitamin C and the Common Cold,* Dr. Pauling has provided some cogent arguments that large doses (5 g daily) of vitamin C are beneficial to health. He makes the following points:

- Several carefully carried out experiments have established that amounts of vitamin C larger than the RDA appear to reduce the severity and frequency of common colds.
- There appears to be evolutionary evidence that large intakes (2–3 g daily) are normal and good for health.
- Vitamin C is nontoxic even at very high dosage levels.
- Vitamin C is certainly better and cheaper than over-the-counter cold remedies, many of which have toxic side effects at relatively low dosages.
- The medical and regulatory establishments have issued biased and misleading statements about the value of vitamin C in reducing the severity and frequency of colds.

The American Medical Association and the National Academy of Sciences keep insisting that there is no evidence for Dr. Pauling's claims.

Pure vitamin C (alternatively called ascorbic acid) is a white, crystalline organic compound with a chemical formula of $C_6H_8O_6$ (F. Wt. 176.12 g mol^{-1}). The structure of ascorbic acid is usually drawn in one of the two following ways:

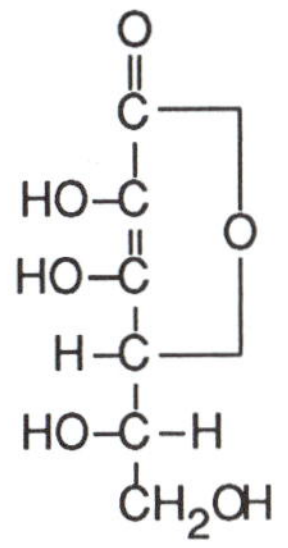

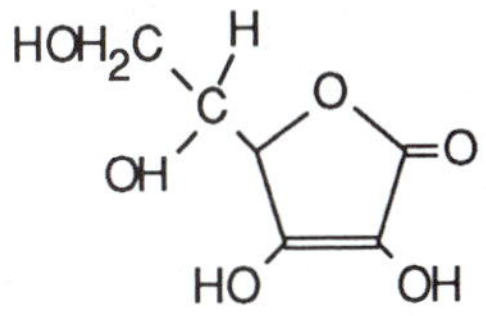

Vitamin C is a weak acid that is very soluble in water (0.3 g mL^{-1}). One of the most important chemical properties of ascorbic acid, particularly with respect to its action as a vitamin, is that it is a reducing agent — i.e., it is relatively easily oxidized. Pure, solid ascorbic acid can be stored for years without much change occurring. However, in solution the vitamin can decompose rapidly due to oxidation by dissolved atmospheric oxygen. The decomposition is accelerated by heat, light, alkalies, oxidative enzymes, and the presence of traces of iron and copper ions. The oxidation is slowed down in an acidic medium and by cold temperatures. The vitamin is commercially available as the solid crystalline acid or as the solid salt sodium ascorbate and is manufactured in the microbiological oxidative fermentation of calcium gluconate by *Acetobacter suboxidans*.

Ascorbic acid is widely distributed in both the plant and animal kingdom. It is synthesized by all known species except the primates (including man), the guinea pig, Indian fruit bats, and the red-vented bulbul bird. These species are unable to synthesize ascorbic acid because their cells do not have the enzyme that catalyzes the oxidation of L-gluconolactone into L-ascorbic acid. A dietary deficiency can result in scurvy, which in primates first appears as symptoms of weakness and fatigue followed by shortage of breath and aching in the bones. As the disease worsens, muscle degeneration and bleeding occur, the gums swell and bleed, and teeth fall out. The final result is convulsions and death. An historical example was the celebrated voyage around the Cape of Good Hope of Vasco da Gama. He lost 100 men out of a 160 man crew! A British doctor, James Lind, tested numerous remedies on sailors and in 1747 discovered that oranges, lemons and limes were curative. American Indians in the sixteenth century knew that pine needle broth and the adrenal glands of slain elk (both good sources of vitamin C) would cure scurvy. Eskimoes traditionally eat certain layers of whale skins to obtain vitamin C. It is now accepted that a daily intake of 10 mg of ascorbic acid will alleviate and cure the clinical symptoms of scurvy in adults. At the present time scurvy is not common in the United States, although it does occur in infants fed diets consisting exclusively of cow's milk (which has little vitamin C) and in aged persons on poor diets. In adults, the occurrence of scurvy is associated with poverty, alcoholism, famine, and nutritional ignorance. The disease is quite prevalent in other parts of the world.

The specific biochemical functions of vitamin C are not at all clear. It is known that the vitamin is required for the formation of collagen (a substance present in cartilage, bone, and teeth) and for cell respiration and enzyme function. There is some evidence that ascorbic acid can act as a biochemical antioxidant by working as a reducing agent to detoxify hydrogen peroxide. In humans, the simultaneous ingestion of vitamin C and iron has been shown to significantly increase iron absorption, thus helping to alleviate the common metal ion deficiency. Little is known concerning the ascorbic acid metabolism of the adult female; however, there appears to be a physiological difference in the retention of vitamin C in males and females.

In the United States, the most common dietary source of vitamin C is the vegetable-fruit group of foods. Good sources are fresh and frozen citrus fruits and juices and green leafy vegetables, such as broccoli, parsley, and spinach (see Table 11.1). Surprisingly, apples, grapes, and peaches are relatively low in vitamin C. It is important to remember that the two chemical characteristics discussed earlier, water solubility and ease of oxidation, can lead directly to significant vitamin C losses. Storage, food handling and preparation, and cooking methods can all have a dramatic effect on the final dietary intake. Commercial food processing — e.g., canning, blanching, freezing, and milling — can also lead to substantial losses. Blanched frozen peas can lose up to 50% of their original vitamin C content. Two-year storage of unfrozen canned fruits and juices (at 27°C) can cause vitamin losses of up to 50%. Long-term storage of potatoes results in significant vitamin loss. The stability of ascorbic acid in potatoes is very important because this vegetable is a significant dietary source in many countries. Slicing or maceration of fresh vegetables releases enzymes (e.g., ascorbate oxidase) that can quickly oxidize vitamin C. A particularly destructive cooking method is that of boiling vegetables in large amounts of water and then throwing away the liquid. Steaming and wok cooking are much better ways of retaining all nutrients, especially the water soluble vitamins. The above types of losses have led nutritionists to be alarmed about the actual vitamin C content in the average diet. The keys to a good vitamin C intake are *fresh, quickly cooked, appropriate foods* — and, of course, vitamin supplements.

The vitamin C content of tissue fluids, foods, blood, and so on, can be accurately determined by suitable methods of chemical analysis. In this series of experiments, you have the opportunity to investigate an excellent redox volumetric method for the analysis of vitamin C and to design your own research experiments in ascorbic acid chemistry.

Background Chemistry

Most of the modern methods for the analysis of vitamin C are based on three important chemical properties of the vitamin: water solubility, ease of oxidation, and fluorescence. The fact that ascorbic acid will emit a visible fluorescence under certain conditions is the basis for a recent spectrometric method. Unfortunately, the instrumentation required is extremely expensive, and the sample preparation rather complicated. One of the simplest and least expensive methods utilizes the ease of oxidation of the vitamin. According to most biochemical texts, ascorbic acid is readily oxidized in the following reaction, producing a product dehydro-L-ascorbic acid ($C_6H_6O_6$):

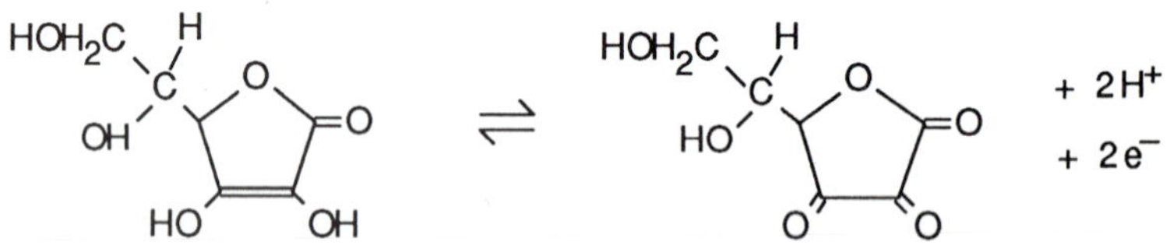

$$C_6H_8O_6 \rightleftharpoons C_6H_6O_6 + 2H^+ + 2e^-, E° = 0.127 \text{ v (at pH 5)}$$

Don't forget that the loss of electrons is defined as *oxidation*. One of the most widely used oxidizing agents for the analysis of vitamin C is an organic dye called 2,6-dichloroindophenol. When this dye acts as an oxidizing agent towards vitamin C, it becomes reduced — i.e., it *gains* electrons from the vitamin C:

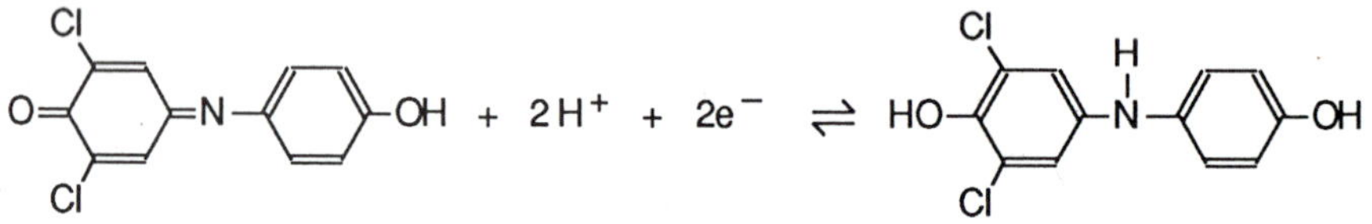

$$\text{2,6-dichloroindophenol} + 2H^+ + 2e^- \rightleftharpoons \text{reduced form (HO–C}_6\text{H}_2\text{Cl}_2\text{–NH–C}_6\text{H}_4\text{–OH)}$$

To simplify things, let us write In instead of 2,6-dichloroindophenol. The above reduction then becomes

$$In + 2H^+ + 2e^- \rightleftharpoons InH_2$$

The overall redox reaction for the oxidation of vitamin C by 2,6-dichloroindophenol is, therefore,

$$C_6H_8O_6 + In \rightleftharpoons C_6H_6O_6 + InH_2$$

The main reasons for the choice of this particular dye as an oxidizing agent are that it will only oxidize vitamin C (and not other substances that might be present) and that it will act as a self-indicator during the analysis. The analysis is carried out using standard volumetric analysis procedures and is done in an acidic solution (pH ~ 1–2). The sample containing vitamin C is usually extracted with an acid solution that is then titrated with a neutral, blue solution of 2,6-dichloroindophenol until the color changes from colorless to pink.

The chemistry of the titration is interesting. It is schematized in Figure 11.1.

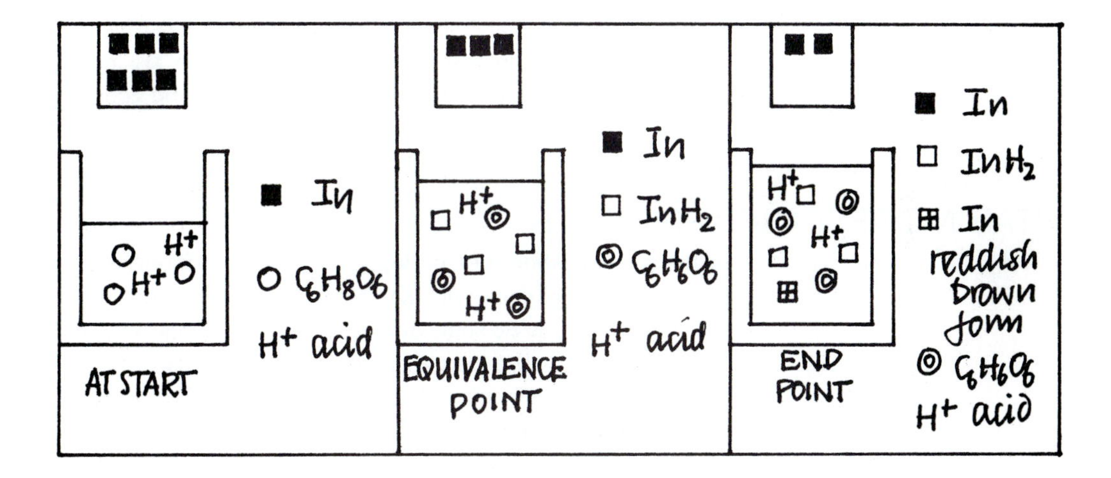

Figure 11.1 Chemical Reactions in the Vitamin C Titration

At the start of the titration, the solution simply contains vitamin C and an acid that is added to preserve the vitamin and give a pH of 1–2. The solution is colorless at this point. The blue titrant (2,6-dichloroindophenol) is now added, and the dye oxidizes some of the vitamin to colorless dehydroascorbic acid in the redox reaction described earlier. Both products of the redox reaction are colorless and, therefore, there will be no color change until *all* of the vitamin C has been oxidized (the equivalence point). A slight excess of titrant is now added to obtain the end point. The excess of dye actually turns a brownish-pink color in the acidic solution. The end point is thus a color change from colorless to brownish pink.

Normally in volumetric analysis, the molarities of the various solutions are carefully calculated from the titration data. However, in the redox analysis of vitamin C, it is usual practice to use a known concentration solution of vitamin C to standardize the dye solution and to report the volume (or drops) of dye that is equivalent to 1.0 mg of vitamin C. A titration of an unknown concentration of vitamin C, together with the above information, can then provide a rapid method of determining the number of milligrams of vitamin C in the sample. Vitamin C is notoriously unstable in solution, and precautions must be taken to stabilize it during the analysis. The culprits are trace amounts of Cu^{2+} and Fe^{3+} ions that act as catalysts in the oxidation of vitamin C by oxygen of the air. Several reagents, such as metaphosphoric acid and EDTA (ethylenediaminetetraacetic acid), have been used to chelate these metal ions, thereby effectively preventing them from acting as catalysts.

The chemical analysis of foods and other "real" samples for vitamin C by the redox titration method is not always completely straightforward. The titration involves a color change that must be discernible by the analyst. If the sample is highly colored (e.g., a purple grape drink) then the analysis cannot be carried out by this particular method. A second problem is one that almost always confronts the analyst in trying to analyze unknown materials. The problem is how much sample to use in the analysis — too little and the analysis will not be sensitive enough, too much and the titration could go on forever! Fortunately, there are many good sources of data concerning the vitamin C content of various fresh and cooked foods, and these can be used to make intelligent decisions about what sample size to use and what dilutions to make. Table 11.1 shows the vitamin C contents of some foods.

VITAMIN C RESEARCH PROJECTS

In the second laboratory period, you have the opportunity to carry out your own research investigations into the chemical and nutritional aspects of vitamin C. It is imperative that you do some thinking about your project before you come to laboratory. Your instructor will expect a one-page proposal that outlines the general goal of the project, together with some indication as to what type of samples you are going to use and what sample preparation methods you will select. All the solutions and apparatus required for vitamin C studies and analysis will be provided, except that you must provide the basic samples if you are going to analyze fruits, vegetables, and the like. You have plenty of time to carry out and report on your project during the laboratory period. Your grade will be based on originality, execution, and reported conclusions. Here are some general suggestions for the types of projects that you might consider.

- Study some of the factors that are important in vitamin C decomposition, particularly Cu^{2+}, Fe^{3+}, heat, and pH. This type of study could be done on pure vitamin C solutions or on fruit juices, etc.
- Carry out a study to determine the optimum conditions for the redox titration of ascorbic acid with 2,6-dichloroindophenol. Are there any interferences in this method? It has been reported that Fe^{2+} will reduce the dye and that Fe^{2+} salts are often present in multivitamin supplements. Reducing sugars (e.g., glucose) are also present in many fruits and vegetables and could perhaps reduce 2,6-dichloroindophenol.
- Determine the effects of various types of food preparation on vitamin C content. A study on potatoes might include peeling, chopping, cutting, slicing, chipping, grating, microwaving, boiling, steaming, etc.
- Dissect a fruit or vegetable and try to determine if there are any differences in vitamin C content in various parts of the fruit or vegetable. Core samples of, say, a potato might reveal cross-sectional vitamin C differences.

- Compare various brands of frozen or canned fruit juices for vitamin C content. Try to ascertain what factors determine the vitamin C content of a commercial juice.
- Compare the vitamin C content of various kinds of related vegetables — e.g., different kinds of peppers or different kinds of citrus fruit.
- Study the factors that control the decomposition of vitamin C in stored fruit juices. The variables you might look at are time, temperature, light, type of fruit, and pH.
- Study the excretion of vitamin C by analyzing urine samples. Large doses (>500 mg) are supposedly not absorbed very efficiently by the body and the vitamin C is excreted into the urine via the kidneys.
- Compare the vitamin C content of different breakfast cereals, particularly with respect to any claims made on the packet about RDA percentages.
- Develop an entirely new method for the determination of vitamin C. It has been suggested that a starch/I_2 solution is a suitable oxidizing agent for ascorbic acid.
- Investigate the stoichiometry of the redox reaction between ascorbic acid and 2,6-dichloroindophenol. The titration depends on having an exact and known stoichiometry. How could you determine this?

NOTE: The projects outlined above are merely suggestions. You are encouraged to design your own investigation. If you need any help, please discuss it with your instructor.

Vitamin C Content of Foods

Food	Portion as Used (FDA Established)	Vitamin C Weight (g)	Vitamin C Portion (mg)	Vitamin C Per 100 g Food (mg)
Vegetables				
Asparagus	cooked, 1/2" diam. at base, 4 spears	60	16	27
Beans, sprouted mung	uncooked, 3 1/2 oz.	100	19	19
Broccoli	cooked, 1 medium stalk, chopped	180	162	90
	yield from 10 oz. frozen pkg, 1 3/8 c	250	143	57
Brussel sprouts	cooked, 7–8 sprouts, 1 c	155	135	87
Cabbage	raw, coarsely shredded, 1 c	70	33	47
	cooked, 1 c	145	48	33
Carrots	cooked, diced 1 c or 1 raw carrot, 6" × 1"	145	6	9
Cauliflower	cooked (flower buds), 1 c	120	66	55
Celery	raw, 8" × 2", 1 stalk	40	4	10
Collards	cooked, 1 c	190	87	46
Corn, sweet	cooked, 5" × 2", 1 ear	140	7	5
Cucumbers	raw, pared, 10 oz., 1 cucumber	207	23	11
Lettuce, iceberg	4 1/2" diam., 1 head	454	29	6
Mushrooms	canned, solid and liquid, 1 c	244	4	2
Okra	cooked, pod 3" × 5/8", 8 pods	85	17	20
Onions	raw, 2 1/2" diam., 1 onion	110	11	10
Parsley	raw, chopped, 1 Tbsp.	4	7	175
Peas, green	cooked, 1 c	160	33	21
	canned, strained baby food, 1 oz.	28	3	11
Peppers, hot red	dried, seeds removed, 1 Tbsp.	15	2	3
Peppers, sweet	raw, seeds removed, 1 pod	74	94	127
Peppers, sweet	cooked, boiled, 1 pod	73	70	96
Potatoes	boiled without skin (about 3# raw), 1 potato	122	20	16
	boiled with skin (about 3# raw), 1 potato	136	22	16
	heated, frozen french fries, 10 pieces	57	12	21
	mashed, milk added, 1 c	195	19	10
	chips, 2" diam., 10 chips	20	3	15
Sauerkraut	canned, solid and liquid, 1 c	235	33	14
Spinach	cooked, 1 c	180	50	28
Squash, summer	cooked, diced, 1 c	210	21	10
Tomatoes	raw, 3" diam, 7 oz., 1 tomato	200	42	21
	catsup, 1 c	273	41	15
Turnips	cooked, 1 c	155	34	22

Table 11.1 Vitamin C Content of Foods

Fruits

Apples	raw, whole (skin, core, etc.), 1 apple	150	3	2
Apple juice	bottled or canned, 1 c	248	2	1
Avocadoes	raw, whole (kernel included), 1 avocado	284	30	11
Bananas	raw, 1 medium banana	175	12	7
Blackberries	raw, 1 c	144	30	21
Blueberries	raw, 1 c	140	20	14
Cantaloupe	raw, 5" diam., 1/2 melon	385	63	16
Cranberry, cocktail juice	canned, 1 c	250	40	16
Fruit, cocktail	canned, in heavy syrup, 1 c	256	5	2
Grapefruit	raw, 1/2 medium fruit	241	44	18
	canned, packed in syrup, 1 c	254	76	30
Grapefruit, juice	fresh, 1 c	246	92	37
	frozen concentrate, undiluted, 6 fl. oz.	207	286	138
Grapes	raw, American slip skin, 1 c	153	3	2
Lemonade	frozen concentrate, 6 fl. oz.	219	66	30
Lemons	raw, 2 1/8" diam., 1 lemon	110	39	35
Lime, juice	fresh, 1 c	246	79	32
Oranges	raw, 3" diam., 1 orange	180	66	37
Orange, juice	fresh, 1 c	248	124	50
	frozen concentrate, 6 fl. oz.	213	360	169
Papayas	raw, 1/2" cubes, 1 c	182	102	56
Peaches	raw, 1 medium peach	114	7	6
	canned, in heavy syrup, 1 c	257	7	3
Pears	raw, 1 medium pear	182	7	4
Pineapples	raw, diced, 1 c	140	24	17
	canned, in heavy syrup, 2 slices + juice	122	8	7
Plums	raw, 2 oz., 1 medium plum	60	3	5
Raisins	seedless, 1/2 oz. pkg.	14	trace	trace
Raspberries	raw, 1 c	123	31	25
Rhubarb	cooked, sugared, 1 c	272	17	6
Strawberries	raw, capped, 1 c	149	88	59
Watermelon	raw, about 2" wedge, 1 wedge	925	30	3

1 c = 1 cup

Table 11.1 (continued) Vitamin C Content of Foods

Pre-Laboratory Quiz

1. What is the adult RDA for vitamin C?

2. What disease is caused by severe vitamin C deficiency?

3. A two-time Nobel prize winner recommends taking large daily doses of vitamin C. What is his name?

4. Give an alternative chemical name for vitamin C.

5. Which species require a dietary source of vitamin C?

6. Two chemical properties of vitamin C are often responsible for the loss of vitamin C in food preparation. What are they?

7. Give a definition of reduction.

8. Which metal ions catalyze the aerial oxidation of vitamin C?

9. Give 4 good food sources of vitamin C.

10. Give the name of a dye that is used as an oxidizing agent in the redox volumetric analysis of vitamin C.

Laboratory Experiments

Flowchart of the Experiments

Section A.	Optimizing the Reaction Conditions for the Determination of Vitamin C

Section B.	Standardization of 2,6-Dichloroindophenol

Section C.	The Analysis of a Commercial Vitamin C Tablet

Section D.	Vitamin C Concentration in Fresh Fruit Juices

Section E.	Analysis of a Breakfast Cereal for Vitamin C

Section F.	Research Project

Requires two three-hour class periods to complete

CAUTION: The acid and base solutions used in this experiment are corrosive. If you get any chemicals on your skin, wash with cold water and then check with your instructor.

Section A. Optimizing the Reaction Conditions for the Determination of Vitamin C

Goal:

To investigate the factors that determine the utility of a redox reaction for the quantitative volumetric analysis of vitamin C.

Experimental Steps:

1. Thoroughly clean two 1 x 12 well strips with distilled water.
2. Use good transfer technique to fill large-drop microburets with 2,6-dichloroindophenol, sodium hydroxide solution, and standard vitamin C solution.

   NOTE: A standard solution of vitamin C has been prepared for you by weighing pharmaceutical-grade crystalline Vitamin C, dissolving it in distilled water, and then making it to an exactly known volume. The solution has a concentration of 0.50 mg mL^{-1} and contains a preservative.

   First, investigate the effect of pH on the redox reaction between vitamin C and 2,6-dichloroindophenol.

3. Drop 1 drop of vitamin C solution in each of 10 wells in a 1 x 12 well strip.
   - What is the approximate pH of this solution?

     HINT: What is the alternative name for vitamin C?

   All the titrations in Section A are approximate in that you are simply trying to establish the appropriate conditions for the redox reactions. This means that you can assume that all the large-drop microburets deliver approximately the same volumes. In later experiments you will need to use quantitative delivery technique.

4. Now carry out a serial titration of the vitamin C solutions with 2,6-dichloroindophenol and stir each well as you go.
   - Record what you see and note how many drops of 2,6-dichloroindophenol are necessary to obtain an endpoint.
5. Drop 1 drop of vitamin C solution in each of ten wells in a 1 x 12 well strip.
6. Drop 1 drop of sodium hydroxide into each of the ten wells containing vitamin C (Step 5) and stir.
   - What is the approximate pH of the vitamin C solutions?
7. Carry out a serial titration of the alkaline vitamin C solutions with 2,6-dichloroindophenol and stir each well as you go.
   - Note how many drops of 2,6-dichloroindophenol are necessary to obtain an endpoint.

- Record what you see occurring to the titration solutions over the next several minutes.

8. Obtain a large-drop microburet of sulfuric acid.
9. Carry out a serial titration of 1 drop of vitamin C solution with 2,6-dichloroindophenol in the presence of 1 drop of sulfuric acid and stir.
    - Record what you see and note how many drops of 2,6-dichloroindophenol are necessary to obtain an endpoint.
    - Compare the titrations that you have just completed.

NOTE: The redox determination of vitamin C by titration with 2,6-dichloroindophenol works well in most situations. However, there is one major interference reaction that gives erroneous results. Samples that contain Fe^{2+} cannot be analyzed easily because 2,6-dichloroindophenol is reduced by Fe^{2+}.

10. Clean a large-drop microburet and fill it with Fe^{2+} solution.
11. Deliver 1 drop of Fe^{2+} solution and 1 drop of sulfuric acid to each of 10 wells of a 1 x 12 well strip and stir.
12. Titrate the Fe^{2+} with 2,6-dichloroindophenol until you obtain an endpoint.
    - Record the number of drops of 2,6-dichloroindophenol that were required to obtain an endpoint.
    - Write a balanced redox reaction for this titration.
    - What types of commercially-available products are likely to contain vitamin C and Fe^{2+}?

Section B. Standardization of 2,6-Dichloroindophenol

Goal:

To standardize a 2,6-dichloroindophenol solution against an exactly known concentration solution of vitamin C using quantitative technique and the optimum conditions investigated in Section A.

NOTE: In this section's experiments, you will be carrying out quantitative volumetric analysis, and it is important that you use the same microburet in the same way for all titration solutions (except the sulfuric acid). Don't forget to use good wash, rinse and transfer technique.

Experimental Steps:

1. Deliver 2 drops of standard vitamin C solution (0.50 mg mL^{-1}) from a clean, large-drop microburet into each of 12 wells of a 1 x 12 well strip.
2. Add 1 drop of sulfuric acid to each well and stir.
3. Use good transfer technique to fill the large-drop microburet used in Step 1 with 2,6-dichloroindophenol.
4. Titrate the vitamin C with the 2,6-dichloroindophenol with the same delivery technique used in Step 1.

- Record the number of drops of 2,6-dichloroindophenol required to obtain an endpoint. (Now you know exactly how many drops of dye are equivalent to 1 drop of vitamin C containing 0.5 mg mL^{-1} vitamin C.)

- Calculate how many drops of 2,6-dichloroindophenol will oxidize the vitamin C in 1 drop of the standard vitamin C solution.

NOTE: The standardization of the 2,6-dichloroindophenol will remain an accurate one provided that you use the dye within the same laboratory period. If more time elapses, then a restandardization is necessary.

5. Now you need to calibrate the large-drop microburet that you used to carry out the titration. This can be done in several ways.

Method 1

If you have access to a balance that weighs to a milligram or better, the calibration is simple.

(1) Place a small piece of plastic or weighing boat on the balance and determine its weight. (2) Fill the large-drop microburet (used in Steps 1 and 3) about 1/2 full with distilled water. (3) Deliver 30 drops to the boat (in exactly the same manner as in the titrations, counting as you go). (4) Reweigh and calculate the weight of 30 drops. (5) Assume a density of 1.00 g mL^{-1} for water and calculate the drop volume (in mL).

- Record the drop volume in your notebook.

Method 2

If you do not have a balance, then use the known volume of the wells in a 1 x 12 well strip.

(1) Fill the large-drop microburet (used in Steps 1 and 3) about 1/2 full with distilled water. (2) Deliver drops to one of the wells in a 1 x 12 well strip (in exactly the same manner as in the titrations and *counting* as you go). (3) Fill the well so that the water is level with the top edge but has a slight bulge in the middle. (4) Assume that each well of a 1 x 12 well strip has a volume of 0.40 mL and calculate the drop volume (in mL).

- Record the drop volume in your notebook.

Once you have calibrated the microburet and have calculated the drop volume, then you can calculate the number of mg of vitamin C in 1 drop of vitamin C solution.

- Calculate the number of mg of vitamin C per drop by multiplying the concentration of the standard vitamin C solution (0.50 mg mL^{-1}) by the drop volume (in mL) and record your answer.

- Now calculate the number of mg of vitamin C oxidized by 1 drop of 2,6-dichloroindophenol and record your answer.

NOTE: Don't forget that the stoichiometry of the redox reaction is 1:1:1:1 and that you have already calculated how many drops of 2,6-dichloroindophenol will oxidize the vitamin C in 1 drop of the standard vitamin C solution.

NOTE: Be careful to save and look after the calibrated microburet because you will need it in the other sections of this laboratory.

Section C. The Analysis of a Commercial Vitamin C Tablet

Goal: *To determine the vitamin C content of a commercially available vitamin C tablet.*

Discussion: An analysis of solid samples usually requires that the weighed sample be dissolved in an appropriate solvent (water) and the resulting solution be made to an exactly known volume. This solution can then be analyzed by the redox volumetric technique.

Your instructor will demonstrate the correct techniques for preparing the solution of a commercial vitamin C tablet and will make available to you portions of the solution.

NOTE: Remember that you are again going to carry out a quantitative volumetric analysis and that you must use good wash, rinse, and titration techniques.

Experimental Steps:

1. Use the calibrated microburet to carry out a serial titration of the unknown vitamin C solution.

   NOTE: Since you don't know how much vitamin C is in the solution, it is perhaps best to do some rough experiments in order to determine how many drops of the solution should be used. Then try to achieve the highest accuracy.

   - Report the concentration of vitamin C in mg mL^{-1}.
   - Calculate the number of milligrams of vitamin C in the tablet.

     NOTE: You will need the volume of the solution that the instructor made at the start of the experiment.

Section D. Vitamin C Concentration in Fresh Fruit Juices

Goal: *To juice several types of fruit, filter a portion of the juice, and carry out a vitamin C analysis.*

Experimental Steps:

1. Assign 1 of your peers (or look for volunteers) the role of professional juicer (any OJs, LJs, LJII s, or GJs).
2. Use a juicer to obtain the juice from oranges, limes, lemons, and/or grapefruit.
3. Organize a bench and class project to obtain the quantitative measurements on the fruit. It would be useful to know the diameter and volume of the fruit and the fruit type. It is critical to know the total volume of the fruit juice produced by a particular fruit.
4. Assign an FJPRP (fruit juice public relations person) to make sure that the appropriate fruit information is disseminated to everyone in the class.
5. Obtain enough juice to carry out a small-scale filtration using the straw method you have used earlier in the course.
6. Filter the juice into a well of the 1 x 12 well strip, saving a little unfiltered juice.
7. Use the calibrated microburet to carry out a serial titration of the filtered juice.

- Calculate and record the vitamin C concentration of the fruit juice (in mg mL^{-1}).
- Calculate and record the total vitamin C content of the fruit.

9. Try a titration on the unfiltered juice.
   - Would you expect any difference?

Section E. Analysis of a Breakfast Cereal for Vitamin C

Goal: *To design and execute an analysis of one of the commercially available cereals available in the laboratory.*

Experimental Steps:

- It's your turn! Carry out the goal stated above.
- Find the number of milligrams of vitamin C per 1 oz. serving for the cereal.
- Compare your analysis with the information presented on the box and the RDAs given in the Introduction section of this module.

Section F. Research Project

Goal: *To design, execute, and report on a research project involving some aspect of the chemistry of vitamin C.*

Discussion: Please see the Background Chemistry section for some ideas and suggestions. As discussed, your instructor may require you to submit a one-page research proposal prior to the laboratory period in which you carry out the project.

Chapter 12

Alcohol Abuse: Chemical Tests for Intoxication

MACDUFF: "What three things does drink especially provoke?"

PORTER: "Marry, sir, nose painting, sleep, and urine.
Lechery, sir, it provokes and unprovokes. It provokes the desire, but takes away the performance: therefore much drink may be said to be an equivocator with lechery; it makes him and it mars him: it sets him on and it takes him off; it persuades him and disheartens him; makes him stand to and not stand to; in conclusion equivocates him in a sleep, and giving him the lye, leaves him."

William Shakespeare

Introduction

Without doubt, the most widely used mind-altering drug in human society is alcohol. We celebrate many of the most memorable events in life, and life itself, with a toast, a prayer, and a drink. Since time immemorial we have drunk in joy and sorrow — at births, funerals, good fortune, and marriages and in communion with God. The moderate use of wine, beer, and spirits has played a significant part in the gastronomic experience. On the other hand, the immoderate abuse of alcohol has led to the destruction of health, families, and even cultures. Alcoholism is now recognized as a disorder of enormous destructive power, affecting more than a third of all American families. The sale of alcoholic beverages generates billions of dollars in federal, state, and local tax revenues. Alcohol abuse, however, costs 30 billion dollars in the lost earning power of alcoholics. The tremendous toll in death, injury, and damage as a result of drunk driving has driven society to recognize the necessity for legal controls on drinking and driving. In the United States, the basis for control and law enforcement has been the implied consent law and the specification of blood alcohol levels as driving limits.

The major types of alcoholic beverages consumed today are beer, wine, and distilled spirits. Almost all alcohol is produced by fermentation processes, and the volume is almost a billion gallons a year in the United States. In the fermentation process microorganisms (primarily yeast) derive energy from the oxidation of sugars and produce alcohol as a waste product. Alcohol, in small amounts, has also been found in the gastrointestinal systems of all mammals, apparently arising as a trace metabolite of glycolysis. The principal active component in alcoholic beverages is ethanol, which has the chemical structure:

```
    H   H
    |   |
H — C — C — O — H
    |   |
    H   H
```

$$C_2H_5OH$$

The overall chemical equation for fermentation was established by Gay-Lussac in the early nineteenth century:

$$C_6H_{12}O_6 \rightarrow 2\,C_2H_5OH + 2\,CO_2$$

Theoretically, 51.1% of the sugar should produce ethanol and 48.9% should produce carbon dioxide. In practice however, the yield of ethanol in molasses or grape fermentation does not exceed 47%. Alcoholic fermentation ceases at about 16% by volume ethanol because of the inhibitory effect of ethanol itself.

There are three common ways of expressing the concentration of ethanol in alcoholic beverages: percent by volume (% by vol.), percent by weight (% by wt.), and proof. *Percent by volume* is the number of milliliters of ethanol per 100 milliliters of beverage. The United States *proof* number is twice the percent by volume and is different from the European proof number. The *percent by weight* is defined as the number of grams of ethanol contained in 100 milliliters of beverage. This definition of percent by weight should really be referred to as a percent by weight by volume; however, in practice it is called percent by weight. A very useful conversion factor is

$$\%\text{ by volume} = \frac{\%\text{ by weight}}{0.79}$$

where 0.79 is the density of ethanol expressed in $g\ mL^{-1}$. Some approximate ranges for the ethanol concentrations in various types of beverages are shown in Table 12.1. Ethanol is also found as a solvent in many pharmaceutical formulations and cosmetic products — e.g., mouthwashes, cough medicines, hair sprays, and deodorants.

Beverage	% by Volume	Proof *
Beer	2–5	4–10
Malt liquor	5–8	10–16
Wine	8–15	16–30
Sherry, port	12–25	24–50
Champagne	18–25	36–50
Whisky	40–50	80–100
Rum	40–50	80–100
Vodka	40–50	80–100
Liqueurs	25–60	50–120

* United States standard

Table 12.1 Ranges of Ethanol Concentration in Beverages

A tremendous amount of information is now available about the absorption, metabolism, and excretion of ethanol in the human body, although research on neurobehavioral effects is still in the early stages. The physiological behavior of ethanol is closely linked to the special chemical properties of the ethanol molecule. Pure ethanol is a colorless liquid with a boiling point of 78.5 °C and a density of 0.7894 g mL^{-1} (at 25 °C). The molecule is quite small (molar mass 46 g mol^{-1}) and contains a hydroxyl group (–OH) and an ethyl group (C_2H_5–). Ethanol and water are miscible (exist as one phase) in all proportions because of the intermolecular hydrogen bonding that can occur between the hydroxyl group of ethanol and the water molecule, as shown below:

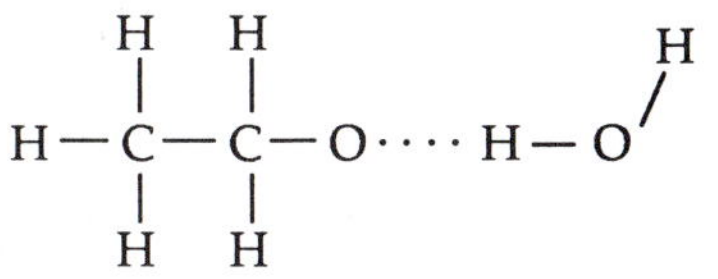

```
    H   H              H
    |   |             /
H — C — C — O · · · · H — O
    |   |
    H   H
```

The two-carbon ethyl group is nonpolar and lends considerable fat solubility to the molecule. The peculiar balance of water solubility and fat (lipid) solubility gives ethanol its unusual physiological properties.

After ingestion ethanol and the other constituents of a beverage flow into the stomach and begin to be absorbed, although the primary route of entry into the bloodstream is from the small intestine. Ethanol passes quite easily in both directions through cell walls and cell membranes. Ethanol also crosses the blood barrier with ease because of its high lipid and water solubility. The rate at which ethanol reaches the brain is controlled by many factors, including the rate of intake, the rate of absorption from the small intestine, the rate of distribution, and the rate of elimination from the body. The rate of absorption from the small intestine appears to be the slowest of these processes and is, therefore, the rate-determining step.

The presence or absence of food in the stomach and the concentration of ethanol in the ingested drink have a major influence on the rate of absorption. The passage of ethanol (and everything else) from the stomach to the small intestine is controlled by the pyloric valve. This valve remains closed until food in the stomach has been acted upon by acids and enzymes secreted by the stomach wall. Eating while drinking will therefore slow down the rate of ethanol absorption from the intestine because it will take longer to arrive. It is now accepted that the concentration of ethanol in a beverage, rather than the type of drink, is also a determining factor in rate of absorption. A 12 oz. can of beer, a 6 oz. glass of wine, and one mixed drink of 1 1/2 oz. distilled spirits all contain about the same amount of ethanol.

Once ethanol is absorbed into the bloodstream, it is rapidly distributed around the body by the flow of blood and the vascular tissue. In any organ with a rich blood supply and dense vascularization, equilibrium is rapidly established between the alcohol in blood and in tissue. For example, diffusion into the brain is extremely rapid, and the placenta is easily penetrated by ethanol. A fetus is exposed to the same concentration of ethanol that is present in the tissues of the mother. The concentration of ethanol in body fluids and organs after equilibration corresponds rather closely to their water content. The mean distribution ratios between whole blood (given an arbitrary value of 1.0) and various fluids, tissues, and organs are shown in Table 12.2.

Organ, Tissue, or Fluid	Mean Ratio*
Brain	1.17
Blood plasma	1.16
Fat	0.019
Liver	0.91
Saliva	1.12
Urine	1.35
Breath	0.00048

* Whole blood is 1.0.

Table 12.2 Distribution Ratios for Fluids, Tissues, and Organs

The degree of intoxication or inebriation is closely related to the concentration of ethanol in the brain. Forensic tests for chemical intoxication are based on the close correlation of the ethanol concentration in the brain and in blood (and in breath, and in urine, etc. — see Table 12.2).

The *blood alcohol level* (BAL) is usually expressed as a % by weight and is defined as the number of grams of ethanol in 100 mL of whole blood. A BAL of 0.10% by weight would indicate that 100 mL of blood contained 0.10 gram (or 100 mg) of ethanol. The progressive effect of ethanol on the brain and the correlation with blood alcohol level can be clearly seen in Table 12.3. The blood alcohol level depends primarily on the amount of ethanol ingested, the time period of ingestion, and the body weight of the consuming person. Several methods of calculating approximate blood alcohol levels from the number of drinks consumed and body weight data have been developed. One simple example is shown in Figure 12.1.

Estimated amount of 80 proof liquor needed to reach approximate given levels of alcohol in the blood.

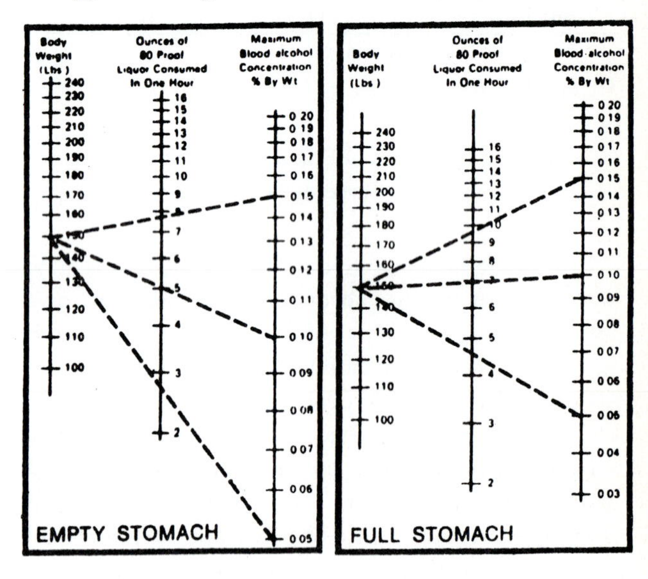

Figure 12.1 Graphs for Calculation of BAL

BAL Range	Affected Area of Brain	Some Symptoms
0.01 – 0.10	frontal lobe	removal of inhibitions, loss of self-control, development of euphoria, altered judgement, dulling of attention
0.10 – 0.20	psychomotor	apraxia, agraphia, ataxia, tremors, loss of skill
0.10 – 0.30	somesthetopsychic	dulled and distorted sensibilities
0.15 – 0.35	cerebellum	disturbance of equilibrium
0.20 – 0.30	visuo-psychic	disturbance of color perception and dimensions, motion, and distance
0.25 – 0.40	diencephalon	apathy, inertia, tremors, sweating, dilation of surface capillaries, stupor, coma
0.40 – 0.50	medulla	depression of respiration, peripheral collapse, death

Table 12.3 BAL and the Effect of Ethanol on the Brain

The elimination of ethanol from the body occurs as a result of oxidation in the liver and by direct excretion. Direct excretion, through breath and urine (and traces in tears, saliva, and perspiration), accounts for only about 2–5% of ethanol removal. More than 95% of ingested ethanol is metabolized in the liver, where the enzyme alcohol dehydrogenase catalyzes the oxidation of ethanol to acetaldeyhde, acetic acid, and eventually to carbon dioxide and water:

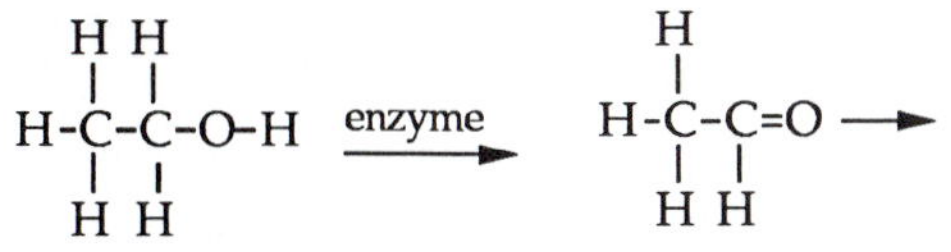

```
  H H                        H
  | |                        |
H-C-C-O-H   enzyme       H-C-C=O  --->
  | |       ------->       | |
  H H                      H H

 ethanol                acetaldehyde
```

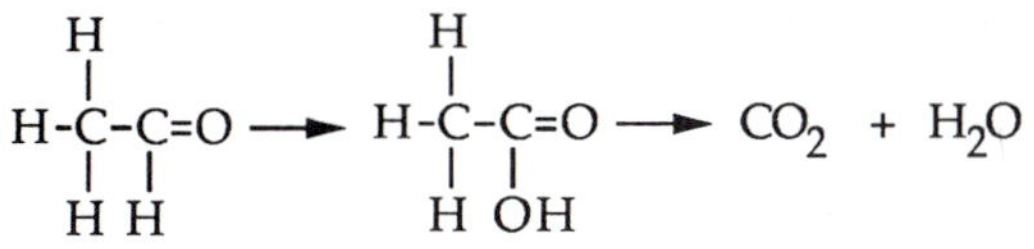

```
  H            H
  |            |
H-C-C=O ---> H-C-C=O ---> CO2 + H2O
  | |          | |
  H H          H OH

acetaldehyde   acetic acid
```

The rate of ethanol elimination is almost entirely controlled by the alcohol dehydrogenase activity, and the activity is regarded as being reasonably constant in humans at a value of 0.015% by weight per hour.

ALCOHOL AND DRIVING

The tragic contribution of alcohol to death, injury, and property damage on the highways has long been recognized. There is now an overwhelming amount of research data which shows that:

- Alcohol is the largest single factor leading to fatal automobile crashes.
- The use of alcohol by drivers and pedestrians leads to 1,000,000 crashes and some 30,000 deaths in the United States each year.
- More than half of the adult population uses the highways, at least occasionally, after drinking.
- The scientific evidence is irrefutable that the problem is primarily one of persons (predominantly men) who have been drinking heavily to an extent rare among drivers and pedestrians not involved in crashes.
- Arrests for driving while intoxicated and related offenses constitute a group whose BALs are so high that the group is almost totally distinguishable in this respect from drivers, drinking or not, who are not involved in crashes. This refutes the assumption that persons arrested for drunken driving are "ordinary drinkers," or persons who have had just a couple of drinks, and who happen to get caught.
- One to four percent of drivers on the road (those with BALs at or above 0.10% by weight) account for about 50–55% of *all* single-vehicle crashes in which drivers are fatally injured.

All these comments have appeared in alcohol and highway safety reports to the United States Congress. Although most of the research on alcohol in relation to transportation has been in the area of automobile crashes and violations, there is abundant evidence to link alcohol with aviation crashes, rail crashes, and pedestrian fatalities.

The antisocial manifestations of alcohol abuse have long been regarded as legal problems to be dealt with through the criminal court process. The control of drinking drivers begins with laws defining the limits of proper conduct. Many states in the United States have established two categories of drinking drivers:

- Driving under the influence (DUI), with a BAL of 0.10% by weight or more
- Driving while the ability to drive is impaired (DWI), with a BAL range from 0.05-0.09% by weight

Establishing legal and illegal blood alcohol levels for the motorist is relatively simple to enact; however, obtaining evidence in the form of blood, breath, or urine samples is much more difficult. Because the drinking driver who has been arrested is very unlikely to consent to provide bodily samples, more than forty-five states have passed *implied consent laws*. These laws state that by the act of driving a motor vehicle, the driver is deemed to give his/her consent for chemical tests for intoxication if he/she is arrested for driving while impaired or under the influence of alcohol. To ensure compliance with constitutional limitations, the implied consent laws generally require an arrest for good cause before a test for intoxication is carried out.

The most important aspect of laws enacted for controlling drunk drivers is that the legal definition of presumed intoxication is stated in terms of specific blood alcohol levels. Enforcement of these laws is totally dependent on obtaining an appropriate sample and on the availability of methods for the chemical analysis of ethanol in the sample. Three types of fluid samples — blood, breath, and urine — are most commonly used in the determination of bodily ethanol concentrations. Established protocols and procedures must be used in obtaining, storing, identifying, preserving, and analyzing a sample because the sample is usually used as evidence in legal proceedings. In practice, urine samples are seldom used for ethanol analysis because the ratio of urine ethanol to blood ethanol is only valid when the urine sample is not diluted by urine present in the bladder prior to intoxication. A urine analysis can only be performed on a sample obtained from a witnessed second urination. Breath samples may be used because it is well established that the ratio of blood alcohol level to alveolar air concentration (at the average temperature of exhaled air, 34 °C) is 2100 to 1. One distinct advantage of a breath sample is that it may be directly analyzed without the necessity of separating the ethanol from the interfering substances that are always present in blood, urine, and tissue samples.

Of the more than 400 quantitative analytical methods of the determination, 3 methods seem to be more popular than all others:

- Distillation of the sample and oxidation of ethanol to acetic acid by potassium dichromate in sulfuric acid solution and subsequent spectrophotometric determination of the blue Cr^{3+} found in the reaction
- Direct or indirect gas chromatographic analysis
- Enzymatic analysis using alcohol dehydrogenase coupled with a spectrophotometric finish

It must be emphasized that all analytical methods require considerable skill, equipment, and facilities costing thousands of dollars. There is currently a considerable international research effort directed towards finding a simple, inexpensive, and accurate method for ethanol determination that could be used by any law enforcement officer. The main thrust of this research appears to be the application of infrared spectroscopy and the development of colorimetric microanalytical sensors.

Background Chemistry

Ethanol is one of the simplest compounds in a class of organic substances called alcohols. All alcohols contain at least one hydroxyl group attached to a carbon atom. Some other examples of alcohols are wood alcohol (methanol), antifreeze (ethylene glycol), and rubbing alcohol (2-propanol). The chemical properties and the physiological effects of alcohols are very dependent on the size, chemical composition, and molecular structure of these compounds. Chemists have invented ingenious pictorial (ideographic) ways of representing molecules that not only specify the elemental composition and mole ratios of elements, but also provide many clues to the explanation of property and reactivity. Table 12.4 shows a variety of ways in which the ethanol molecule is represented in the scientific literature. Also included in the table are various representations of the water molecule in order for you to make a comparison of the two molecules and begin to think about intermolecular interactions between them. As you progress in your career as a scientist, you will encounter many examples of these types of structural representations. Please take a little time to learn how to "read" the information coded in these molecular abstractions.

An ethanol molecule can be thought of as consisting of two quite different parts within the whole covalent structure. One part, the ethyl group (C_2H_5), is nonpolar and lipophilic (liking fats), and the other part, the hydroxyl group (OH), is polar and hydrophilic (liking water). The electronegativity difference between oxygen and the hydrogen atom attached to it is sufficient to cause the oxygen atom to carry a slight negative charge (δ^-) and the hydrogen atom to have a slight positive charge (δ^+). Ethanol is therefore a polar covalent compound with a dipole moment of 1.69 Debye. Ethanol is less polar than water, which has a dipole moment of 1.87 Debye, but considerably more polar than methane with zero dipole moment. The hydroxyl group looks suspiciously like that of a base — e.g., sodium hydroxide (NaOH) — however, it must be emphasized that the hydroxyl group in ethanol does not dissociate. The carbon and oxygen atoms have about the same electronegativity, and consequently, the whole molecule is firmly held together by covalent bonding.

Many of the important chemical and physiological properties of ethanol can be explained by the presence of both lipophilic and hydrophilic groups within the same small molecule. Mixtures of ethanol and water are miscible (form one phase) in all proportions because both molecules are small and both can form intermolecular hydrogen bonds. Both ethanol and water are neutral (uncharged) and appear to move with ease through bilayer cell membranes, including nerve cell membranes that make up the blood-brain barrier. Figure 12.2 illustrates a section of a cell membrane consisting of large phospholipid molecules oriented in a bilayer. The small water and ethanol molecules are shown moving through the membrane.

A comparison of the boiling points of pure ethanol (78.5 °C) and water (100 °C) clearly demonstrates that the smaller dipole moment of ethanol leads to weaker hydrogen bonding in pure ethanol than in water. Ethanol is therefore more volatile than water, in spite of the fact that the hydrogen bonding in both liquids is responsible

	Ethanol	Water	Description
1.	C_2H_6O	H_2O	The molecular formula that gives the elemental composition and mole ratios of elements.
2.	CH_3CH_2OH	HOH	More information about which atoms are bonded together.
3.	OH	no equivalent	An abbreviation called a skeleton structure.
4.	H-C-C-O-H (with H above and below each C)	H-O-H	A more detailed picture showing specific covalent bonds between all the atoms.
5.			More detail. An attempt is made to show 3-dimensionality with bond angles, shape, and lone pairs.
6.	H-C-C-O-H (with H above and below each C; δ- on O, δ+ on H)	δ- O, H δ+, H δ+	This type of structure is used to picture the dipolar nature of molecule.
7.			These are the so-called ball and stick models that show atoms as colored balls of appropriate size. Bond angles are easily seen.
8.			These space-filling models are recognized to be the most realistic representations of molecules. Size and shape are the emphasis.
9.			Intermolecular interactions and dynamics are represented by showing several molecules. A "snapshot" of the gas phase shows molecules quite far apart.
10.			An instantaneous picture of the liquid state showing intermolecular hydrogen bonding between close molecules.
11.			The solid state showing the extraordinarily highly ordered structure of ice and the disordered glasslike state of solid ethanol.
12.			Pictorial representation of intermolecular hydrogen bonding in a homogeneous solution of water and ethanol.

Table 12.4 Various Representations of Ethanol and Water

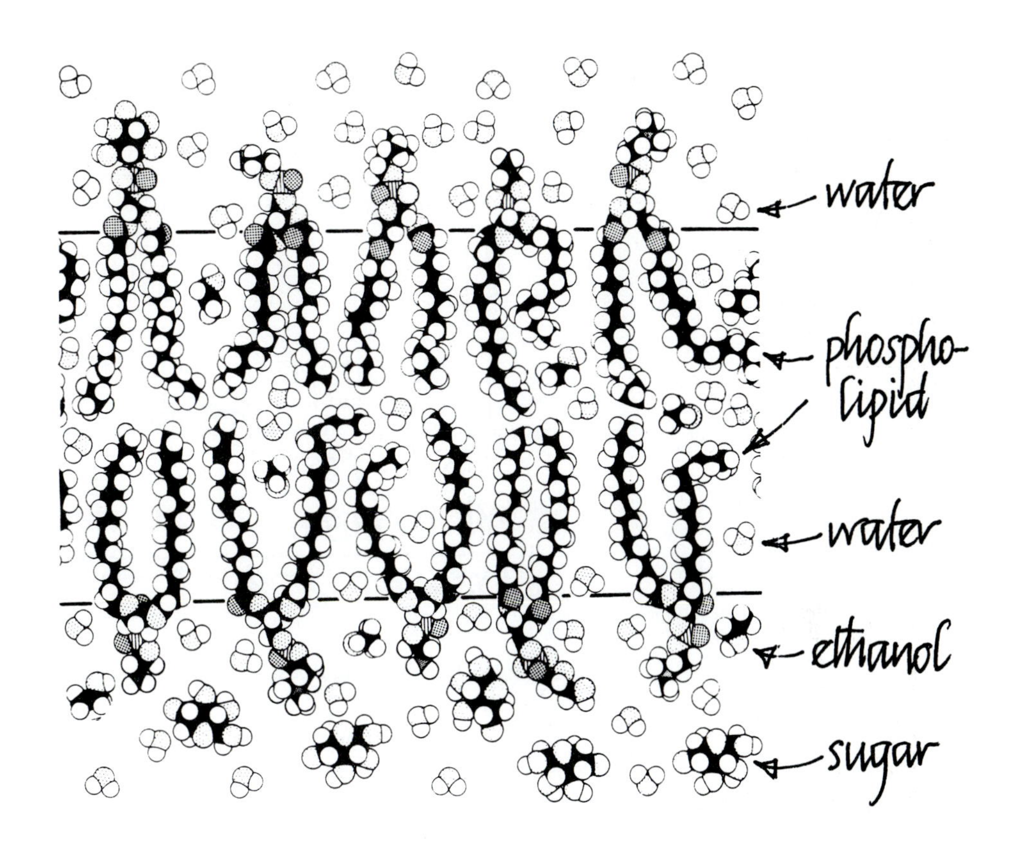

Figure 12.2 Water and Ethanol Moving Through Cell Membrane

for the unusually high boiling points. The greater volatility of ethanol is the basis for most of the ancient methods of separating ethanol from water in the concentration of ethanol in the production of spirits and liqueurs — e.g., whisky, vodka, gin, and brandy. The distillation of an aqueous ethanol solution is most conveniently described by a boiling point versus composition diagram as shown in Figure 12.3. Intermolecular hydrogen bonding produces negative deviations from Raoults law and results in the formation of a minimum boiling azeotrope system.

Consider the distillation of a dilute, aqueous ethanol solution — e.g., a blood sample or a beverage sample. A sample of the solution composition shown in Figure 12.3 will boil at temperature T_1 and produce vapor that is richer in ethanol than water (composition C_1). Repeated reboiling and condensation (in a process called fractional distillation) will produce a liquid with a constant boiling point of 78.3 °C and a constant composition of 4.8% water and 95.2% ethanol. Absolutely pure ethanol can only be produced by adding an entrainer called benzene and by continuing the distillation through several azeotropes.

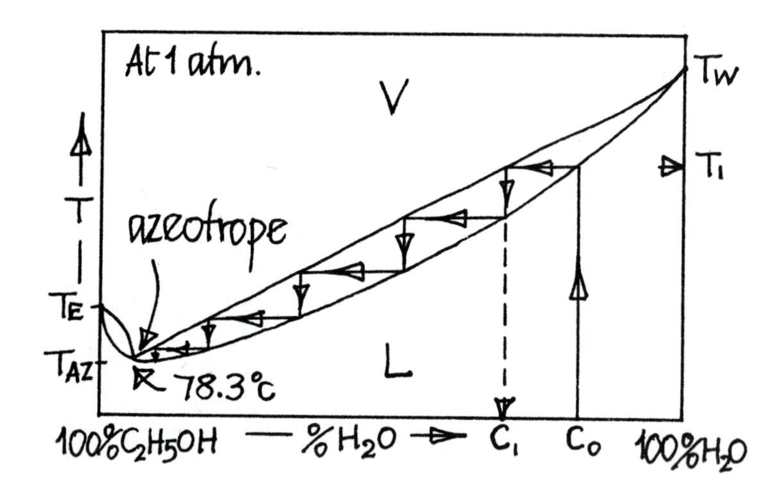

Figure 12.3 Boiling Point Versus Composition Diagram for Ethanol

One of the most common methods of determining the ethanol concentration in a sample is to oxidize the ethanol with a suitable oxidizing agent. The analysis is then completed by a quantitative measurement of the amount of unreacted oxidizing agent, or the amount of reacted product, or both. A strongly acidic solution of potassium dichromate ($K_2Cr_2O_7$) effectively oxidizes ethanol to acetic acid according to the redox equation:

$$2K_2Cr_2O_7 + 8H_2SO_4 + 3C_2H_5OH \xrightarrow{\text{heat}} 2Cr_2(SO_4)_3 + 2K_2SO_4 + 3CH_3COOH + 11H_2O$$

In this reaction, orange dichromate ion ($Cr_2O_7^{2-}$) oxidizes ethanol to acetic acid and is reduced to blue-green chromium (III) ion (Cr^{3+}). An ethanol determination is usually carried out by first preparing and then oxidizing a series of standard solutions containing exactly known ethanol concentrations. After oxidation, the standards range in color from orange to a pale blue color in going from the lowest to the highest ethanol concentration. A spectrophotometric analysis of the chromium (III) species is often used to finish the ethanol determination. However, a much simpler method is to utilize the set of oxidized ethanol standards as a colorimetric comparison set. A visual comparison of an unknown sample with the colorimetric set is sufficient to effect an analysis.

Unfortunately, there are several serious methodological problems that must be overcome in order to guarantee accurate results in ethanol determinations. One major problem is that physiological fluids — e.g., blood and urine — and beverage samples almost always contain a relatively high concentration of substances that are easily oxidized by an acidic dichromate solution. These oxidizable substances grossly interfere in the analysis and must be separated from ethanol before the oxidation is carried out. A distillation of the original sample accomplishes a separation of volatile ethanol from the nonvolatile interfering substances. A second problem is that the dichromate oxidation is powerful enough to oxidize virtually any other volatile compounds, such as methanol and acetone, that might be present in a sample. If the sample is suspected of containing volatile compounds other than ethanol, then a more selective method of analysis must be employed. Gas chromatography and alcohol dehydrogenase enzymatic methods are commonly used in these situations.

The quantitative distillation of all the ethanol from a physiological or beverage sample is a rather time-consuming method requiring expensive, fragile, specially designed glassware and careful technique. The apparatus used in distillation is shown in Figure 12.4 and consists of a distillation flask, a distillation head, a cold water condenser and a receiving flask.

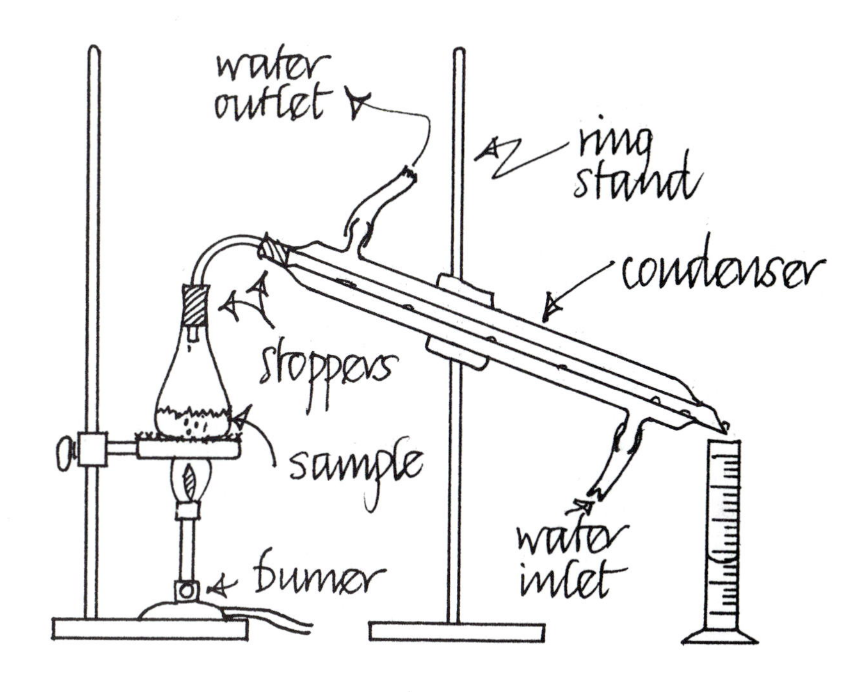

Figure 12.4 Distillation Apparatus

An aliquot (portion) of the sample is delivered to the distillation flask and is brought to boiling by an electric heating mantle (or Bunsen burner). Vapor is condensed back to liquid in the condenser. All the liquid distillate is collected and diluted to a known volume in a volumetric flask. A suitable aliquot is then removed for analysis by the oxidation method.

In this laboratory module, a simpler, less expensive microdistillation technique has been devised for the separation of ethanol from interfering substances in an unknown sample. A drop of the sample, diluted when necessary, is quantitatively delivered to the inside of a small plastic cap where it adheres and is kept in place by surface tension forces. The cap is placed onto a small glass vial containing a small volume of the dichromate oxidizing mixture. The vial(s) is placed in an oven at 80 °C for about 1 hour. During this time the ethanol distills over into the oxidant and is converted to acetic acid. Standards and many unknown samples can be microdistilled and oxidized simultaneously. Spectrophotometric, volumetric, or comparison analysis of the final colored solutions is then an easy matter.

Pre-Laboratory Quiz

1. What is the chemical formula for ethanol?

2. A certain brand of whisky is 80 proof. Calculate the ethanol concentration as a % by volume and % by weight.

3. What 3 factors control the rate at which ethanol reaches the brain?

4. Explain why eating slows down the rate of absorption of ethanol.

5. Define blood alcohol level (BAL).

6. Use the graphs provided in Figure 12.1 to calculate your BAL if you were to drink 4 cans of regular beer on an empty stomach.

7. What is the rate of elimination of ethanol from the human body?

8. Give the 2 commonly defined BALs for driving under the influence and driving while impaired.

9. Explain why ethanol is a polar covalent compound.

10. Give the chemical equation for the oxidation of ethanol by a strongly acid dichromate solution.

Laboratory Experiments

Flowchart of the Experiments

Section A.	Preparation of a Set of Colorimetric Standards for the Determination of Ethanol
Section B.	The Analysis of Unknown Samples Containing Ethanol
Section C.	An Exploration of the Relationship between Chemical Structure and the Chemical and Physiological Behavior of Alcohols
Section D.	Classical Distillation of Samples Containing Ethanol
Section E.	Drinking and Driving . . . A Sad Story

Requires one three-hour class period to complete

CAUTION: In this laboratory module you will be using a solution that contains potassium dichromate dissolved in 9 M sulfuric acid solution. This solution is highly corrosive and a powerful oxidizing agent. Even though you will be using very small volumes of this solution, you should take great care when you transfer and heat these solutions. Wear your safety goggles (over your eyes!) when carrying out any procedure involving potassium dichromate–sulfuric acid solutions. If you get this solution on your hands, wash immediately with cold water and report to your instructor. Do not pour the potassium dichromate–sulfuric acid solution down the sink. Waste containers are provided for any solutions containing potassium dichromate–sulfuric acid. These wastes will be recycled in an environmentally safe manner.

Section A. Preparation of a Set of Colorimetric Standards for the Determination of Ethanol

Goal:

To prepare a set of 10 colorimetric standards by oxidizing standard ethanol solutions that have been evaporated by a special microdistillation technique.

Discussion:

Wear your safety goggles! Note that you should carry out Section A up to and including Step 23. Then before putting the standards in the oven in Step 24, carry out Steps 1–11 of Section B. This allows you to heat both standards and samples simultaneously in the oven.

Experimental Steps:

1. Clean 10 small glass vials and drain them upside down on a paper towel. Clean 10 vial caps and use twisted paper towel or a cotton swab to remove any remaining moisture from the vials and from the inside of the caps. Set the caps aside for a moment.
2. Turn a 96-well tray (RB) upside down to create a convenient rack support for the vials.
3. Place the clean vials into the upside-down tray (at the edge where they fit perfectly, fortuitously).
4. Number the vials 1 to 10, towards the top of each vial, with a permanent marker.
5. Use good transfer technique to fill a cutoff (large-drop) microburet with potassium dichromate–sulfuric acid solution.
6. Use standard vertical delivery technique to deliver 6 drops of the potassium dichromate–sulfuric acid solution to each of the vials. Set the tray with vials aside.
7. Clean a 1 x 12 well strip and slap it on a towel to remove any remaining water. Use a cotton swab to make sure the wells are dry.
8. Clean the same microburet and use good transfer technique to fill it with 1.00% by weight ethanol solution.
9. Deliver 1 drop ethanol solution to the first well of the 1 x 12 well strip, 2 drops to the second well, and so on, until 10 drops in well number 10 have been delivered.

10. Clean the microburet and make the dilutions indicated in the following table:

Well No.	Drops of 1% by Weight Ethanol Solution	Drops of Water	% by Weight Diluted Solution
1	1	9	
2	2	8	
3	3	7	
4	4	6	
5	5	5	
6	6	4	
7	7	3	
8	8	2	
9	9	1	
10	10	0	

11. Stir the solutions with a microstirrer.
    - Calculate the % by weight ethanol concentrations for each of the diluted solutions.
12. Place the clean dry caps (from Step 1) upside down on the plastic surface.
13. Pull a *small-drop* microburet (or use the one provided).
14. Wash the small-drop microburet and expel any remaining water.
15. Suck up the diluted ethanol solution from the first well.
16. As illustrated below, use standard delivery technique to deliver 1 drop to the inside space of the cap. Gently lift the cap and push it all the way into vial 1.

    NOTE: Throughout the following sequence of steps it is important to use care in moving the vials in order to keep the drops from sliding into the dichromate solution.

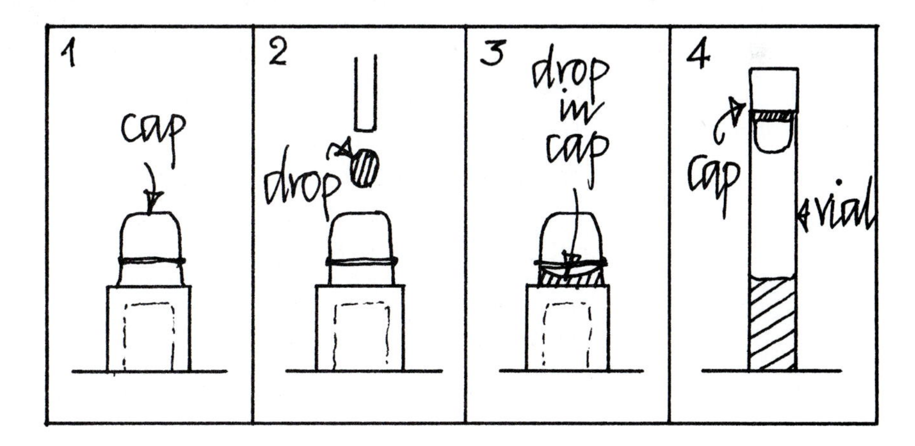

17. Place the vial back in the rack.
18. Wash the small-drop microburet and expel any remaining water.
19. Suck up the ethanol solution from the second well and quantitatively deliver 1 drop to another cap.

20. Gently push the cap onto vial 2. Place it back in the rack.
21. Continue in the same manner until you have delivered a drop of each diluted ethanol solution to a cap and capped the appropriate vial.

    NOTE: If you wish to analyze unknown samples simultaneously with your ethanol standards, then go straight to Section B. It is recommended that you do this because there is less chance of error in a quantitative analysis when both calibration standards and samples are subject to exactly the same experimental conditions.
22. Carefully take the 10 vials (plus any unknown sample vials) into your left hand (right if you are left-handed) and wrap a rubber band around them, forming a bundle.
23. Place the bundle into a marked styrofoam cup, keeping the vials vertical.
24. Place the cup in an oven at 80 °C for 1 hour.

    NOTE: You now have an hour to complete Section C of the laboratory module. Note the time and go to Section C.

After one hour:

25. Remove the cup from the oven.
26. Carefully remove the bundle from the cup and swirl carefully to homogenize the color.

    CAUTION: Do not shake the dichromate solution into the caps!
27. Carefully unbundle the vials and place them in a styrofoam stand or rack against a good white background.
    - Compare any unknown samples with the ethanol standards.

Section B. The Analysis of Unknown Samples Containing Ethanol

Goals:

(1) To microdistill and oxidize various unknown samples. (2) To determine the ethanol concentration by color comparison with a set of colorimetric standards.

Discussion:

Your instructor will tell you about the fluid or beverage samples that are available for analysis. The labelled samples can be found at Reagent Central.

Experimental Steps:

1. Clean 2 small glass vials and caps for each unknown sample you wish to analyze. The vials must be the same type as those used for the set of colorimetric standards prepared in Section A. Remove any remaining water with a twisted microtowel or cotton swab.
2. Place the vials in a rack made by turning a 96-well tray upside down. Use a permanent marker to identify each vial with an appropriate sample mark.
3. Clean the large-drop microburet that you used to deliver drops of dichromate solution in Section A. It must be the same microburet.
4. Use good transfer technique to fill it with potassium dichromate–sulfuric acid solution.
5. Deliver 6 drops to each vial in the rack.
6. Clean the microburet and a 1 x 12 well strip. Slap the strip on a paper towel to remove any remaining water.
7. Use good quantitative technique to dilute the unknown sample in order to ensure that 1 drop (from the small-drop microburet) of the diluted sample will contain an ethanol concentration within the range of the set of colorimetric standards.

NOTE: A typical American beer sample (decarbonated) should be diluted 5-fold. A urine analysis sample (or blood sample) should not be diluted; in fact, use 2 or 3 drops (from the small-drop microburet) of these types of samples.

8. Clean the small-drop microburet. Suck up the diluted sample and deliver 1 drop into each of 2 caps.

   NOTE: Using two caps will enable you to do duplicate analyses on each unknown.

9. If you wish to see what would happen if the sample were not microdistilled, then deliver 1 drop of the diluted sample *directly* into a vial containing the potassium dichromate–sulfuric acid solution.

10. Gently place the caps on the appropriate vials and bundle the vials with the standards (see Step 23, Section A) in a rubber band. Make sure to keep the vials vertical!

11. Place the bundle into a marked styrofoam cup.

12. Put the cup in an oven at 80 °C for about 1 hour.

13. Remove the cup and carefully unbundle the vials. Place the vials in the rack.

14. Compare the unknown samples, one at a time, with the set of colorimetric standards and decide on a match.

    NOTE: In colorimetric comparisons like this, it is critical to have a good background, angle of view, and lighting. If the sample appears to be in between two standards, then interpolate.

    - Calculate the % by weight ethanol concentration in the original sample by multiplying the % by weight of the matched standard by the dilution factor. For example, if a beer sample is diluted 5-fold and 1 drop of the diluted sample is used in the analysis, then the % by weight match must be multiplied by 5.
    - Record your ethanol analyses in your laboratory record.

15. Remove the caps from the vials and wash well with water. Carefully clean the inside of the caps (where the drop was placed) with a wet cotton swab.

16. Pour the dichromate-acid solution from both standards and samples into the labelled waste container provided.

17. Wash your hands. Wash the vials with water and return them to the rack.

Section C. An Exploration of the Relationship between Chemical Structure and the Chemical and Physiological Behavior of Alcohols

Goals:

(1) To examine literature data for a series of alcohols. (2) To explore some of the relationships that link chemical structure and the properties of these physiologically active compounds.

Discussion:

Table 12.5 contains literature data and chemical structures for a series of primary, aliphatic alcohols. You will need to consult the data in Table 12.5 and perhaps also refer to the Background Chemistry as you work through this section.

First look at the melting point data for the alcohols.

- What causes the melting points of alcohols to increase from 1-propanol to 1-decanol?
- Suggest some reasons why solid ethanol melts at a temperature 116 °C *lower* than ice.

  HINT: Think about the structure of ice.
- Make a graph of boiling point versus number of carbon atoms in an alcohol.
- Give an explanation for the obvious relationship between boiling point and carbon number.
- Why does ethanol have a lower boiling point than water?
- Which of the alcohols is least volatile?
- Make a graph of molar solubility of alcohol versus carbon number and a graph of $\log_{10}$ molar solubility of alcohol versus carbon number.
- Explain, with the aid of simple pictures, why the solubility of alcohols in water dramatically decreases from 1-butanol to 1-decanol.
- By what factor does the molar solubility change as the alcohol carbon number increases by 1?
- Comment on the mathematical relationship between water solubility and structure of an alcohol.

It is interesting to note that the very long chain alcohol called cetyl alcohol ($C_{16}H_{33}OH$) is so insoluble in water that it forms a monolayer on the surface of the water (see illustration below). The hydroxyl group hydrogen bonds to the water below, and the rest of the molecule sticks up into the air and is held up by the cooperative van der Waals forces between the lipophilic $C_{16}H_{33}$ chains. Cetyl alcohol has been successfully used in natural ponds, lakes, and reservoirs as an agent to restrict the loss of water by evaporation.

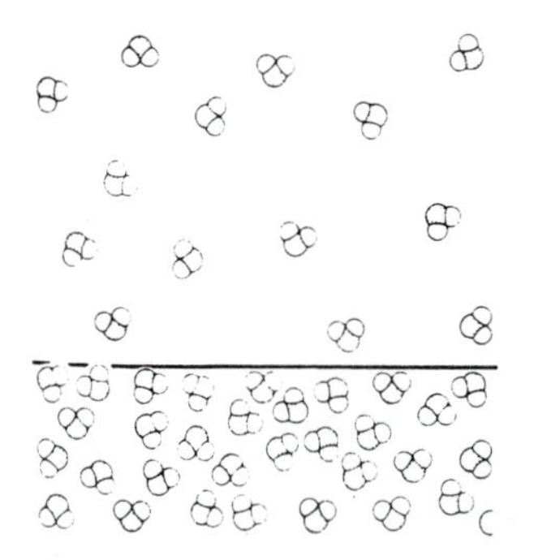

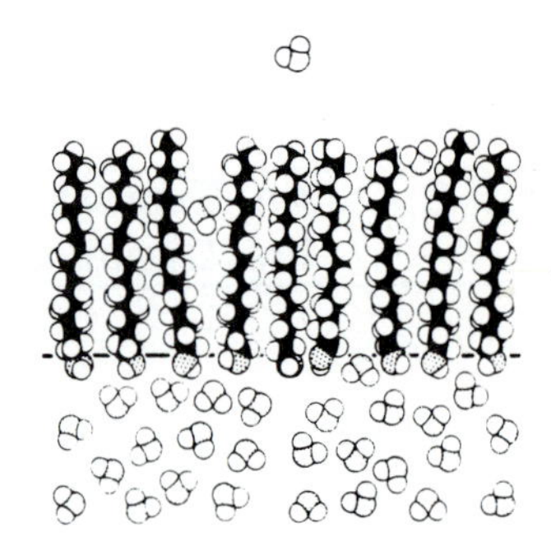

Name	Chemical Formula	Abbreviated Structure	Space-Filling Model	Molar Mass ($g\ mol^{-1}$)	Melting Point (°C)	Boiling Point[a]	Molar Solubility of Alcohol[b]	K_d in C_6H_6[c]	Azeotropic Data: % by Weight Water	Azeotropic Data: Boiling Point
Methanol	CH_3OH			32.0	-94	65	∞	0.015	does not form	
Ethanol	CH_3CH_2OH			46.1	-116	78.5	∞	0.046	4.0	78.3
1-propanol	$CH_3(CH_2)_2OH$			60.1	-126	97	∞	0.23	28.3	87.0
1-butanol	$CH_3(CH_2)_3OH$			74.1	-90	118	0.97	0.80	42.5	92.7
1-pentanol	$CH_3(CH_2)_4OH$			88.2	-79	138	0.25	3.90	54.4	95.8
1-hexanol	$CH_3(CH_2)_5OH$			102.2	-47	158	0.06	large	75.0	97.8
1-heptanol	$CH_3(CH_2)_6OH$			116.2	-35	177	0.015	large	82.0	98.3
1-octanol	$CH_3(CH_2)_7OH$			130.2	-17	195	0.004	large	90.0	99.4
1-nonanol	$CH_3(CH_2)_8OH$			144.3	20	212	0.001	large	—	—
1-decanol	$CH_3(CH_2)_9OH$			158.3	7	229	0.00024	large	—	—

Table 12.5 Literature Data and Chemical Structures for a Series of Primary Alcohols

(a) - (°C at 760 mm)
(b) - In H_2O at 25°C
(c) - 0.1 M salt at 25°C

Data on the solubility of *water* in the alcohol layer, although not shown in Table 12.5, are rather interesting. It has been found that the solubility of water in the alcohol layer remains small and fairly constant in going from 1-pentanol to 1-decanol.

- Explain why the trend in the solubility of water in alcohols is so very different from the trend of the solubility of alcohols in water.

  HINT: Use simple pictures to explain the intermolecular bonding.

Look at the column in Table 12.5 labelled K_d in C_6H_6. These data are obtained by shaking together a volume of an aqueous alcohol solution with an equal volume of the organic solvent benzene C_6H_6. The partition coefficient K_d of the alcohol is

$$K_d = \frac{\text{molar concentration of alcohol in benzene layer}}{\text{molar concentration of alcohol in aqueous layer}}$$

- Explain why the value of Kd increases as the length of the alcohol carbon chain increases.

Many of the alcohols listed in Table 12.5 have been found to be present in free or in ester form in fruits and in fermented beverages. All the alcohols have been found to be toxic. The physiological effects vary depending on the number of carbons in the alcohol chain and on the toxicity of the breakdown products produced in the body. The biological behavior of methanol is very different from ethanol and the longer chain alcohols. Methanol toxicity, with the symptoms of acute visual disturbance (and eventual blindness) and severe acidosis has been known for over a hundred years. The *in vivo* biochemistry of methanol is extremely complicated; however, it is probable that methanol is partially oxidized to formaldehyde (HCHO) in various organs, including the eye. Formaldehyde is an extremely reactive compound that is known to destroy the retina and optic nerve! The higher alcohols, 1-butanol through 1-decanol, have all been found to be extremely toxic and to cause a strong narcotic effect. The narcosis appears to parallel the lipid solubility of the alcohol.

- Write a short paragraph on the effect of the chemical structure on the physiological distribution, elimination, and effects of alcohol.
- Draw a boiling point versus composition diagram for 1-pentanol.
- Explain briefly what happens when a 75% by weight water and 25% 1-pentanol solution is fractionally distilled.

Section D. Classical Distillation of Samples Containing Ethanol

Goals: *(1) To set up a classical distillation apparatus. (2) To distill all the ethanol from a sample containing ethanol and nonvolatile oxidizable substances.*

Experimental Steps:

1. Clean and dry two 100 mL beakers or cups.
2. Pour about 20 mL of the sample into one of the cups and pour the sample from one cup to the other until most of the carbon dioxide has been removed.

3. Set up the distillation apparatus as shown.

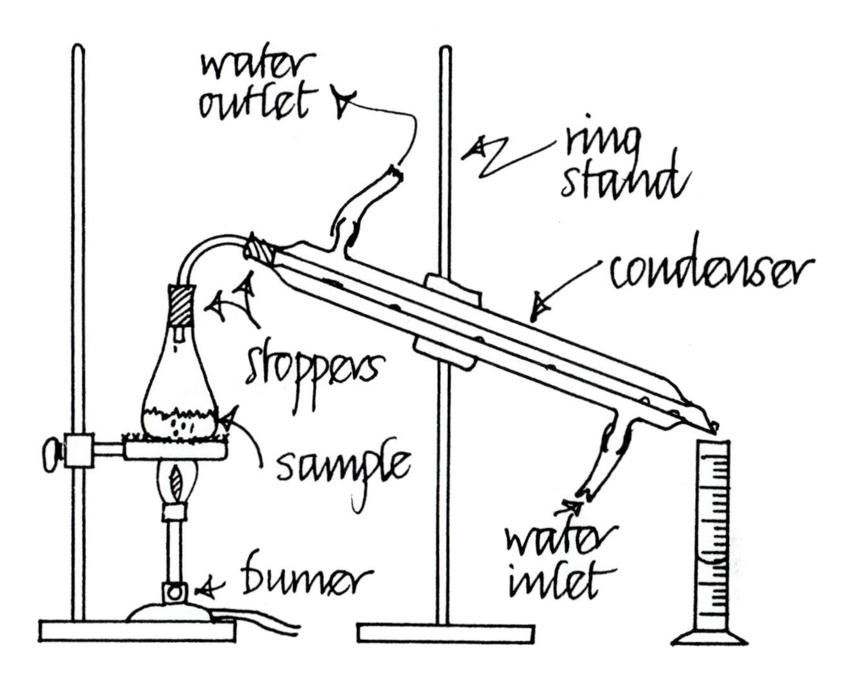

NOTE: Make sure that the cold water enters the condenser at the lowest condenser inlet and leaves by the top one.

4. Pipet 5.00 mL of the decarbonated sample into a clean 125 mL Erlenmeyer flask.
5. Add approximately 50 mL of water to the flask.
6. Place the stopper into the Erlenmeyer flask and push the other stopper into the condenser.

   NOTE: Make sure that the stoppers are pushed firmly into the flask and condenser or they may come loose during the distillation, and ethanol vapor will be lost.

7. Turn the cold water on so that it is flowing smoothly through the condenser.
8. Place a clean, dry, 100 mL graduated cylinder (at a tilt if necessary) directly underneath the condenser spout.
9. Place the Bunsen burner underneath the Erlenmeyer flask and gently boil the sample.
10. Collect 30–40 mL of distillate in the graduated cylinder.
11. Pour all of the distillate into a 100 mL volumetric flask. Carefully wash the graduated cylinder out several times with water from a wash bottle and pour the washings into the volumetric flask.
12. Carefully add water to the 100 mL mark, place the stopper on the flask, and shake vigorously.

    - Analyze the solution by an appropriate method.

Section E. Drinking and Driving . . . A Sad Story

In 1988, I was leading a discussion of this laboratory module with several of my instructors, all of whom were to teach three laboratory sections in the week following the meeting. I noticed that one of the instructors remained very quiet throughout the meeting. Towards the end of the meeting, she began to cry. I asked her what was wrong; she hesitated for a few minutes. She took out her wallet, removed a small photograph, and placed it on the table in front of us. She choked back her tears and told us that her three-month-old daughter had been killed by a drunk driver. I will never forget the intense feeling of shock and sadness that everyone in the room experienced.

Please, don't drink and drive, or drink and fly, or drink and . . . .

Chapter 13

Kinetic Blues

"Equilibrium is when all of the fast things have happened and all of the slow things have not."

Quote attributed to Richard Feynman

Introduction

There are two fundamentally different ways of studying chemical reactions: thermodynamics and kinetics. The *thermodynamic* approach is static and involves the study of the initial and final states of a chemical system. Thermodynamics tells us nothing about the dynamics of a chemical reaction. One of the most useful things to know about a chemical reaction is the rate at which products are formed from reactants. Chemical *kinetics* is the study of the various factors that control the rates of reactions and the study of mechanisms by which reactants become products. The speeds of chemical reactions can vary tremendously, from extremely fast — e.g., in explosions and acid-base reactions — to very slow, which is seen in the geochemical weathering of rocks. These very fast or very slow reactions are difficult to study experimentally. However, even these extreme kinetics are beginning to yield to sophisticated techniques. Great progress has been made in many scientific fields — e.g., molecular biology, medicine, and atmospheric chemistry — due to a knowledge of the fundamental reaction dynamics.

The *rate* of a chemical reaction tells us how the concentration of a reactant or product changes over time. Kinetic studies of a chemical reaction almost always start with experimental measurements made in the laboratory. Various hypotheses are then proposed in an attempt to explain the experimental data. This approach clearly shows that the rate of a chemical reaction generally depends on the following factors:

- The temperature of the system
- The concentrations of the various chemical species present in the system
- The presence of catalysts
- The rate at which reactants can mix or diffuse together

Usually, the experimental investigations are designed to produce quantitative information about all of the above factors. This information can be used to predict how the rate of the reaction will change when the reactions conditions are varied. Kinetic data are also of great interest because it provides the most general and powerful method of determining the mechanism of a reaction. A *mechanism* is defined as all of the actual elementary steps involving molecules that take place simultaneously or consecutively and that added together give the observed, overall reaction.

Generally, it is extremely difficult to identify the exact mechanism of a chemical reaction. Often, the two major sources of difficulty are the extraordinarily fast rate of many elementary steps and the problem of identifying transient chemical species. The most useful approach to identifying a mechanism is to invent plausible schemes that seem to be consistent with the kinetic data. These schemes can then be tested in more sophisticated experiments that are specifically designed to probe for certain steps or individual species. It must always be kept in mind that the suggested schemes are merely that — suggestions!

The application of the principles of kinetic analysis is best discussed by studying a well-known reaction. Consider the reaction between nitric oxide (NO) and dioxygen (O_2) to produce brown nitrogen dioxide (NO_2):

$$2\,NO_{(g)} + O_{2(g)} \rightarrow 2\,NO_{2(g)}$$

Known concentrations of the two reactants NO and O_2 are pumped into a thermostatted reaction container kept at a known temperature. The rate of the reaction is determined either by measuring the decrease in the concentration of NO or O_2 or by measuring the increase in concentration of the product NO_2 over time.

A convenient way of studying the effect of concentration on reaction rate is to determine the initial rate (at the start of the reaction) as a function of the initial concentration of one reagent while keeping the concentration of all other reactants constant. The results from a series of experiments carried out in this way, at some known, constant temperature, are expressed in the form of a *rate law*. The rate law for the above reaction is experimentally found to be

$$\text{Rate} = k\,[NO]^2\,[O_2]$$

where the proportionality constant k is called the *rate constant*. The experimentally determined exponents for the concentration of each species — i.e., 2 for [NO] and 1 for $[O_2]$ — describe the *order* of the reaction for that reactant. The reaction is said to be second order in NO, first order in O_2, and is third order overall. It is important to emphasize that these experimentally determined orders have nothing to do with the stoichiometry of the balanced equation for the reaction. The fact that this reaction is second order in NO and first order in O_2 and that these orders are the same numbers as the coefficients in the reaction is entirely fortuitous! The fact that the reaction is second order in NO simply means that a doubling of the NO concentration will result in a four-fold increase in the reaction rate, provided that the O_2 concentration and the temperature are kept constant. Once the rate law is known and the rate constant k has been measured (at a particular temperature), then the rate of reaction for any concentration of reactants can be calculated.

The rates of most simple chemical reactions increase as the temperature rises. This rate change occurs because an increase in temperature increases the fraction of molecules having high kinetic energies, and these very energetic molecules are the ones most likely to react upon collision. The expression that relates the dependence of the rate constant of a reaction on temperature was first introduced by Arrhenius (a Swedish chemist) and is called the *Arrhenius equation*:

$$k = Ae^{-\frac{E_a}{RT}}$$

where E_a is the activation energy of the reaction (in kJ mol^{-1}), R is the gas constant (8.31 $JK^{-1}mol^{-1}$), T is the absolute temperature (K), e is the base in natural logarithms, and A is the frequency factor. The Arrhenius equation can be expressed in a more convenient form by taking the natural logarithm of both sides,

$$\ln k = \ln A - \frac{E_a}{RT}$$

and converting to base 10 logarithms:

$$2.30 \log_{10} k = 2.30 \log_{10} A - \frac{E_a}{RT}$$

and dividing by 2.30 gives:

$$\log_{10} k = \log_{10} A - \frac{E_a}{2.30\,RT}$$

The activation energy E_a is the minimum amount of energy required for the reaction to occur. Molecules that do not acquire this energy will not react. The potential energy profile of a reaction shows that the activation energy represents an energy barrier that must be overcome in order for products to be formed. Potential energy profiles for exothermic and endothermic reactions are shown in Figure 13.1. The

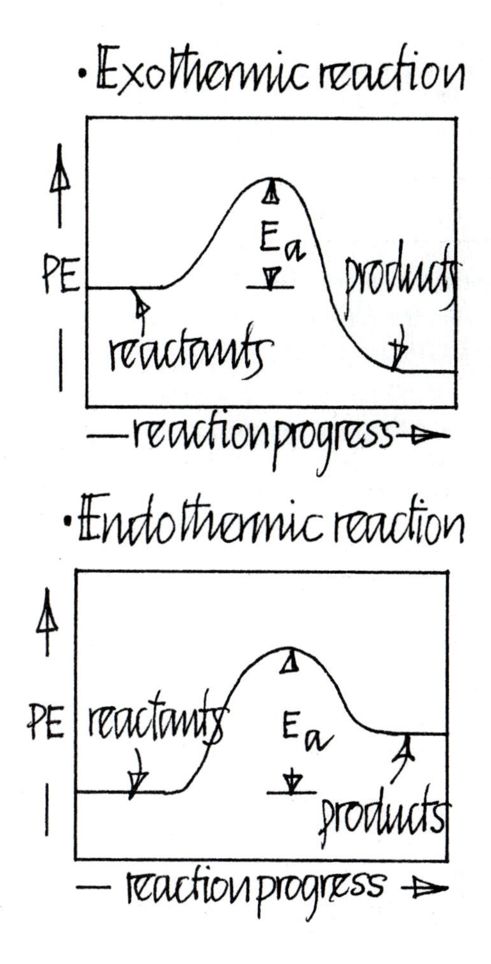

Figure 13.1 Potential Energy Profiles for Exothermic and Endothermic Reactions

activation energy for a reaction can be determined experimentally by carrying out the reaction at different temperatures and by measuring the rate constant at each temperature. A plot of $\log_{10} k$ versus $1/T$ has a slope of $-E_a/2.30R$; hence, E_a can be calculated.

Different types of chemical reactions have dramatically different activation energies, depending on the stability of the reactant molecules. The gas phase reaction

$$2\,NO_{(g)} + O_{2(g)} \rightarrow 2\,NO_{2(g)}$$

discussed earlier, is very interesting and rather atypical. This reaction *slows down* as the temperature increases! Application of the Arrhenius equation to this reaction would give the strange result of a negative value for the activation energy. The reasons for this apparent anomaly have been much debated, and the concensus is that the mechanism for this reaction involves an elementary step that is very sensitive to temperature increases.

The determination of the rate law and the temperature dependence of a chemical reaction represents the starting point for possible mechanistic interpretations of the reaction. The overall third-order kinetics for the reaction

$$2\,NO_{(g)} + O_{2(g)} \rightarrow 2\,NO_{2(g)}$$

originally led scientists to speculate that this reaction involved the simultaneous collision of two NO molecules and one O_2 molecule, i.e., a termolecular collision. Termolecular collisions are statistically rare, and this fact did not seem to be consistent with the large value for the rate constant. A much more plausible mechanism is the suggestion that the formation of NO_2 occurs by a two-step process, the first of which involves the fast dimerization of NO,

$$2\,NO_{(g)} \rightleftharpoons N_2O_{2(g)} \quad \text{dimerization}$$

followed by a slower reaction between the dimer and O_2:

$$N_2O_2 + O_{2(g)} \rightarrow 2\,NO_{2(g)}$$

The slowest reaction in any sequence of reactions determines the rate of the overall reaction and is called the *rate-determining step*. The rate of the slow second reaction is

$$\text{Rate } NO_2 \text{ formation} = k\,[O_2]\,[N_2O_2]$$

Now, for the fast equilibrium reaction, the equilibrium constant K_{eq} is

$$K_{eq} = \frac{[N_2O_2]}{[NO]^2}$$

and, rearranging,

$$[N_2O_2] = K_{eq}\,[NO]^2$$

Substituting $[N_2O_2]$ back into the rate law for NO_2 formation results in

$$\text{Rate } NO_2 \text{ formation} = k\,K_{eq}\,[NO]^2\,[O_2]$$

This last rate law is identical with the experimentally determined rate law, giving support for the suggested two-step mechanism. Further evidence for the two-step mechanism is the unusual temperature dependence reported earlier. The interpretation of the slowdown of the overall reaction with increasing temperature is that the dimer N_2O_2 is unstable at higher temperatures. Another way of saying this is that the equilibrium constant K_{eq} for the dimerization reaction decreases with increasing temperature.

It is important to emphasize that a proposed mechanism is simply a reasonable model that is apparently consistent with the experimental kinetic data. Often, several plausible mechanistic routes for a reaction are possible, and the determination of the "true" mechanism then requires many more experimental investigations.

Background Chemistry

In the series of experiments in this chapter, you have the opportunity to revisit "The System" (see Chapter 1). In "The System" you were presented with a vial containing an aqueous solution of sodium hydroxide, a reducing sugar, and a dye (methylene blue), and you were asked to deduce what was happening. It would be very helpful for you to review your laboratory notes for "The System" — the goal in this series of experiments is to investigate the chemical kinetics of "The System" reactions. One of the most interesting aspects of the chemistry of "The System" is that the chemical kinetics have never been quantitatively studied, in spite of the fact that the phenomenon has been known for a long time.

The blue dye used in "The System" is called methylene blue, and it has the following structure:

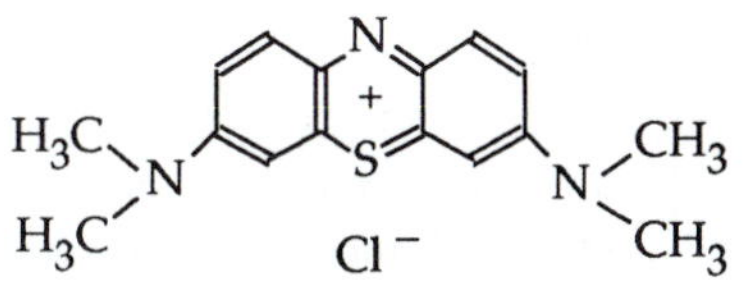

$$C_{16}H_{18}N_3SCl \bullet 3\,H_2O$$

$$338.88\ \text{g mol}^{-1}$$

(The correct chemical name is a real mouthful: phenothiazin-5-ium, 3, 7-bisdimethyl-amino chloride trihydrate.) Methylene blue was first synthesized by Caro in 1876 and was used as a fabric dye because of its beautiful deep blue color. By 1890 the dye was found to be an invaluable biological staining reagent, as well as a very useful indicator of biochemical reduction. The action of alkaline glucose solutions on methylene blue was also discovered in about 1890 and was successfully used as a method of sugar analysis. The use of methylene blue as a catalyst in the oxidation of sugars by air was first recorded by H. A. Spoehr in a research paper in the *Journal of the American Chemical Society* in 1924. Spoehr reported that methylene blue is reduced by alkaline glucose solutions and that when air is drawn through such a colorless solution, the blue color again appears, and carbon dioxide is formed. He abandoned the study of methylene blue as a catalyst because he thought he had discovered a much better catalyst, sodium ferro-pyrophosphate.

The general redox properties of methylene blue also appear to be involved in many other very useful applications of the dye. Methylene blue was one of the first anthelmintics (worming agent) and was one of the first reported antimalarial drugs. Modified derivatives of the dye form a very large class of animal tranquilizers, e.g., chlorpromazine. Methylene blue is still used in the treatment of nitrate intoxication of cattle. Hundreds of other uses for this remarkable dye have appeared in the research literature over the past 50 years. One of the most interesting applications of "The System" reactions is reported in some recent Japanese patents. Various chemical formulations of methylene blue and reducing agents, e.g., glucose, are being used as oxygen color indicators for packaged foods. These formulations (in the colorless form) are coated or printed onto plastic packaging materials used as containers for foods that are packed under vacuum or under nitrogen. The formulation turns blue if oxygen leaks into the package.

The sugar used in "The System" solution is fructose, commonly known as fruit sugar. Fructose has the structure

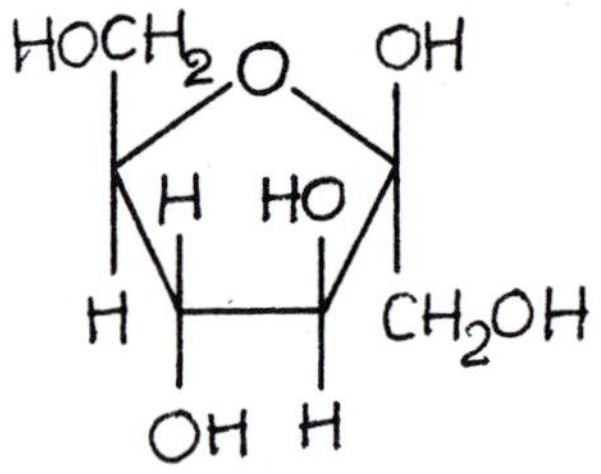

$C_6H_{12}O_6$

180.2 g mol^{-1}

and occurs in a large number of fruits and honey and as the sole sugar in bull and human semen. Fructose is extremely soluble in water and, in fact, is rather difficult to obtain in the pure anhydrous crystalline form. As soon as solid fructose is dissolved in water, the molecule undergoes various rearrangements, e.g., mutarotation and enolization. In alkaline solution, such as in "The System" solution, the changes are very complicated and not very well understood. The simplest approach is to assume that an acid-base reaction can occur between sodium hydroxide and fructose, giving rise to an anion of some type. The anion is then capable of acting as a reducing agent towards the methylene blue dye. It is interesting to note that all the reducing sugars — e.g., glucose, mannose, lactose, etc. — appear to work in a similar manner in "The System," although there are definite differences in the various rates of reactions.

"The System" offers a fascinating, novel series of reactions for the study of chemical kinetics and for the interpretation of mechanistic routes for reactions. In the field of chemical kinetics, it is the experimental facts that form the entire basis for the mechanistic possibilities. In this series of experiments, there are many opportunities for individual contributions and suggestions, and there are many new interpretations to be discovered. Please don't hesitate to discuss with your instructor any ideas and new possibilities you think of during the laboratory.

ADDITIONAL READING

1. Campbell, J.A., "Kinetics — Early and Often," *J. Chem. Ed.* **40**, 1963, p. 578.
   An excellent discussion of the catalytic oxidation of glucose in alkaline solution.

2. Spoehr, H.A., "The Oxidation of Carbohydrates with Air," *J. Am. Chem. Soc.* **46**, 1924, p. 1494.
   This is the original reference in which the methylene blue oxidation of sugars is described.

3. Clark, W.M., Cohen, B., Gibbs, H.D., "Studies on Oxidation-Reduction, VIII. Methylene Blue," *Public Health Reports* **40**, 1925, p. 1131.

Laboratory Experiments

Flowchart of the Experiments

Section A.	Experimental Evidence: A Review of "The System"

Section B.	A Possible Reaction Mechanism for "The System"

Section C.	Further Investigations of the Mechanism and a Comparison of Rates of Reaction

Section D.	The Rate-Determining Step and the Rate Law for the Overall Reaction

Section E.	Determination of the Order of Reaction for Fructose and Hydroxide Ion

Section F.	Determination of the Energy of Activation

Requires one three-hour class period to complete

Section A. Experimental Evidence: A Review of "The System"

Goals:

(1) To assemble "The System" by mixing fructose, sodium hydroxide, and methylene blue solutions. (2) To carry out experimental investigations on "The System" that could lead to the formulation of a reasonable reaction mechanism for the oxidation of fructose by methylene blue.

Discussion:

The reagents for this sequence of experiments are 0.2 M fructose, 0.5 M sodium hydroxide (NaOH), and 0.02 wt % methylene blue (MB) in 50% by volume ethanol/water. You will find these reagents in labelled microburets in a 24-well tray at your place. You may use the microburets to deliver drops and assume that the drop volume is approximately the same. It is important, however, to use a consistent delivery technique throughout. You will need a watch or clock with the capability of timing events in seconds. View the solutions against a white background.

Experimental Steps:

1. Clean a small vial and cap. This is best done by using a microburet filled with distilled water. Suck up any waste liquid and expel to a waste cup.
2. Deliver 5 drops 0.2 M fructose, 5 drops 0.5 M NaOH, and 3 drops of methylene blue (MB) to the vial. Stir with a microstirrer.
   - Watch carefully and time from the stirring until something happens. Use your hand lens.
   - Describe what you see in detail in your laboratory notebook.
3. Place the cap onto the vial. Give the vial one sharp shake.
   - How long does it take for the solution to become colored?
   - How long does it take for the solution to lose its color?
   - What do you see at the interface between the solution and the air?
4. Give the vial several sharp shakes.
   - Does the solution go more intensely colored the more you shake it?
5. Investigate the effect of heat and cold on the reactions. A simple method of doing this is to use wells in the 24-well tray to prepare a cold (perhaps an ice bath) and a warm (about 40–50°C) water bath.
   - Does cooling the vial influence both the blueing and the deblueing reactions?
6. If "The System" shows signs of deterioration, clean the vial and prepare another solution (as in Step 2).
7. Shake the vial vigorously and while the solution is blue, place it in the intense light from a slide or overhead projector (or in bright sunshine).
   - Describe what happens.
   - Will the system still cycle?
8. Review your notes from this section and from "The System" (see Chapter 1).

Section B. A Possible Reaction Mechanism for "The System"

Goals:

(1) To examine a possible reaction mechanism for the oxidation of fructose by methylene blue. (2) To correlate the mechanism with the experimental observations made in the last section.

Discussion:

A review of the evidence from the last section indicates that there are five fundamental processes occurring in "The System."

(1) Initially, the fructose reacts with sodium hydroxide to produce some unknown anion.

This reaction can be regarded as an acid-base reaction between the weak acid fructose and the strong base sodium hydroxide.

(2) The unknown anion reacts with blue methylene blue to produce colorless methylene white and some unknown colorless products.

(3) Vigorous shaking of "The System" results in air mixing with, and dissolving in, the colorless solution.

(4) Dissolved gas, probably dioxygen (O_2) from air, oxidizes methylene white back to methylene blue. The solution may then cycle through the deblueing process, over and over again, provided that fructose, base, and dissolved dioxygen are present in "The System."

(5) Intense light, in the presence of base, changes blue methylene blue into some other, much redder substance.

The processes described above can be incorporated into a more concise reaction scheme by giving symbols to the various ingredients and by writing reasonable chemical reactions for each process. Let

- FH = fructose
- F^- = an anion formed by the reaction of base with fructose
- OH^- = sodium hydroxide, a strong base
- B = the blue form of methylene blue
- W = the colorless form called methylene white
- $O_{2(g)}$ = gaseous dioxygen in air
- $O_{2(aq)}$ = dioxygen dissolved in the solution
- P = product(s) formed in the overall scheme
- R = reddish-purple substance formed by the action of intense light on methylene blue.

The following scheme is a possible reaction mechanism for "The System." At this point, it is important to note that the scheme does not include any kinetic information.

$$FH + OH^- \longrightarrow F^- + H_2O$$
$$F^- + B \longrightarrow W + P$$
$$O_{2(g)} \longrightarrow O_{2(aq)}$$
$$O_{2(aq)} + W \longrightarrow B$$
$$B \xrightarrow{\text{light}} R$$

- What preliminary evidence do you have that some of the above reactions go at different rates than others? Explain.
- Which of the reactions seems to be particularly slow under normal conditions, i.e., room conditions of temperature, pressure, etc.?

Section C. Further Investigations of the Mechanism and a Comparison of Rates of Reaction

Goal: *To explore in further detail the possible reaction mechanism suggested in Section B.*

Discussion: The accumulated experimental evidence should allow you to make an evaluation of the relative rates of individual reactions in the suggested mechanism.

Experimental Steps:

D4 Fructos

B6 NaOH

1. Clean a 1 x 12 well strip.
2. Use large-drop microburets to deliver all the solutions used in this series of experiments.

   NOTE: You will need a watch again.

   - Design a simple experiment to prove that the deblueing agent can only be a product formed by the reaction between fructose and sodium hydroxide.
   - Write a rate law for the formation of the deblueing agent, F^-.
3. Clean the 1 x 12 well strip. Deliver 2 drops of 0.5 M NaOH to a well, followed by 2 drops of 0.2 M fructose. Stir and immediately add 1 drop MB solution. Stir quickly with microstirrer and time the deblueing reaction.
   - Note the time. 40s
4. Carry out Step 3 again, except this time wait 3 minutes *before* adding the MB solution.

- Time the deblueing reaction and record the time. 30 s
- What does the information from the last two steps tell you about the rate of formation of deblueing agent? Explain.

5. Hand-pull a thin stem pipet and cut off the pulled part, leaving about 4 cm of the thin part.

   NOTE: This microburet can be used as an efficient aeration device in this and later experiments.

6. Squeeze the microburet bulb, suck up the solution from one of the wells, expel it, suck it up again, and expel it back.

   - Record what happens in your notebook.
   - What do the results of this experiment tell you about the rates of the following reactions?

$$O_{2(g)} \rightarrow O_{2(aq)}$$

$$O_{2(aq)} + W \rightarrow B$$

7. Repeat Step 3 and allow the deblueing reaction to occur. Use a clean microburet to gently add — i.e., without aerating — some cold tap water to the colorless solution in the well. Stir very gently, *without* aerating the solution.

   - What do you think is causing the change? Cooler temp

   The oxygen dissolution reaction and the oxidation of methylene white reaction (see Step 6) require that dioxygen gas be used up by the cycling system. A dioxygen uptake experiment should provide very strong support for this part of the mechanism.

8. Obtain a clean, thin-stem pipet from Reagent Central.

   NOTE: The pipet can be used both as a reaction vessel and as a manometer to measure pressure changes.

9. Fill a well of a 24-well tray about 3/4 full with a solution that is a mixture of equal volumes of 0.5 M NaOH and 0.2 M fructose. Add 10 drops of MB. Stir. Suck the mixed solution into the thin-stem bulb. The bulb should be about 3/4 full.

10. Now carefully bend the stem over and dip the end in water. Press the bulb slightly and suck a very small amount of water into the stem so that the water plug stays almost at the tip.

11. Place the thin stem of the pipet in a straw clamp so that the bulb is just below the clamp.

12. Place the clamp in a straw stand, using a 96-well RB tray as a base.

13. Mark the stem at the top of the water plug with a black marker pen.

14. Tilt the stem about 45° and hold the thin stem at the clamp tightly between your index finger and thumb. Give the bulb a series of sharp taps with your finger in order to vigorously shake the solution in the bulb.

15. Stop shaking the solution and wait until the solution deblues. Measure the change in pressure by measuring how far the water plug has moved in the stem.

16. Repeat Steps 13 and 14 several times (perhaps 7 or 8 times).

   - If you have time, plot a graph of change in position of the water plug versus the shake number.
   - What does the graph tell you about the uptake of dioxygen?

NOTE: All of the above experiments provide further strong support for the mechanism suggested earlier.

Section D. The Rate-Determining Step and the Rate Law for the Overall Reaction

Goal: *To be able to choose the rate-determining step in the proposed mechanism.*

Discussion: Once the rate-determining step is identified, a rate law for the overall reaction can be obtained and can then be tested in kinetic studies. Note that a plausible mechanism thus far appears to be:

$$FH + OH^- \rightarrow F^- + H_2O$$

$$F^- + B \rightarrow W + P$$

$$O_{2(g)} \rightarrow O_{2(aq)}$$

$$O_{2(aq)} + W \rightarrow B$$

$$B \xrightarrow{\text{light}} R$$

The initial reaction in the mechanism requires some time to come to equilibrium. Once the sequence is allowed to set up, then the initial time period does not matter. We can write the first reaction as a typically fast acid-base equilibrium reaction, i.e.,

$$FH + OH^- \rightleftharpoons F^- + H_2O$$

provided that enough time is allowed for the reaction to come to equilibrium. There are four remaining reactions to consider in choosing the rate-determining step. The last reaction,

$$B \xrightarrow{\text{light}} R$$

which involves the photochemical destruction of the blue species, is apparently far too slow in ordinary room light to have any effect on the overall reaction.

- Which of the three remaining reactions is the slowest?

The slowest reaction is called the rate-determining step because it determines the rate of the overall reaction. The overall reaction may be obtained by adding up all the reactions in the mechanism, cancelling species common to both sides, and leaving out species common to the solvent water. The overall reaction is

$$FH + O_{2(g)} \rightarrow P$$

which is the oxidation of fructose by atmospheric oxygen catalyzed by methylene blue in aqueous alkaline solution.

- Why is methylene blue called a catalyst for the overall reaction?
- What is an alternative definition of a catalyst?

The rate law for the deblueing reaction is

$$\text{rate} = \frac{-d[B]}{dt} = k[B][F^-]$$

where k is the rate constant for the deblueing reaction and [B], $[F^-]$ are the molar concentrations of B and F^-, respectively.

Since F^- is part of an equilibrium reaction,

$$FH + OH^- \rightleftharpoons F^- + H_2O$$

for which we can write an equilibrium constant:

$$K_{eq} = \frac{[F^-]}{[FH][OH^-]}$$

Thus

$$[F^-] = K_{eq}[FH][OH^-]$$

And substituting for $[F^-]$ in the above rate law,

$$\text{Rate} = \frac{-d[B]}{dt} = k\,K_{eq}[B][FH][OH^-]$$

Then combining the constants k and K_{eq}, the final rate law for the rate-determining step and, therefore, the overall reaction is

$$\text{Rate} = \frac{-d[B]}{dt} = k_f[B][FH][OH^-]$$

The interesting conclusion is that the rate is first order in methylene blue, fructose, and hydroxide ion and zero order in oxygen.

Section E. Determination of the Order of Reaction for Fructose and Hydroxide Ion

Goal:

To determine the order of reaction for fructose and for hydroxide ion in the rate-determining step of the catalytic oxidation of fructose by atmospheric oxygen.

Experimental Steps:

1. Make sure that the microburets are filled with the appropriate reagents.

   NOTE: You will need 0.5 M NaOH, 0.2 M fructose and 0.02 wt % MB solutions. The styrofoam cup is filled almost to the top with water that is equilibrated at room temperature. You will need scissors and a hand lens. Obtain a slim straw and thin-stem piece from Reagent Central and clean several wells in the 24-well tray.

2. First, carry out a practice run so that you can become familiar with the technique: Prepare a solution by delivering 10 drops of 0.5 M NaOH, 10 drops of 0.2 M fructose and 5 drops MB to a clean well of a 24-well tray. Stir and wait until the solution deblues.

3. While you are waiting, cut the slim straw into several pieces, each about 5 cm long. Place one of these pieces on the plastic surface in front of you.

   NOTE: Make sure that tweezers are close by and that you have a watch ready.

4. Use the microburet with the long pulled-out stem to suck up and aerate the mixed solution. Expel it back to the well and suck it up again, expel and suck it up again. Quickly transfer some solution (about 1–1.5 cm) to the piece of slim straw. Start timing and then lift the straw (keeping it horizontal) and drop it into the water in the styrofoam cup.

   - Monitor the temperature of the water and be careful to maintain a constant temperature.

   - Record the deblueing time and subsequent deblueing times in the table in Step 7.

   NOTE: One of the reasons for employing this observation technique is that the surface blueing reaction occurs only in a vertical plane at the ends of the cylinder of solution. This observation technique should allow you to make good measurements of the deblueing time. Another reason for the technique is that the large mass of water in the cup quickly brings the solution in the straw to a constant temperature. You need to determine the temperature of the water during these experiments to make sure that it doesn't change.

5. When you feel comfortable with the aeration and timing techniques, prepare another mixed solution (as in Step 2) and carry out an accurate deblueing time measurement.

6. Repeat Steps 2 through 4, this time using a mixed solution containing 5 drops 0.2 M fructose, 5 drops distilled water, 10 drops 0.5 M NaOH and 5 drops MB solution.

   - Remember to record the deblueing time.

7. Repeat Steps 2 through 5, this time using a mixed solution containing 10 drops 0.2 M fructose, 5 drops distilled water, 5 drops 0.5 M NaOH, and 5 drops MB solution.

- Fill in the following data table and transfer the information to your notebook.

.2M .5M

Experiment #	Mixed Solution	[Fructose]	[OH⁻]	Time (sec)	Rate of Deblueing
1	Step 2 solution	10	10	7	0.14
2	Step 6 solution	5	10	90	0.01
3	Step 7 solution	10	5	30	0.03

- You are now going to use the data to calculate orders of reaction.

CALCULATION OF THE ORDER OF REACTION FOR FRUCTOSE AND HYDROXIDE ION

The concentration of fructose and OH^- (in mol L^{-1}) is easily calculated from the dilution. (Remember, the total volume is 25 drops in each instance.) Rate is inversely proportional to time. If the deblueing reaction took 50 seconds, then the rate = 1/50 = 0.02 s^{-1}.

Now the rate law for the deblueing reaction is

$$\text{Rate of deblueing} = K_f \times [\text{fructose}]^a \times [B]^b \times [OH^-]^c$$

where a, b, and c are the experimental orders of reaction with respect to fructose, methylene blue, and hydroxide. What we wish to do is find a, b, and c from the experimental rate and concentration data in the above table. The rate of deblueing in Experiment 1 is

$$\text{Rate 1} = K_f \times [\text{fructose}]_1{}^a \times [B]^b \times [OH^-]_1{}^c$$

The rate of deblueing in Experiment 2 is

$$\text{Rate 2} = K_f \times [\text{fructose}]_2{}^a \times [B]^b \times [OH^-]_2{}^c$$

Dividing the rates,

$$\frac{\text{Rate 1}}{\text{Rate 2}} = \frac{K_f}{K_f} \times \frac{[\text{fructose}]_1{}^a}{[\text{fructose}]_2{}^a} \times \frac{[B]^b}{[B]^b} \times \frac{[OH^-]_1{}^c}{[OH^-]_2{}^c}$$

Since the $[OH^-]$ in Experiments 1 and 2 is the same and since in all three experiments the [B] is kept the same, then

$$\frac{\text{Rate 1}}{\text{Rate 2}} = \frac{[\text{fructose}]_1{}^a}{[\text{fructose}]_2{}^a}$$

and putting in the values for rate 1, rate 2, $[\text{fructose}]_1$, and $[\text{fructose}]_2$, the order of reaction with respect to fructose can be obtained.

- Use the data from Experiments 1 and 3 (from the table in Step 7) to find the order of reaction with respect to OH^-.

10. Compare the experimentally obtained orders of reaction with those deduced for the proposed mechanism (see Section D).

Section F. Determination of the Energy of Activation

Goal: *To quantitatively examine the effect of temperature on the rate of the deblueing reaction.*

Discussion: The temperature dependence data can be used to plot an Arrhenius curve from which the energy of activation for the catalytic oxidation of fructose can be calculated.

Experimental Steps:

1. Construct a constant temperature reaction apparatus from straws, caps, and a styrofoam cup as shown below:

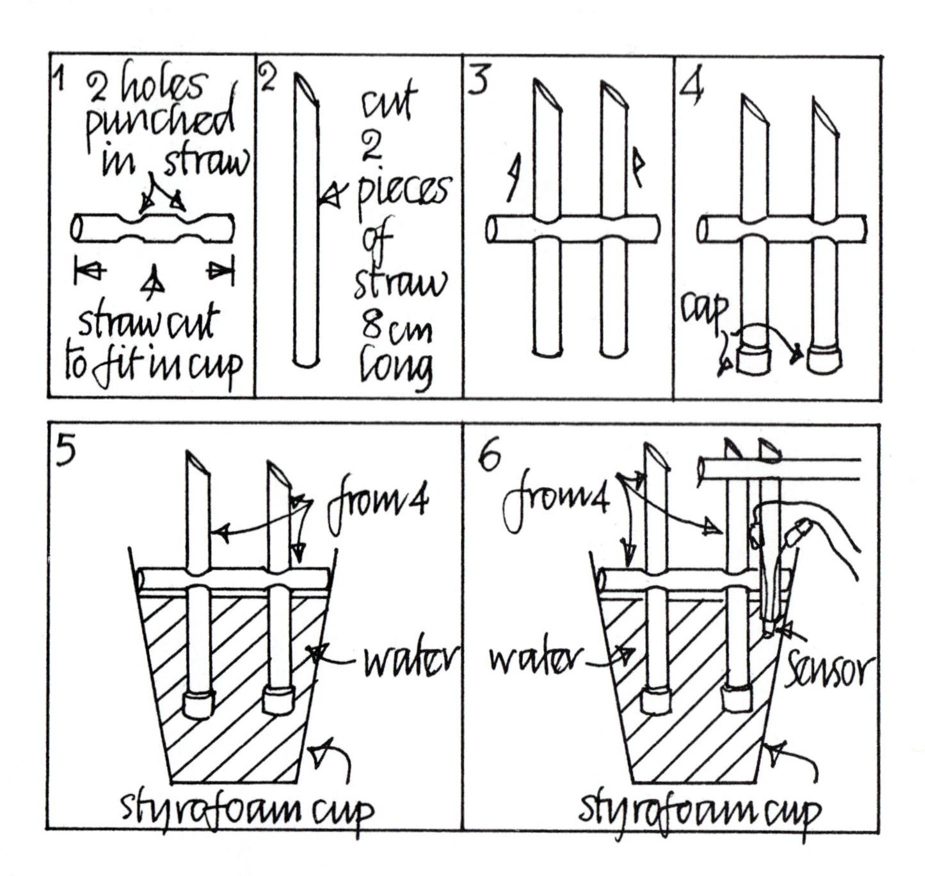

2. Fill the styrofoam cup almost full with warm water (at a temperature of about 30 °C).
3. Make sure that the temperature measurement system (sensor and DMM or thermometer) is working and that the sensor is actually in the water.

NOTE: See Chapter 5 for the details of temperature measurement using sensor and DMM.

4. Fill another cup with ice.

   NOTE: You will be using the ice to change the temperature of the water in the constant-temperature bath.

5. Clean a well of a 24-well tray. Prepare a mixed solution containing 10 drops 0.2 M fructose, 10 drops 0.5 M NaOH, and 5 drops MB solution. Stir and allow to deblue.

6. Use a thin-stem pipet to transfer about 1/2 of the solution to each of the two capped straws in the bath. Allow 1 minute for temperature equilibration.

7. Place the pipet all the way to the bottom of one of the capped straws. Suck the solution up so that it becomes vigorously aerated. Expel it back to the straw. Suck it up again to aerate and expel back.

   NOTE: This procedure must be done quickly to aerate the solution properly.

8. Start timing and time the deblueing reaction.

   - Record the time.

9. Repeat the measurement on the duplicate solution sample in the other straw.

   - Record the temperature of the bath as well as the deblueing time for the duplicate solution.

   NOTE: Timing the deblueing reaction in the above experiment may have been difficult because the reaction was so fast.

10. Cool the water in the bath about 5 °C by adding ice and stirring with a straw. When the temperature has equilibrated, repeat steps 5 through 7.

    - Record the deblueing times and temperature.

11. Cool the bath another 5 °C and repeat Steps 5 through 7.

    - Measure the deblueing times at several temperatures, down to a bath temperature of about 10 °C.

      NOTE: At this low temperature the reaction may take a very long time.

    - Prepare a table of data, as follows:

Bath Temp (°C)	Bath Temp (K)	$\frac{1}{T}$ (K^{-1})	Deblueing Time (sec)		Average	Rate	Log_{10} Rate
			run 1	run 2			

- Plot a graph of $\log_{10}$ rate (vertical axis) versus $1/T$ K^{-1} (horizontal axis). The graph will be in the fourth quadrant.
- Draw the best straight line through the points.
- Determine the slope of the line. The slope is negative and

$$\text{slope} = -\frac{E_a}{2.30 \times R}$$

where E_a = energy of activation (in joules)

and R = the gas constant, 8.314 J K^{-1} mol^{-1}

The 2.30 factor comes from the conversion of natural logarithms to logarithm to base 10. (See the Background Chemistry section.)

NOTE: Check with your instructor if you have trouble understanding how to carry out this graphing.

12. Calculate the energy of activation for the deblueing reaction.
    - What unit is the slope of the line expressed in?
    - What unit is the energy of activation expressed in?

Chapter 14

Acid-Base Equilibria

"How inappropriate to call this planet Earth, when clearly it is Ocean."

Arthur C. Clarke

Introduction

Life on earth flourishes in a water environment in which the acidity is maintained within a very narrow range by dynamic acid-base reactions. A variety of interrelated reactions bind and release hydrogen ions and, in the process, keep the pH not too far from seven. The pH of natural fluids is a useful index of the status of equilibrium reactions in which water participates. The reaction of dissolved carbon dioxide with water is one of the most important processes that release hydrogen ions:

$$CO_{2(g)} \rightleftharpoons CO_{2(aq)}$$

$$CO_{2(aq)} + H_2O \rightleftharpoons H_2CO_3$$

$$H_2CO_3 \rightleftharpoons H^+ + HCO_3^-$$

$$HCO_3^- \rightleftharpoons H^+ + CO_3^{2-}$$

Pure water in contact with air having an average CO_2 content has a pH that stays about 5.65. In man, the hydrogen ion concentration in blood is maintained very close to 4×10^{-8} M (pH 7.4), which is a requirement for the optimal performance of many enzyme systems. The ability of natural systems to maintain hydrogen ion concentrations at precise low levels is a direct consequence of the "feedback" properties of acid-base equilibria. These equilibria are dynamic, reversible, and fast. Any challenge to the natural system by abnormal or unusual changes is immediately detected and effectively neutralized.

The effect of an outside influence on a system at equilibrium may be predicted by means of *Le Chatelier's principle*, which states that if an equilibrium is upset, the system responds in a direction that will reestablish equilibrium. It is important to emphasize that the principle holds even in very complicated natural systems in which there are many connected chemical and biological equilibria. Equilibria are *connected* if a component of one reaction (either reactant or product) also takes part in another reaction. A good example is the carbon dioxide–water system introduced earlier. An increase in the atmospheric carbon dioxide concentration ($CO_{2(g)}$) will lead to an increase in dissolved carbon dioxide ($CO_{2(aq)}$), which will increase the carbonic acid (H_2CO_3) concentration and will eventually lead to increased acidity (H^+ concentration). Similarly, any reaction that results in the removal of carbonate (CO_3^{2-}) or bicarbonate (HCO_3^-) will have a ripple effect in the system and lead to increased acidity and an increased dissolution of gaseous carbon dioxide. Homeostasis (i.e., maintenance), which occurs as a result of Le Chatelier's principle, is often called *buffering action*. In chemistry, acid-base systems with these maintenance properties are called *buffer systems*.

Earth is the water planet. Water is the key to all maintenance of life. It is the liquid that mediates all terrestrial exchanges of hydrogen ions between acids and bases. Water, therefore, provides the reference point for all measures of acidity and alkalinity. In this series of laboratory experiments, you have the opportunity to investigate most of the important aspects of acid-base equilibria, including Le Chatelier's principle, strengths of acids and bases, and buffering action.

Background Chemistry

Many theories have been developed to interpret the nature of acids and bases. One of the simplest and most useful is the Brønsted-Lowry theory. In this theory an acid is defined as a proton (H^+) donor and a base is defined as a proton acceptor. Let us consider the nature of water as a solvent. In liquid water, the water molecules collide billions of times per second, and a few of these collisions are powerful enough to break covalent bonds. A reaction occurs that may be written as

$$H_2O + H_2O \rightleftharpoons H_3O^+ + OH^-$$

in which the products are hydronium ions (H_3O^+) and hydroxide ions (OH^-). The double arrow $\rightleftharpoons$ is used to show that the reaction is dynamic and reversible. Water molecules are continuously breaking up, and hydronium ions and hydroxide ions are continuously combining to produce water molecules. The rate of the forward reaction

$$H_2O + H_2O \rightarrow H_3O^+ + OH^-$$

is equal to the rate of the reverse reaction

$$H_2O + H_2O \leftarrow H_3O^+ + OH^-$$

and, therefore, the concentration of the reactants and products does not change with time. The water system is said to be at *equilibrium*.

It is interesting to note that in the above equilibrium reaction, water is acting both as an acid (proton donor) and as a base (proton acceptor). In *pure* water, the self-dissociation produces (at 25 °C) equal concentrations of H_3O^+ and OH^-, which are 10^{-7} M. The law of mass action can be applied to any reaction at equilibrium, and when applied to the self-dissociation of water, gives

$$K_c = \frac{[H_3O^+][OH^-]}{[H_2O]^2}$$

where K_c is called the equilibrium constant for the reaction. Now the molar concentration of H_2O in water is 55.6 M and can be regarded as constant. We can simplify the above expression to

$$K_w = [H_3O^+][OH^-]$$

where K_w is called the *ion-product constant* of water. Since at 25 °C,

$$[H_3O^+] = [OH^-] = 10^{-7}\,M$$

Then

$$K_W = 10^{-7} \times 10^{-7} = 10^{-14}$$

Of course, all equilibrium constants are temperature dependent. The value for K_W at 25 °C, i.e., 1.00×10^{-14}, is used throughout this laboratory module. It is convenient to use logarithmic scales for the expression of $[H^+]$ and often for equilibrium constants:

$$pH = -\log_{10} [H^+]$$

$$pK = -\log_{10} K$$

and generally,

$$pX = -\log_{10} X$$

Thus, the pH of pure water at 25 °C is 7.00, and the pK_W for water at 25 °C is 14.00.

We can now consider what happens when an acid, e.g., acetic acid (CH_3COOH), is added to water. Since CH_3COOH is an acid, protons are donated by CH_3COOH and are accepted by the solvent water, which acts as a base:

$$CH_3COOH + H_2O \rightleftharpoons CH_3COO^- + H_3O^+$$

The equilibrium constant for this reaction is called an acid dissociation constant K_a, and applying the law of mass action,

$$K_a = \frac{[CH_3COO^-][H_3O^+]}{[CH_3COOH]}$$

The value of K_a is 1.75×10^{-5} at 25 °C. This very small value means that the equilibrium lies very much to the left. Very few molecules of CH_3COOH dissociate to give CH_3COO^- (acetate ion) and H_3O^+. Acids that have K_a values smaller than 10^{-2} are said to be *weak acids*. Acids with K_a values greater than 10^{-2} (usually >> 10^{-2}) are said to be *strong acids*. Hydrochloric acid is an example of a strong acid in aqueous solution:

$$HCl + H_2O \rightleftharpoons Cl^- + H_3O^+$$

where the acid dissociation constant is so large ($\sim 10^{+4}$) that the equilibrium lies almost completely to the right, and the reaction is normally written as

$$HCl + H_2O \rightarrow Cl^- + H_3O^+$$

The addition of acids to water, whether weak or strong, will always result in an increase in $[H_3O^+]$ and, therefore, a *decrease* in pH (since $pH = -\log_{10} [H^+]$). The pH of acidic solutions is always less than 7.00.

Bases in aqueous solution can also be characterized as weak or strong depending on the value of the base dissociation constant K_b. The addition of the base ammonia (NH_3) to water results in

$$NH_3 + H_2O \rightleftharpoons NH_4^+ + OH^-$$

where the solvent water acts as an acid and

$$K_b = \frac{[NH_4^+][OH^-]}{[NH_3]} = 1.75 \times 10^{-5} \text{ (at 25 °C)}$$

Ammonia is a *weak* base and is not dissociated to any great extent in aqueous solution. Sodium hydroxide (NaOH) is an example of a strong base that is completely dissociated in aqueous solution. The pH of basic solutions is always greater than 7.00.

All the reactions discussed so far and all the reactions you will investigate in this module involve water as the solvent. All the equilibrium constants are, therefore, valid only for *aqueous* solutions and are dependent on the proton-donating and proton-accepting ability of the water molecule. It is very convenient to employ a simpler way of writing acid-base equilibria in aqueous solution than that used thus far. It may be assumed that all ions in an aqueous solution are hydrated, including the H_3O^+ ion. Thus, self-dissociation of water may be written as

$$H_2O \rightleftharpoons H^+ + OH^-$$

and the dissociation of a weak acid HA may be written as

$$HA \rightleftharpoons H^+ + A^-$$

Unfortunately, the dissociation of weak bases must still be written with H_2O as a reactant in order to maintain the species balance, e.g.,

$$NH_3 + H_2O \rightleftharpoons NH_4^+ + OH^-$$

All the acids discussed thus far have been capable of donating *one* proton and are called *monoprotic acids*. There are many acids that can donate two or more protons, e.g., H_3PO_4 is an example of a triprotic acid:

$$H_3PO_4 \rightleftharpoons H^+ + H_2PO_4^- \qquad K_1$$

$$H_2PO_4^- \rightleftharpoons H^+ + HPO_4^{2-} \qquad K_2$$

$$HPO_4^{2-} \rightleftharpoons H^+ + PO_4^{3-} \qquad K_3$$

These polyprotic acids donate protons in a stepwise manner. The first acid dissociation constant is usally larger than the second acid dissociation constant, and so on. In the H_3PO_4 dissociation shown above, $K_1 > K_2 > K_3$.

Acid-base equilibria are often investigated by carrying out titration reactions. An *acid-base titration* is the process of incremental addition of a solution of acid (or base) to a solution of base (or acid) such that the extent of the neutralization reaction can be monitored. A titration is really a very simple way of obtaining a tremendous amount of useful information about an acid-base system.

One of the most important criteria for the use of titration methods in the study of equilibria is that equilibrium be achieved rapidly. Proton transfer reactions, especially in water, are usually very fast, with half-lives of less than milliseconds. The rapid rate of chemical reactions involving protons is in part due to the rapid rate at which protons travel through an aqueous solution. It has been proposed that H^+ moves through an aqueous solution by a "jump" or "proton-hopping" mechanism (see Figure 14.1).

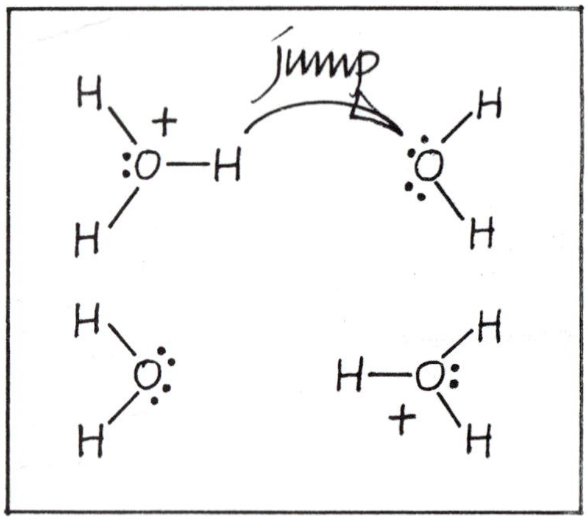

and on, and on . . . .

Figure 14.1 Proton-Hopping Mechanism in Water

Once the acid-base system has achieved equilibrium, then a wide variety of quantitative monitoring techniques can easily be used to probe the state of the system. The master variable in acid-base systems is $[H^+]$ (or pH), and the most common method of analysis is to examine the change of concentration of other species (at equilibrium) as a function of changes in $[H^+]$.

In this module you will experimentally investigate a wide variety of acid-base reactions and compare your results with computer simulations carried out on the main frame computer system at Colorado State University (a Cyber 205). Graphical presentation of the data in the form of titration curves reveals a rich source of useful information about acids and bases and their reactions.

Laboratory Experiments

Flowchart of the Experiments

Section A.	Conductimetry and the Strength of Acids and Bases

Section B.	Acid-Base Equilibria and Indicator Dyes

Section C.	Determination of the K_a Values of Weak Acid Indicators

Section D.	pH Measurement with Indicator Color Probes

Section E.	The Study of Acid-Base Equilibria by Graphical Interpretation of Titration Data Part 1. Titration of a Strong Acid with a Strong Base Part 2. Titration of a Weak Acid with a Strong Base Part 3. The Measurement of the K_a of a Weak Acid Part 4. The Henderson-Hasselbalch Equation Part 5. Titration Curves and Buffer Solutions Part 6. The Dilution of Buffers Part 7. The Selection of Weak Acids to Make Buffer Solutions Part 8. The Titration of Weak Bases with Strong Acids

Section F.	The Titration of Polyprotic Acids

Section G.	Calculations on Diprotic and Triprotic Acids

Requires two three-hour class periods to complete

Throughout this module you will be investigating acid-base equilibria by analyzing the progress of acid-base reactions in volumetric titrations. The quality of the results from these experiments will depend on the care with which you carry out solution transfers, dilutions, and drop additions. Large-drop microburets (described in Chapter 3, Section C) are satisfactory for most of the titrations. Microburets may be calibrated (when necessary) by means of an analytical balance or volumetrically.

CAUTION: Most acids and bases are corrosive. If you get any of these solutions on your skin, wash well with cold water and inform your instructor.

Section A. Conductimetry and the Strength of Acids and Bases

Goals:

(1) To be able to interpret the conductivity experiments demonstrated by your instructor.
(2) To be able to be identify strong and weak acids and bases by means of K_aK_b, pK_a, and pK_b values.

Discussion:

Your instructor will demonstrate several conductivity experiments for you. Record your observations of these experiments in your notebook. A discussion of two of the experiments follows.

1. The ability of a solution to conduct electricity depends on the concentration of *ions* in the solution. A 0.1 M HCl solution is a good conductor because the HCl molecules are completely dissociated into ions:

$$HCl + H_2O \rightarrow H_3O^+ + Cl^-$$

which may be conveniently abbreviated

$$HCL \rightarrow H^+ + Cl^-$$

because water is the solvent in aqueous solutions. The dissociation occurs extremely rapidly to instantaneously give a solution that is 0.1 M H^+ and 0.1 M Cl^-. Acids that dissociate completely are called *strong acids,* and the K_a value is very large (> 10^4 for HCl).

2. The 0.1 M CH_3COOH (acetic acid) solution is a poor conductor because the solution contains mostly undissociated CH_3COOH molecules. Acetic acid dissociates rapidly, but only to a small extent:

$$CH_3COOH + H_2O \rightleftharpoons H_3O^+ + CH_3COO^-$$

This reaction may be abbreviated

$$CH_3COOH \rightleftharpoons H^+ + CH_3COO^-$$

The K_a of CH_3COOH is small (1.8×10^{-5}); thus, the solution contains only small concentrations of H^+ and CH_3COO^- (acetate ion). Acids that do not appreciably dissociate in aqueous solution are called *weak acids.* Weak acids have K_a values that are smaller than about 10^{-2}.

Your instructor will also demonstrate conductance experiments with NaOH (strong base) and NH_3 (weak base) solutions and carry out an acid-base reaction between NH_3 and CH_3COOH. Try to come up with your own interpretation of what is occurring in these experiments. Use the following guidelines:

- Try to interpret the results from the conductance experiments in the form of chemical equilibria reactions.
- Predict what would happen in a conductivity experiment if an H_2SO_4 solution was titrated with the strong base $Ba(OH)_2$.

HINT: First, write a balanced chemical equation and think about the products that are formed.

- Use your calculator to find the pK_a and pK_b values for all the acids and bases used in this section.

Section B. Acid-Base Equilibria and Indicator Dyes

Goal:

To investigate several aspects of acid-base equilibria by means of a colored dye called bromocresol green.

Discussion:

Bromocresol green is one of a class of organic compounds that can exist in different colored forms depending on whether the compound is *protonated* or *unprotonated*. (The terms protonated and unprotonated refer to the acid form and the conjugate base form of the indicator.) You have already done experiments on a naturally occurring dye (red cabbage extract) in an earlier module. The chemical structure of most of these organic indicators is rather complex; e.g., bromocresol green has the formula $C_{21}H_{14}O_5Br_4S$ (molar mass: 698 g mol^{-1}). However, these structures can be symbolized in a simple and convenient way by writing the protonated, acid form of the dye as HIn. The dissociation of the acid indicator can thus be written

$$HIn \rightleftharpoons H^+ + In^-$$

for which

$$K_a = \frac{[H^+][In^-]}{[HIn]}$$

The protonated form HIn usually has a different color than the unprotonated form In^-.

Experimental Steps:

1. Clean a 1 x 12 well strip.
2. Locate a microburet of 0.03% bromocresol green indicator solution. Deliver 1 drop to each of 6 wells.

3. Use a clean large-drop microburet to quantitatively dilute 1 drop of 0.05 M HCl to 0.01 M HCl.

   NOTE: Use good wash and transfer techniques!

4. Add 1 drop of 0.01 M HCl to the first well. Stir.

   - What observation would lead you to conclude that the time for mixing to occur is much longer than the time required for the acid-base reaction?
   - You have just added a strong acid (HCl) — i.e., a solution of H^+ and Cl^- ions (each of which is 0.01 M concentration) — to the dye system that was originally at equilibrium. On the basis of Le Chatelier's principle, explain the color change.
   - What is the approximate pH of the mixed solution?

5. Add 1 drop of 0.01 M NaOH to the indicator in the third well.

   - Describe what happens and explain the change according to Le Chatelier's principle.

6. Add 3 drops of pH 4.0 solution (from Reagent Central) to wells 1, 2, and 3. Stir.

   NOTE: The pH 4.0 solution is a buffer solution (you will be exploring buffers in a later section) that reacts with strong acids *and* strong bases to produce a final solution that has a pH of approximately 4, i.e., $[H^+] \approx 10^{-4}$ M.

   - Explain the color changes you observed in Step 6.
   - Design and execute a simple experiment to show that the indicator equilibrium reaction is *reversible.*
   - Record the results in your notebook.

Section C. Determination of the K_a Values of Weak Acid Indicators

Goal: *To carry out a determination of the acid dissociation equilibrium constant of several colored, weak acid indicators.*

Discussion: One of the simplest ways to measure K_a values of weak acid indicators is to place the weak electrolyte in a series of solutions of exactly known pH values and allow any acid-base reactions to come to equilibrium. Once equilibrium is achieved — and of course, with acid-base reactions this happens rapidly — the appropriate color changes reveal the relative concentration of HIn and In^- in the various solutions.

Experimental Steps:

1. Clean a 96-well tray (preferably with flat-bottomed wells).
2. Locate the 12 reagent bottles containing solutions of known pH 1.0 through 12.0. Deliver 2 drops of pH 1.0 to the first well in row A, 2 drops of pH 2.0 to the second well, 2 drops of pH 3.0 to the third well, and so on. Repeat the procedure in rows B, C, and D. (You now have 4 rows of 12 solutions going from pH 1.0 on the left to pH 12.0 on the right.)
3. Use a microburet to deliver 1 drop of methyl orange to each well of row A.

4. Similarly, deliver 1 drop of bromocresol green to the wells of row B, 1 drop of bromothymol blue to the wells of row C, and 1 drop of phenophthalein to row D. Stir the solutions with a microstirrer.

   - Record the solution colors in a table in your laboratory notebook.

   In the above experiment, each weak acid indicator is "locked" in a certain equilibrium position that is dependent on the $[H^+]$ of the solution. The $[H^+]$ varies between 10^{-1} M (pH 1.0) and 10^{-12} M (pH 12.0) in each row. The K_a for an indicator is a constant for a particular indicator (at constant temperature):

$$HIn \rightleftharpoons H^+ + In^-$$

$$K_a = \frac{[H^+][In^-]}{[HIn]}$$

   Rearranging the K_a expression gives

$$\frac{K_a}{[H^+]} = \frac{[In^-]}{[HIn]}$$

   which shows that at different $[H^+]$'s, the $[In^-]/[HIn]$ ratio must vary. Since HIn and In^- are different colors, then at some $[H^+]$ the $[In^-]/[HIn]$ ratio will be 1.0, i.e.,

$$[In^-] \approx [HIn]$$

   and the color of the solution will be intermediate between the color of HIn and the color of In^-. Experimentally, this means a color change from the color of HIn to that of In^- in going from left to right in a row. Now, when $[HIn] \approx [In^-]$, then

$$\frac{K_a}{[H^+]} \approx 1.0$$

   and therefore $K_a \approx [H^+]$ at that place in the row.

6. Determine the K_a and pK_a values of the 4 weak acid indicators and report them in your notebook.

   - Why are these values only approximate?
   - How could they be measured more accurately using the same type of method?

   It is interesting to note that the human eye-brain system, with normal color vision, can easily discriminate a 100-fold change in the $[In^-]/[HIn]$ ratio. An indicator solution will appear to have the "pure" color of HIn at

$$\frac{[HIn]}{[In^-]} = \frac{10}{1}$$

Thus,

$$\frac{K_a}{[H^+]} = \frac{10}{1}$$

Similarly, the solution will be perceived to have the "pure" color of In^- at

$$\frac{[HIn]}{[In^-]} = \frac{1}{10}$$

Thus,

$$\frac{K_a}{[H^+]} = \frac{1}{10}$$

Thus, the pH transition range of an indicator from one "pure" color to another is

$$\text{pH transition range} = pK_a \pm 1.0$$

Table 14.1 lists some common acid-base indicators, along with pH transition ranges and colors in acidic and basic solutions. Using the table, calculate the pH transition range for the 4 weak acid indicators that you have investigated in this section.

Section D. pH Measurement with Indicator Color Probes

Goal: *To use a solution of a combination of indicators (a* universal *indicator) as a color probe for sensing [H⁺] and, therefore, pH.*

Discussion: The author and Peter Markow have tested over 5000 combinations of acid-base indicators and have selected a universal indicator that may be used to obtain an easily seen color change, over small intervals, from pH 1.0 to pH 12.0. You will be preparing a set of solutions of different pH and color in order to be able to use the set as a *colorimetric pH meter* for measuring the pH of various sample solutions.

Experimental Steps:

1. Clean a 1 x 12 well strip.
2. Use a microburet to deliver 1 drop of the Thompson-Markow universal indicator solution to each well.
3. Add 3 drops of pH 1.0 solution to the first well, 3 drops of pH 2.0 solution to the second well, and so on. Stir.
   - Describe the color of each pH value in your lab book.

   NOTE: You now have 12 solutions of known pH, each of which has a different color. This set of solutions can now be used to measure the pH of any unknown solution by a simple color comparison.
4. Measure the pH of 5 of the colorless solutions located in the 24-well tray, e.g., 0.05 M CH_3COOH, etc.

   NOTE: To measure the pH, drop 3 drops of the unknown solution into a well of another 1 x 12 well strip. Add 1 drop of universal indicator and stir.

Name	pH Transition Range	Acid Color	Basic Color
Malachite green	0 – 2.0	yellow	green
Cresol red	0.2 – 1.8	red	yellow
Metacresol purple	1.2 – 2.8	red	yellow
Thymol blue	1.2 – 2.8	red	yellow
Orange IV	1.3 – 3.2	red	yellow
Methyl yellow	2.9 – 4.0	red	yellow
Bromophenol blue	3.0 – 4.6	yellow	purple
Congo red	3.0 – 5.0	blue	red
Methyl Orange		red	yellow
Bromocresol green		yellow	blue
Methyl red	4.4 – 6.2	red	yellow
Chlorophenol red	4.8 – 6.4	yellow	purple
Litmus	4.5 – 8.3	red	blue
Paranitro phenol	5.0 – 7.0	colorless	yellow
Bromocresol purple	5.2 – 6.8	yellow	purple
Bromothymol blue		yellow	blue
Neutral red	6.8 – 8.0	bluish red	yellow
Phenol red	6.4 – 8.2	yellow	red
Cresol red	7.0 – 8.8	yellow	purple
Metacresol purple	7.4 – 9.0	yellow	purple
Curcumin	7.8 – 9.2	yellow	brown
Thymol blue	8.0 – 9.6	yellow	blue
Phenolphthalein		colorless	pink
Thymolphthalein	9.3 – 10.5	colorless	blue
Alizarin yellow R	10.0 – 12.1	yellow	brown-red
Curcumin	10.2 – 11.8	red	orange
Malachite green	11.5 – 14.0	blue	colorless
Clayton yellow	12.2 – 13.2	yellow	amber
Red cabbage extract	2.5 – 4.0	red	pale violet
Red cabbage extract	6.0 – 8.5	pale violet	blue
Red cabbage extract	9.0 – 10.5	blue	green
Red cabbage extract	10.5 – 14.0	green	yellow

Table 14.1 Some Common Acid-Base Indicators

D3 D6 D5 D4
HCl HNO_3 H_2SO_4 HCl
.05M 1.0M 1.0M 1.0M
pH 2.5 1.5 1.5 1.5

5. Place the pH meter strip and the second strip together and view against a white background. Slide the strips relative to each other and decide on a match.

   NOTE: If the color of the unknown lies between 2 wells of the pH meter strip, then try to interpolate the pH value.

   - Report the pH values in your notebook.

6. Keep the pH meter strip for the next series of experiments. Evaporation of the solutions may be stopped by sealing the strip with a piece of transparent sticky tape.

   NOTE: The accuracy of pH measurement by this method depends on your ability to discriminate color hue differences. If you have difficulty with color discrimination, consult your instructor.

Section E. The Study of Acid-Base Equilibria by Graphical Interpretation of Titration Data

Goal: *To investigate various quantitative aspects of acid-base chemistry by interpreting experimental titration curves and by comparing the data with computer-simulated titrations.*

Discussion: Titration curve data provide a simple, powerful basis for both experimental and theoretical analysis of reaction stoichiometry (the measurement of concentration of acid, base, and salt solutions); the determination of equilibrium constants of acids and bases; the determination of molar mass; the buffer concept; the nature of polyprotic acid systems; and the use of Henderson-Hasselbalch expressions.

Section E is subdivided into Parts 1 through 8. Each part is an investigation of a particular type of titration and the information that can be derived from graphical presentations of titration data. Recall from the Background Chemistry section of this chapter that an acid-base titration is the process of incremental addition of a solution of acid (or base) to a solution of base (or acid) such that the extent of the neutralization reaction can be monitored. Each point in the titration represents the state of the acid-base system after equilibrium has been reached. The reason is that acid-base reactions are very fast, and equilibrium is usually achieved in less than a millisecond — provided that the reaction mixture is stirred! The change in $[H^+]$ that occurs during the titration may be measured by color comparison with the colorimetric pH meter.

As you work through each of the experimental sections in this chapter, you will need to refer to Figures 14.2 through 14.18. All these figures are located in sequential order at the end of this chapter. Feel free to remove them from the book.

Section E. Part 1. Titration of a Strong Acid with a Strong Base

Experimental Steps:

1. Clean a 96-well tray (flat-bottomed wells).
2. Deliver 1 drop of universal indicator to each well of row A.

C4

.01 NaOH

	pH
1	2.5
2	3
3	3.4
4	3.7
5	3.9
6	4.2
7	10
8	10
9	10
10	10
11	10
12	10

NOTE: Use a clean large-drop microburet for all titrations. First, use good tranfer technique to rinse and half-fill the microburet with 0.05 M HCl.

3. Quantitatively deliver (using standard technique) 1 drop of 0.05 M HCl to each well of row A.

4. Wash, rinse, and half-fill the *same* microburet with 0.01 M NaOH.

5. Carry out a *serial* titration of the 0.05 M HCl with 0.01 M NaOH — i.e., add 1 drop NaOH to the first well, 2 drops NaOH to the second well, and so on. Stir.

6. Use the pH meter strip (from the last section) to measure the pH values for the solution in each well.

   7-12 - light purple no difference in color

   - Record the pH values.
   - On the graph paper provided in Figure 14.2, plot a graph of pH versus drops of added 0.01 M NaOH — the axes are already drawn for you. Draw a *smooth* curve through the data points using a sharp pencil.

   NOTE: You might want to look at Figure 14.3 to get an idea of the form of the curve. Figure 14.3 is a computer-simulated titration curve for the titration that you have just carried out experimentally. This type of graph is called a *titration curve.* The simulated curve was computer generated by solving the exact mathematical equations for the acid-base equilibria involved in this titration. The pK_W of water was given a value of 14.00. The dilution of the solutions by the added indicator was taken into account in computing the pH values in the titration. If you wish to know more about the computer program, consult your instructor.

   - Compare your graph with Figure 14.3.
   - What chemical reaction is taking place in the titration you just performed?
   - What is the net ionic equation for the reaction?

   Since all drops from the same microburet are the same volume, the addition of 1 drop of 0.01 M NaOH to 1 drop of 0.05 M HCl will neutralize 1/5th of the HCl. This occurs because of the stiochiometry of the reaction. All the HCl will have been neutralized by the addition of 5 drops of base. This point in the titration is called the *equivalence point.* The computer data show that the equivalence point occurs at exactly pH 7.00 and 5.00 drops of 0.01 M NaOH.

7. Look at your experimental titraton curve.

   NOTE: The equivalence point is not easy to locate on experimental curves. One way is to draw a vertical line at the 5 drop point. You can see that the difficulty arises because of the very large pH rise (i.e., approximately 8 pH units, a change in $[H^+]$ of 100,000,000!) that occurs close to the equivalence point. The very large pH rise near the equivalence point is not always seen in a strong acid–strong base titration.

8. Look at Figure 14.4.

Seven titration curves are shown for strong acid (HCl) with strong base (NaOH). For each curve the concentration of acid and base are the same, e.g., the curve labelled 1 M is the curve for the titration of 5 drops of 1 M HCl with 1 M NaOH.

- Why is the pH rise for the titration of 10^{-5} M HCl with 10^{-5} M NaOH very small?
- On Figure 14.5 (which is a copy of Figure 14.4), draw in the pH transition ranges for the folowing single indicators: methyl orange, bromocresol green, bromothymol blue, and phenolphthalein. Use the pK_a values that you calculated from the experimental data obtained in the earlier part of the laboratory.

  NOTE: Remember that the *range* is $pK_a \pm 1$!
- What indicator would be suitable for a titration of 0.001 M HNO_3 with 0.001 M KOH solution? Give reasons for your selection.

9. Figure 14.6 shows 4 titration curves of *monoprotic strong acids* titrated with 0.10 M NaOH.
   - As a homework assignment, calculate the *molarity* of each acid, given

   Curve A = 25.00 mL of acid of unknown concentration titrated with 0.10 M NaOH
   Curve B = 5.00 mL of acid of unknown concentration titrated with 0.10 M NaOH
   Curve C = 10.00 mL of acid of unknown concentration titrated with 0.10 M NaOH
   Curve D = 25.00 mL of acid of unknown concentration titrated with 0.10 M NaOH

pH
10

Section E. Part 2. Titration of a Weak Acid with a Strong Base

Experimental Steps:

pH
1 10
2 11
1 11
11 11
12 12

1. Deliver 1 drop of universal indicator to each well in row B of the 96-well tray. C1
2. Use good wash, rinse, and transfer technique to deliver 1 drop of 0.05 M CH_3COOH to each well of row B. C3
3. Carry out a serial titration of the acetic acid with 0.01 M NaOH. Stir.
4. Match colors with the pH meter strip and measure the solution pH values.
5. Plot the titration curve on the same graph paper (Figure 14.2) as the previous titration.
   - Draw a smooth titration curve. The computer-simulated curve is shown in Figure 14.7.
   - What are the chemical reaction and net ionic reaction for the neutralization reaction that takes place during the titration?

   NOTE: Data selected from the computer simulation show that the equivalence point pH is 8.31 and that it occurs after the addition of exactly 5 drops of 0.01 M NaOH.
6. Compare this pH value with that obtained at the equivalence point in the strong acid–strong base titration.

   The solution at the equivalence point in the acetic acid titration is a sodium acetate solution. Sodium acetate (CH_3COONa) is a salt that is completely dissociated in aqueous solution into acetate ions CH_3COO^- and sodium ions Na^+:

$$CH_3COONa \rightarrow CH_3COO^- + Na^+$$

The acetate ion is a weak base (i.e., it can accept a proton):

$$CH_3COO^- + H_2O \rightleftharpoons CH_3COOH + OH^-$$

Thus, the solution at the equivalence point is slightly basic (pH 8.31) because of the hydroxide ions OH^- produced in the last reaction.

Section E. Part 3. The Measurement of the K_a of a Weak Acid from the Titration Curve

Experimental Steps:

1. Look at Figure 14.7. The part of the curve before the equivalence point gives the pH of partially neutralized acetic acid solutions.

   - What is the limiting reagent?

   The solutions contain unreacted acetic acid and acetate ions from the neutralization reaction

$$CH_3COOH \rightleftharpoons CH_3COO^- + H^+$$

$$CH_3COONa \rightarrow CH_3COO^- + Na^+$$

$$H_2O \rightleftharpoons H^+ + OH^-$$

The K_a expression for acetic acid and K_w for water are

$$K_a = 1.75 \times 10^{-5} = \frac{[CH_3COO^-][H^+]}{[CH_3COOH]}$$

$$K_w = 1.0 \times 10^{-14} = [H^+][OH^-]$$

At a point in the titration halfway from the start to the equivalence point (at 2.5 drops NaOH), 1/2 of the CH_3COOH has been neutralized. Thus, if we neglect the $[H^+]$ from water, then

$$[CH_3COOH] \approx [CH_3COO^-]$$

Substituting this into the K_a expression,

$$K_a = \frac{[CH_3COO^-]\ [H^+]}{[CH_3COOH]}$$

$$K_a \approx [H^+]$$

Or $pK_a \approx pH$ *at the midpoint*!

- To find the pK_a of CH_3COOH from Figure 14.7, draw a vertical line from 2.5 drops to intersect the curve. From the intersection draw a horizontal line to the pH axis, and $pH = pK_a$.

- What is the K_a of CH_3COOH?

This K_a value is typical of the K_a values for many organic carboxylic acids that have the structure

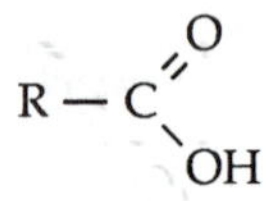

$$R-C\begin{matrix}{}^{\nearrow O} \\ {}_{\searrow OH}\end{matrix}$$

The hydrogen atom that dissociates is the H atom attached to the oxygen atom of the carboxylate group.

Section E. Part 4. The Henderson-Hasselbalch Equation

Discussion:

Again, consult Figure 14.7. A very useful relationship that links the pH of solutions in titrations of weak acids with pK_a is the *Henderson-Hasselbalch equation,*

$$pH = pK_a + \log_{10} \frac{[A^-]}{[HA]}$$

for a weak acid HA.

If the pK_a of the weak acid is known — e.g., the value obtained for CH_3COOH in Part 3 — then for

$$CH_3COOH \rightleftharpoons CH_3COO^- + H^+$$

$$K_a = \frac{[CH_3COO^-][H^+]}{[CH_3COOH]}$$

Taking $\log_{10}$ of the K_a expression

$$\log_{10} K_a = \log_{10}[H^+] + \log_{10}[CH_3COO^-] - \log_{10}[CH_3COOH]$$

and changing the sign throughout results in

$$-\log_{10} K_a = -\log_{10}[H^+] - \log_{10}[CH_3COO^-] + \log_{10}[CH_3COOH]$$

Since $pK_a = -\log_{10} K_a$ and $pH = -\log_{10}[H^+]$

then

$$pH = pK_a + \log_{10} \frac{[CH_3COO^-]}{[CH_3COOH]}$$

The last term in the above equation is simply the logarithm of the ratio of [CH_3COO^-], formed in the neutralization, to the unreacted [CH_3COOH]. You may show that this is correct by consulting Figure 14.7 and filling in the following table.

Drops of 0.01 M NaOH Added	$\frac{[CH_3COO^-]}{[CH_3COOH]}$	pH from Curve	pH Calculated from H-H Equation
1	1/4	4.1	4.95
2	1/2	4.6	5.2
3	3/4	4.9	5.45
4	1	5.3	5.7

NOTE: Use the pK_a value for CH_3COOH that you obtained in Part 3.

Section E. Part 5. Titration Curves and Buffer Solutions

Discussion:

Look at Figure 14.7, particularly the part of the curve before the equivalence point.

Note that the pH change that occurs on the addition of strong base NaOH is quite small. Partially neutralized solutions of weak acids or weak bases (see later) that resist change in pH upon the addition of small amounts of strong acid or strong base are called *buffer solutions.* The part of the titration curve where buffering occurs is called the *buffer region.* This buffering action occurs because the solutions contain relatively large concentrations of weak acid CH_3COOH and its salt, CH_3COONa (produced in the neutralization reaction). The addition of OH^- from the added strong base NaOH results in

$$CH_3COOH + OH^- \rightarrow CH_3COO^- + H_2O$$

and the OH^- is converted in water. The pH of the solution will change very little. The addition of H^+ from a strong acid to a buffer solution of this type results in

$$CH_3COO^- + H^+ \rightarrow CH_3COOH$$

and produces the *weak* acid CH_3COOH. The pH of the solution will change, but not by very much. In general, acidic buffer solutions are prepared by titrating appropriate weak acids with strong bases until the desired pH is obtained.

The slope of the titration curve in Figure 14.7 is a quantitative measure of the buffering ability, or *buffer capacity,* of the solution. Titration curves with small values of slope — i.e., "flat" regions — mean that those solutions have good buffer capacity. The steep part of the titration curve — i.e., close to the equivalence point — means that these solutions have small or no buffer capacity.

Section E. Part 6. The Dilution of Buffers

Experimental Steps:

1. Look at Figure 14.8. The 5 titration curves are for
   (a) 1.0 M CH_3COOH with 1.0 M NaOH
   (b) 0.1 M CH_3COOH with 0.1 M NaOH
   (c) 0.01 M CH_3COOH with 0.01 M NaOH
   (d) 0.001 M CH_3COOH with 0.001 M NaOH
   (e) 0.0001 M CH_3COOH with 0.0001 M NaOH
   - Draw a vertical line at 2 drops of added NaOH to intersect all 5 curves.
   - What is the value of the $\frac{[CH_3COO^-]}{[CH_3COOH]}$ ratio?
   - What are the pH values of the 5 solutions?

   NOTE: These data show that buffer solutions can be diluted and still maintain approximately the same pH value. However, the buffer will collapse with excessive dilution.

Section E. Part 7. The Selection of Weak Acids to Make Buffer Solutions of Desired pH

Experimental Steps:

1. Look at Figures 14.9 and 14.10 which show titration curves for a series of weak acids with different K_a values.

   NOTE: All the curves are for 0.1 M weak acid solutions titrated with 0.1 M strong base. The K_a values range from very large (strong acid) to 10^{-10} (very weak acid) as shown on the figures. The buffer regions with good buffer capacity are in the range $pK_a \pm 1$.
   - Draw 2 vertical lines on Figure 14.9 to show the regions of good buffer capacity.

   NOTE: One interesting feature of the titration curves in Figures 14.9 and 14.10 is that all the curves give a small-slope, good buffer region *after* the equivalence point. All the solutions contain the salt of the weak acid and excess of strong base NaOH. Solutions that contain strong bases (or strong acids — e.g., see titration curves on Figure 14.4) are also regarded as buffers. The buffering action arises as a consequence of the high concentration of water in aqueous solutions (~ 55.6 M H_2O) and the nature of the pH scale. These buffers are called *pseudobuffers*.

Section E. Part 8. The Titration of Weak Bases with Strong Acids

Experimental Steps:

1. Deliver 1 drop of universal indicator into each well of row C of the 96-well tray. Retain the serial titrations in rows A and B for comparison.

2. Carry out a serial titration of the weak base 0.05 M NH_3 with 0.01 M HCl. The diluted HCl can be made by careful quantitative dilution of 0.05 M HCl. Stir.

3. Measure the pH values of the solutions.
   - Plot the titration curve on Figure 14.2.

4. Compare with the computer-simulated curve (Figure 14.11).

   NOTE: Data from the computer calculation give an equivalence point pH of 5.70 at 5.0 drops of added 0.01 M HCl.

5. Look at Figure 14.11.
   - Determine the K_b and pK_b of ammonia and the K_a and pK_a of the ammonium ion.
   - What is the chemical reaction in this titration?

6. Look at Figure 14.12. The titration curves are for the following titrations:

   (a) 1.0 M NH_3 titrated with 1.0 M HCl

   (b) 0.1 M NH_3 titrated with 0.1 M HCl

   (c) 0.01 M NH_3 titrated with 0.01 M HCl

   (d) 0.001 M NH_3 titrated with 0.001 M HCl

   (e) 0.0001 M NH_3 titrated with 0.0001 M HCl

   - Draw in lines indicating buffer regions.
   - What is the effect of dilution on the buffer solutions?

   NOTE: Figures 14.13 and 14.14 show the effect of K_b of the weak base on titration curves. All the titraton curves are for 0.1 M base with 0.1 M monoprotic strong acid.

7. Select an indicator that would work well for a titration of 0.1 M weak base with a K_b of 1.0×10^{-6} with an 0.1 M HCl solution.
   - Are titrations of bases with K_b values of less than 10^{-8} feasible in aqueous solutions? Give reasons for your answer.
   - Suppose you wish to make a buffer solution of pH 8.2 for a clinical physiological experiment. Describe how you would do this.

     HINT: Look at Figure 14.13.

Section F. The Titration of Polyprotic Acids

Goal: *To apply the principles of graphical interpretation of titration data to investigate polyprotic acid equilibria.*

Experimental Steps:

1. Deliver 1 drop of universal indicator to each well of row D of the 96-well tray.

2. Use a large-drop microburet to deliver 1 drop of 0.04 M H_3PO_4 (phosphoric acid) to each well in row D.

3. Carry out a serial titration of the phosphoric acid with 0.01 M NaOH. Stir.

   - Determine the solution pH values and plot the titration curve on the same graph as the computer-simulated titration curve (Figure 14.15).

4. Look at Figure 14.15.

   There appear to be only 2 equivalence points, even though phosphoric acid H_3PO_4 is a *triprotic acid*! The pH at the 2 equivalence points can be found from the computer data. The first equivalence point pH is 4.83 and occurs after 4.0 drops of NaOH have been added. The second equivalence point pH is 9.36 after a total of 8.0 added drops of NaOH.

5. On the computer-simulated titration curve in Figure 14.15, label the start of the titration as point A, the first equivalence point C, and the second equivalence point D.

   At point A, no NaOH has been added, and the solution contains phosphoric acid (H_3PO_4) and dihydrogen phosphate ions $H_2PO_4^-$ from

$$H_3PO_4 \rightleftharpoons H^+ + H_2PO_4^-$$

   The first dissociation constant K_1 of H_3PO_4 has a value of 7.02×10^{-3}:

$$K_1 = \frac{[H^+][H_2PO_4]}{[H_3PO_4]} = 7.02 \times 10^{-3}$$

   which shows that H_3PO_4 is a borderline weak/strong acid. Thus, the first dissociation is significant, and the pH of the solution is quite low (~ 2). Now as NaOH is added in the titration, some of the H_3PO_4 is neutralized according to

$$H_3PO_4 + NaOH \rightarrow NaH_2PO_4 + H_2O$$

   - Which ionic species are present in solution at any point between A and C (excluding indicator ions)?

   At point C, all the H_3PO_4 has been titrated, and the solution contains only NaH_2PO_4 (completely dissociated, of course, because it is a salt).

   - What indicator would be *most* suitable for the location of the first equivalence point?

   As the titration proceeds beyond C, the weak acid $H_2PO_4^-$ is neutralized:

$$NaH_2PO_4 + NaOH \rightarrow Na_2HPO_4 + H_2O$$

   and NaOH is the limiting reagent.

Chapter
15

Redox Equilibria and Electrochemistry

"It is better to wear out than to rust out."

Bishop Richard Cumberland
1631-1718

Introduction

Life on this planet derives from the photosynthetic process by which plant cells trap and use solar energy. The deceptively simple photosynthetic equation

$$H_2O + CO_2 \xrightarrow{\text{light}} \text{carbohydrates} + O_2$$

occurs via a complex cascade of biochemical reactions. The critical events are a light-activated transfer of an electron to an acceptor in the chloroplast, a flow of electrons from water through an electron transport chain, and the resulting formation of carbohydrates from carbon dioxide. The key reactions that involve electron transfer are called redox reactions. In evolutionary terms the change from anaerobic life to aerobic (oxygen-using) life was an important step because it allowed a rich energy source to be used. The redox reactions of oxygen with glucose, mediated by the transport proteins myoglobin and hemoglobin, provide an efficient way of generating the energy for life processes.

Most of the energy used by our industrial society comes from the combustion of fossil fuels, such as gas, coal, and oil. Fossil fuels are the end products of millions of years of sun-trapping reaction by plants. The burning of fossil fuels is a redox process by which electrons are transferred from dioxygen to water and carbon dioxide. As you can see, the common thread in the production of useful energy by plant and animal life involves reactions that involve the gain and loss of electrons.

The study of electron transfer processes in natural and synthetic systems has became an extremely important subfield of chemistry and is called electrochemistry. The field of electrochemistry is broadening rapidly as research reveals the incredibly varied part played by redox reactions in such diverse disciplines as neurophysiology, meteorology, and geochemistry. Indeed, it is worth quoting an axiom of electrochemists — "There are only two kinds of reactions: redox and the rest!" In this series of experiments, you have the opportunity to explore the fundamentals of solution electrochemistry. The experiments have been designed such that the basic principles of redox equilibria can be investigated. You may then apply the principles to your own field of endeavor.

Background Chemistry

Chemical reactions that involve the movement of electrons between chemical species are called *redox reactions* (an abbreviation of reduction-oxidation reaction). Redox reactions can be identified, and the movement of electrons can be followed, by using a sort of formal "bookkeeping" of the number of electrons associated with atoms. This accounting of electrons involves assigning arbitrary *oxidation numbers* to individual atoms within a chemical species. Redox reactions are reactions in which there are changes in oxidation numbers. Every redox reaction can be divided into two *half-reactions*: one that involves a gain of electrons and one that involves a loss of electrons. The gain of electrons is called *reduction* and the loss of electrons is called *oxidation*.

Consider the following interesting chemical reaction in which a copper coin is dropped into a dilute solution of silver nitrate ($AgNO_3$). Beautiful, shining needles of metallic silver crystals grow slowly from the copper surface, especially if the solution is not disturbed. The solution also slowly changes from colorless to a pale blue color, indicating that Cu^{2+} ions are being produced. The overall chemical equation is

$$Cu_{(s)} + 2AgNO_3 \rightleftharpoons Cu(NO_3)_2 + 2Ag_{(s)}$$

for which the net ionic equation is

$$Cu_{(s)} + 2Ag^+ \rightleftharpoons Cu^{2+} + 2Ag_{(s)}$$

This reaction is a redox reaction because the oxidation number of copper is changing from 0 to +2 ($Cu_{(s)}$ to Cu^{2+}) and that of silver is changing from +1 to 0 (Ag^+ to $Ag_{(s)}$). The NO_3^- ion remains unchanged during the reaction and is called a *spectator ion*. The reaction can be divided into two half-reactions:

$$\text{Reduction:} \quad 2Ag^+ + 2e^- \rightleftharpoons 2Ag_{(s)}$$

$$\text{Oxidation:} \quad Cu_{(s)} \rightleftharpoons Cu^{2+} + 2e^-$$

which add together to give the redox reaction. Redox reactions in which the reactants are observed to react and produce products — i.e., proceed from left to right — are called *spontaneous* reactions. The reverse reaction,

$$Cu^{2+} + 2Ag_{(s)} \rightleftharpoons Cu_{(s)} + 2Ag^+$$

will not occur and is said to be a nonspontaneous reaction. The energy given out in the spontaneous reaction in which the copper coin reduced Ag^+ to silver crystals is lost as heat to the solution. The redox reaction will eventually come to equilibrium, and no more heat will be given out.

Redox reactions in which electrons are completely lost by one species and completely accepted by another are very useful because the two half-reactions can often be physically separated. The electrons that are transferred may then be allowed to flow through external wires in a circuit and be made to do useful work. *Electrochemistry* is the study of redox reactions that either produce or utilize electrical energy (moving electrons and/or ions) in devices called electrochemical cells.

In the redox reaction

$$Cu_{(s)} + 2Ag^{+} \rightleftharpoons Cu^{2+} + 2Ag_{(s)}$$

the two half-reactions can actually be separated by placing the reactants in different compartments, partitioned by some type of porous medium that prevents mixing, but not ion flow. The compartments, called *half-cells*, each contain a metal electrode in contact with its own metal ion, as shown in Figure 15.1.

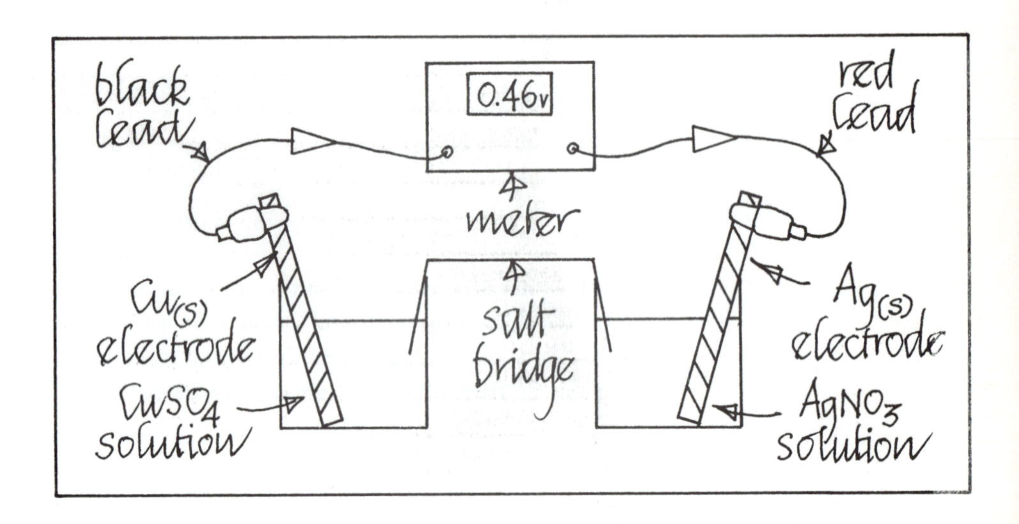

Figure 15.1 An Electrochemical Cell Made up of Two Half-Cells

An external connection between the two electrodes completes the circuit, and electrons will flow from the copper electrode through the external wire and meter and into the silver electrode. The copper electrode will dissolve, forming Cu^{2+} ions in solution, and Ag^{+} ions will pick up electrons at the surface of the silver electrode and be deposited as silver atoms. The electrode at which oxidation takes place (the copper electrode) is called the *anode*, and the electrode at which reduction takes place (the silver electrode) is called the *cathode*. The combination of the two half-cells is called an *electrochemical cell*.

In the electrochemical cell under discussion, it is a fact that oxidation,

$$Cu_{(s)} \rightleftharpoons Cu^{2+} + 2e^-$$

occurs at the copper electrode (anode) and that reduction,

$$2Ag^+ + 2e^- \rightleftharpoons 2Ag_{(s)}$$

occurs at the silver electrode (cathode). The relative tendency of a particular species to give up or accept electrons is manifested as a measurable electrical force, or voltage (*potential*), between the two electrodes. This force may be considered as being the sum of two potentials called *half-cell potentials* or *single-electrode potentials*.

The tendency of a species to give up or accept electrons can only be compared relative to another species. In order to obtain consistent electrochemical data, it is necessary to compare all single electrodes to a standard reference electrode. The universal reference electrode, chosen by international agreement, is the *standard hydrogen electrode* (SHE), which is shown in the diagram in Figure 15.2.

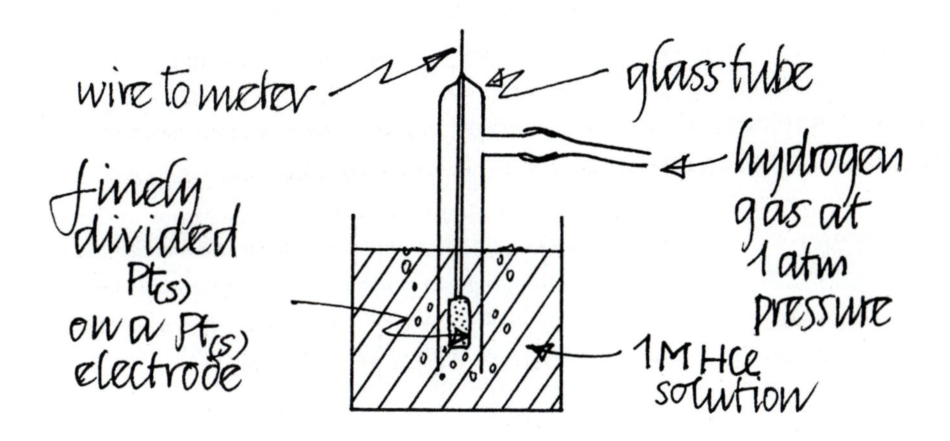

Figure 15.2 A Standard Hydrogen Electrode

The half-reaction at the SHE,

$$2H^+ + 2e^- \rightleftharpoons H_{2(g)}$$

is arbitrarily written, for the moment, as a reduction.

An arbitrary assignment of *zero* electrode potential (0.00 V) is given to the SHE, and all other electrode potentials are referred to it. It is now customary to report *single* electrode potentials in tables, and it must be remembered that these single half-cell potentials are really measured in combination with a SHE at 0.00 V.

The single-electrode potential value is dependent on the concentration of the ion surrounding the electrode and on the temperature. Standard conditions of 1 M concentration and 298 K (25 °C) have been chosen, and by international agreement all standard electrode potentials are reported as *standard reduction potentials* (E°). Some examples follow.

$$Al^{3+} + 3e^- \rightleftharpoons Al_{(s)} \qquad E^\circ = -1.66\ V$$

$$Cu^{2+} + 2e^- \rightleftharpoons Cu_{(s)} \qquad E^\circ = 0.34\ V$$

$$2H^+ + 2e^- \rightleftharpoons H_{2(g)} \qquad E^\circ = 0.00\ V$$

$$Ag^+ + e^- \rightleftharpoons Ag_{(s)} \qquad E^\circ = 0.80\ V$$

A useful way of thinking about these E° values is to remember that the more *positive* the E° value, the more that reaction goes to the right. The E° value for oxidation reactions is obtained simply by changing the *sign* of the appropriate reduction reaction, e.g.,

$$Cu_{(s)} \rightleftharpoons Cu^{2+} + 2e^- \qquad E^\circ = -0.34\ V$$

One very important practical consideration in the measurement of cell potentials is that the cell reaction must be carried out under standard conditions. A simple wire connection between the two electrodes would allow the electrons to flow and the redox reaction to go to completion. In the process the concentrations of ions in each half-cell would change dramatically, and the cell voltage would drop to zero, its equilibrium value. Cell potential measurements are, therefore, usually made with instruments that have very high resistance in order to minimize the flow of electrons during the measurement.

Another important consideration is the electrical connection that must be made between the two half-cell solutions before the cell voltage measurement can be made. The connection is called a *salt bridge.* The salt bridge allows electrical neutrality to be maintained in each half-cell. In the voltage measurement, a few electrons must flow from the anode, through the meter, and into the cathode. Cations are generated in the anode half-cell solution, and to maintain a charge balance, anions flow from the salt bridge. Cations are consumed in the cathode half-cell, and to maintain charge balance, cations flow from the salt bridge. The net result is simply a flow of inert electrolyte from the salt bridge into the cell.

Standard reduction potential values refer, of course, to voltages measured under standard conditions. In the case of nonstandard conditions, the reduction potential value will be different from those reported at 1 M and 25 °C. The quantitative relationship between cell potential and concentration (and temperature) is called the *Nernst law*:

$$E = E° - \frac{RT}{nF} \ln Q$$

where E° is the standard potential, Q is the reaction quotient, F is the Faraday constant, n is the number of moles of electrons transferred in the cell reaction, R is the gas constant, and T is the absolute temperature. The Nernst law may be rewritten, after substituting for all constants and putting T = 298 K, as

$$E = E° - \frac{0.0592}{n} \log_{10} Q$$

The Nernst law is valid both for the potentials of half-reactions and for overall redox reactions. For a reaction that has come to equilibrium, the cell voltage must be *zero*, i.e.,

$$E = 0 \text{ and } Q = K$$

where K is the equilibrium constant for the reaction. At equilibrium the Nernst law becomes

$$E = 0 = E° - \frac{0.0592}{n} \log_{10} K$$

and, rearranging,

$$\log_{10} K = \frac{nE°}{0.0592}$$

The value of the equilibrium constant for a redox reaction (at 25 °C) can be calculated using the above relationship.

An electrochemical cell in which a spontaneous redox reaction can occur is called a *voltaic cell*, or battery. Voltaic cells are useful energy storage devices in which a suitable redox system is packed into an appropriate container. The voltage produced by a voltaic cell is, of course, dependent upon the particular redox reaction used, the concentration of materials, and the design of the package. A battery in use is a device in which chemical energy is converted into flowing electrons in a redox reaction that eventually winds down to an equilibrium state. At equilibrium, the battery is dead and the cell voltage is zero.

An electrochemical cell in which a nonspontaneous redox reaction is made to occur by means of an external power source is called an *electrolysis cell*. Automobile batteries that are being recharged are examples of electrolysis cells. When a run-down, lead-acid automobile battery is recharged, an outside voltage source pushes the nonspontaneous reaction to completion. Electrolysis cells are used extensively in the industrial production of a wide variety of useful chemicals such as chlorine, sodium hydroxide, aluminum, and in metal finishing and plating.

Unfortunately, spontaneous electrochemistry is not always a useful process. The corrosion of iron and steel is estimated to cost over 80 billion dollars a year in the United States. We see the results in automobiles and other consumer products, as well as in bridges, buildings, and in storage tanks. Many of the typical results of corrosion are best explained in terms of electrochemical mechanisms. Iron, which shows evidence of corrosion, is often found to have anodic areas at which

$$Fe_{(s)} \rightleftharpoons Fe^{2+} + 2e^-$$

and cathodic areas at which atmospheric dioxygen is reduced to OH^- ions:

$$O_{2(g)} + 2H_2O + 4e^- \rightleftharpoons 4OH^-$$

Iron (II) hydroxide ($Fe(OH)_2$) precipitates as an insoluble solid and is subsequently oxidized to a loose, flaky deposit of hydrated iron oxide called rust. The rusting process is spontaneous under a wide variety of conditions. The major factors appear to be the presence of dioxygen, water, microimpurities, and dissimilar metals in contact with each other.

Laboratory Experiments

Flowchart of the Experiments

Section A. Redox Reaction Investigations

Section B. A Small-Scale Electrochemical Cell

Section C. An Electrochemical Series from Cell Data

Section D. Electrographic Analysis of Metals

Section E. Nernst's Law and Potentiometric Redox Titrations

Part 1. The Ce^{4+} - Fe^{2+} Titration
Part 2. The Fe^{2+} - $Cr_2O_7^{2-}$ Titration

Section F. Lead-Acid Automobile Battery

Requires one three-hour class period to complete

The overall goal in this series of experiments in electrochemistry is to be able to identify, write, balance, and apply common reduction-oxidation (electron transfer) reactions. You should also be able to construct a variety of electrochemical cells, both voltaic and electrolytic, and be able to use standard electrochemical instrumentation and conventions to measure and calculate standard reduction potentials. Finally, you should be able to apply the Nernst law to potentiometric titrations with microelectrode sensing of redox reaction potentials.

CAUTION: Solutions of many transition metal ions, e.g., Pb^{2+}, Cu^{2+}, and Sn^{2+}, are toxic. Even though you will be using only very small amounts of these solutions, it is important to exercise caution when using these chemicals. It is also important to observe the hazardous waste regulations in this experiment. All transition metal ion solutions and waste will be collected and disposed of by the instructors and technical staff. Acids and bases, e.g., H_2SO_4 and NH_3, are also corrosive and toxic. Please use these solutions with caution. If you get any of these chemicals on your hands, wash well with cold water and report to your instructor.

Section A. Redox Reaction Investigations

Goals:

(1) To study a redox reaction involving copper and zinc species. (2) To describe and characterize some of the electrochemical properties of the reaction.

Experimental Steps:

1. Clean your plastic reaction surface.
2. Obtain 4 pieces each of copper metal ($Cu_{(s)}$) and zinc metal ($Zn_{(s)}$) from Reagent Central. Use tweezers to transfer the metal pieces to the surface.
3. Drop 2 drops of 1 M $CuSO_4$ solution onto the *top* of a piece of $Zn_{(s)}$.
4. Drop 2 drops of 1 M $ZnSO_4$ solution onto the *top* of a piece of $Cu_{(s)}$.
5. Study each of the solution-metal interfaces with your hand lens.
   - Describe and record what you see.

NOTE: The discoloration of the $Zn_{(s)}$ is due to the formation of finely divided $Cu_{(s)}$ at the zinc metal surface. The chemical reaction is a *spontaneous* reaction in which an electron transfer is taking place, and it is called a reduction-oxidation, or *redox*, reaction:

$$Cu^{2+} + Zn_{(s)} \rightleftharpoons Cu_{(s)} + Zn^{2+}$$

- What half-reactions (one a *reduction*, the other an *oxidation*) can be written for this redox reaction?

$Cu^{2+} \rightarrow Cu$
$Zn \rightarrow Zn^{2+}$

- How many electrons are being transferred from $Zn_{(s)}$ to Cu^{2+}?
- Which is the reducing *agent* and which is the oxidizing *agent* in this titration?
- Is this redox reaction a homogeneous or a heterogeneous reaction?
- Why does the $Cu_{(s)}$ produced in the redox reaction look very different from the copper metal piece?
- Do you think that the redox reaction has a large or small equilibrium constant?

6. Design a simple experiment to prove that Cu^{2+} disappears in the reaction and is replaced by Zn^{2+} ions in solution.
7. Look at the piece of copper metal with $ZnSO_4$ on it.
   - Describe the surface of the metal.

   NOTE: A good way to do this is to remove the $ZnSO_4$ drops with a cotton swab and compare the copper surface with another copper piece.

   The nonspontaneous reaction, i.e., the redox reaction that did *not* occur, is

$$Cu_{(s)} + Zn^{2+} \rightleftharpoons Cu^{2+} + Zn_{(s)}$$

   - Does this reaction have a small or a large equilibrium constant?
8. Look at the spontaneous reaction again.
   - What are the *oxidation numbers* for $Cu_{(s)}$, $Zn_{(s)}$, Cu^{2+}, and Zn^{2+} and for sulfur and oxygen in $SO_4{}^{2-}$?
9. Leave the metal pieces on the plastic surface because you will be using them in the next section.

Section B. A Small-Scale Electrochemical Cell

Goals: *(1) To construct a small-scale electrochemical cell using the redox system investigated in Section A. (2) To learn how to use a multimeter to measure electrochemical cell potentials and be able to use the appropriate sign conventions to calculate standard reduction potentials from cell potentials.*

Experimental Steps:

1. Obtain a multimeter from your instructor. Set up the meter to measure DC voltage. Use the two needle probes (red and black).

   NOTE: If you don't know how to do this, check with your instructor.
2. Place a small rectangular piece of filter paper (already cut for you) onto the plastic surface.
3. Place a piece of $Cu_{(s)}$ and a piece of $Zn_{(s)}$ onto the paper about 3 cm apart.

The diagram below illustrates the locations of the solutions delivered in Steps 4–6.

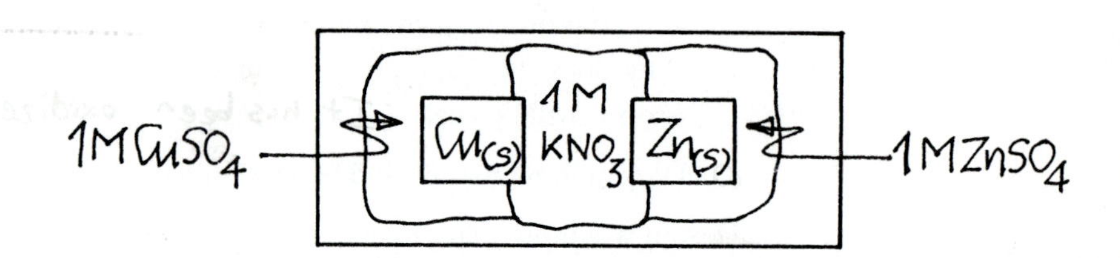

4. Deliver 2 drops of 1 M $CuSO_4$ to the paper just at the edge of the copper metal so that the solution soaks under the metal piece.
5. Deliver 2 drops of 1 M $ZnSO_4$ to the paper so that it soaks under the zinc metal piece.
6. Deliver 2 drops 1 M KNO_3 to the dry paper in between the wet circles.

   NOTE: The KNO_3 solution will spread out and run into the part of the paper wet with $CuSO_4$ and $ZnSO_4$.

7. Switch the multimeter on and touch the *red* probe to the piece of copper metal and the *black* probe to the piece of zinc metal.
   - Read and record the voltage. Don't forget the sign!
8. Reverse the probes.
   - Record what happens to the sign and switch the meter off.

The voltage (potential) that you have just recorded is the electrochemical cell voltage of a cell, which is made up of two half-cells joined electrically by wires (the probes) and the salt bridge solution (KNO_3). The sign conventions in these types of cell voltage measurements are sometimes confusing and are not easy to understand. The multimeter has two probes: red and black. The red is the positive terminal; the black is the negative terminal. The probes are touched to the electrodes in the cell, and the measured voltage will have a *positive sign if the black probe is on the anode and the red probe is on the cathode.* If the sign of the measured voltage is negative, then the reverse of the above is true: The black probe is on the cathode, and the red probe is on the anode.

REMEMBER: The anode is the electrode at which oxidation occurs. The cathode is the electrode at which reduction occurs.

The voltage read in Step 7 should be positive. Since the black probe is on $Zn_{(s)}$, this must be the anode.

- Write the half-reaction occurring at $Zn_{(s)}$.
- Write the half-reaction occurring at the cathode ($Cu_{(s)}$).

The standard reduction potential E° for

$$Cu^{2+} + 2e^- \rightleftharpoons Cu_{(s)}$$

is 0.34 V (measured against a standard hydrogen electrode).

This value can now be used to calculate the standard reduction potential for

$$Zn^{2+} + 2e^- \rightleftharpoons Zn_{(s)}$$

At the anode the half-reaction is

$$Zn_{(s)} \rightleftharpoons Zn^{2+} + 2e^- , \quad E^\circ_{ox}$$

and at the cathode the half-reaction is

$$Cu^{2+} + 2e^- \rightleftharpoons Cu_{(s)} , \quad E^\circ_{red} = 0.34 \text{ V}$$

The sum of these two half-reactions is the cell reaction

$$Cu^{2+} + Zn_{(s)} \rightleftharpoons Cu_{(s)} + Zn^{2+} , \quad E^\circ_{cell} = E^\circ_{ox} + E^\circ_{red}$$

- Calculate E°_{ox}, the standard oxidation potential for

$$Zn_{(s)} \rightleftharpoons Zn^{2+} + 2e^-$$

and then reverse the sign to obtain the standard reduction potential for

$$Zn^{2+} + 2e^- \rightleftharpoons Zn_{(s)}$$

- You might want to practice your skills by writing the half-reactions for the case of the measured voltage from Step 8.
- Calculate the equilibrium constant for the spontaneous redox reaction

$$Cu^{2+} + Zn_{(s)} \rightleftharpoons Cu_{(s)} + Zn^{2+}$$

by using the relationship

$$\log_{10} K = \frac{nE^\circ_{cell}}{0.0592} \quad \text{(at 25°C)}$$

where K is the equilibrium constant, n is the number of electrons transferred in the redox reaction, and E°_{cell} is the standard cell potential.

NOTE: Equilibrium constants for redox reactions cannot be found in the usual reference literature (as can equilibrium constant for acid-base reactions, etc.). Redox reaction equilibrium constants are generally calculated from standard reduction potential data.

Section C. An Electrochemical Series from Cell Data

Goals:

(1) To construct a series of electrochemical cells. (2) To obtain cell potentials for the calculation of standard reduction potentials.

Discussion:

The standard potential data obtained in this section may then be used to interpret corrosion phenomena, design voltaic and electrolytic cells, and arrange reduction reactions into an electrochemical series.

Experimental Steps:

1. Obtain a 9 cm filter paper and make a cell template similar to the template shown in the diagram below.

2. Write on each sector the atomic symbols of the metals, as shown.
3. Place the cutout paper into the lid of a clean, dry plastic petri dish.
4. Use your tweezers to transfer the appropriate metal pieces (e.g., $Ag_{(s)}$ wire, $Cu_{(s)}$ squares, etc.) to each sector.

   NOTE: Make sure they are arranged in the order shown.
5. Drop 2 drops of the appropriate metal ion solutions onto the paper at the edge of each metal piece so that each metal is contacting the solution (as in Section B).
6. Drop 1 M KNO_3 salt bridge solution into the middle so that it soaks outwards and contacts all the other wet areas. Try not to use too much!
7. Switch the multimeter to measure DC voltage and start making measurements with the red probe on $Ag_{(s)}$ and the black on $Cu_{(s)}$. Don't forget the sign!

.388

8. Now keep the red probe on $Ag_{(s)}$ and make measurements by moving the black probe around the circle in a clockwise manner.

   - Record voltages and signs.

9. Move the red probe to $Cu_{(s)}$ and continue around with the black probe. When you have finished, switch the meter off.

   NOTE: The measurements made on $Mg_{(s)}$ and $Al_{(s)}$ will be *very* unstable. Just try to obtain an approximate voltage.

   - Make a table of the cells and their cell voltages in your notebook.
   - Use the principles that you learned in Section B to calculate the standard reduction potentials for each half-cell, given that the E° for

     $$Cu^{2+} + 2e^- \rightleftharpoons Cu_{(s)}$$

     is 0.34 V.

   Note that E° for a half-reaction is not dependent on the coefficients, provided, of course, that the reaction is balanced, i.e., E° for

   $$Ag^+ + e^- \rightleftharpoons Ag_{(s)}$$

   is the same as E° for

   $$2Ag^+ + 2e^- \rightleftharpoons 2Ag_{(s)}$$

10. Arrange the standard reduction potentials and corresponding half-reactions in order, starting with the most negative at the top of the table and ending with the most positive at the bottom.

    - Which is the strongest oxidizing agent in the table?
    - Which is the strongest reducing agent in the table?

    The literature value of E° for the half-reaction,

    $$Fe^{2+} + 2e^- \rightleftharpoons Fe_{(s)}$$

    is −0.44 V.

    - Compare your value for steel.
    - Why are they so different?
    - If you wanted to design a battery from relatively common materials that produced a voltage of about 1 V, which cell reaction would you choose?
    - Why were the measurements on $Mg_{(s)}$ and $Al_{(s)}$ so difficult?

11. Clean up by placing the filter paper and metal pieces in the appropriate waste containers.

Section D. Electrographic Analysis of Metals

Goal:

To carry out a chemical analysis of an unknown metal sample by making the sample the anode of an electrolytic cell.

Experimental Steps:

1. Obtain from Reagent Central the following items: a petri dish, a piece of aluminum foil and a piece of filter paper (both already cut for you), a 9 V battery, and leads with clips.
2. Place the foil into the petri dish.
3. Place the filter paper onto the foil and add 2 or 3 drops of 1 M KNO_3 to the paper.
4. Place a clean nickel (the coin!) onto the paper.
5. Clip an electrical lead to the foil and clip the other end of it to the negative terminal of the battery.
6. Clip the other lead to the positive terminal of the battery and then touch the other end to the coin for no longer than 3 seconds.
7. Remove the coin, lift up the paper, and place it on a small piece of microtowel. Drop 1 or 2 drops of dimethylglyoxime reagent onto the paper.
   - Observe the result with your hand lens.

   NOTE: You may carry out similar experiments with other kinds of scrap metals. You must be able to carry out a qualitative test for the metal ion produced in the electrolytic cell. A good test for iron is to add 2 drops of H_2O_2 followed by 1 drop of NH_4CNS. A red-brown coloration indicates Fe^{3+}.

   - Which metal is the cathode and which is the anode in this experiment?
   - What half-reaction is occurring at the coin?
   - Do you think that the electrolysis could be carried out with a standard 1.5 V battery?
   - Whose picture is on a nickel?

Section E. Nernst's Law and Potentiometric Redox Titrations

Goals:

(1) To be able to use potentiometric techniques to analyze the progress of redox titrations. (2) To graphically interpret the titration data using Nernst's Law.

Section E. Part 1. The Ce^{4+} – Fe^{2+} Titration

Experimental Steps:

1. Clean a 1 x 12 well strip. Shake to remove any water.
2. Using a clean 24-well tray, obtain a small volume of 0.06 M Fe^{2+} solution in one well and, in another well, a small volume of 0.01 M Ce^{4+} solution.

3. Use good wash, rinse, fill, and transfer technique to fill large-drop microburets with Fe^{2+} and Ce^{4+} solutions.

4. Deliver 1 drop of Fe^{2+} to each well of the 1 x 12 strip.

5. Carry out a serial titration of the Fe^{2+} by adding 1 drop of Ce^{4+} to the first well, 2 drops Ce^{4+} to the second well, and so on. Stir.

6. Clip a piece of pencil lead into the red spring-clip probe of the multimeter and a piece of copper wire into the black spring-clip probe. Switch the multimeter to read DC voltage.

   NOTE: The spring clips slip onto the needle probes. The pencil lead is the sensing microelectrode, and the copper wire is the reference microelectrode.

7. Hold a spring clip with electrode in each hand and dip the electrodes into the solution in the first well. Record the voltage. Do not touch the electrodes together.

8. Dip the electrodes in each of the wells for the titration.
   - Record the voltages.

9. Switch the meter off.

10. Now add 1 drop of ferroin indicator to each of the wells and stir.
    - Record the number of drops of Ce^{4+} required to produce an end point color change.
    - Plot a graph of solution voltage (vertical axis) versus number of drops of Ce^{4+}. Draw the best smooth line through the points. (The graph is called a *potentiometric titration curve.*)

    NOTE: The redox reaction occurring during the titration is

$$Fe^{2+} + Ce^{4+} \rightleftharpoons Fe^{3+} + Ce^{3+}$$

The Ce^{4+} (cerium(IV) ion) is the oxidizing agent that oxidizes Fe^{2+} to Fe^{3+}. The equilibrium constant for the redox reaction may be calculated from the standard reduction potentials:

$$Fe^{3+} + e^{-} \rightleftharpoons Fe^{2+} \qquad E° = 0.77\ V$$

$$Ce^{4+} + e^{-} \rightleftharpoons Ce^{3+} \qquad E° = 1.44\ V$$

At each point in the titration, equilibrium is achieved because the redox reaction is fast — this titration is analogous to the acid-base titrations you studied earlier. The concentrations of the various species *are* changing during the titration. At the beginning the $[Fe^{2+}]$ is 0.06 M, and at the equivalence point, it is very small because it has all reacted with the added Ce^{4+}. The nonstandard concentrations (they are not 1 M!)

require that the Nernst law be used to see how the voltage of the titration solution changes. The Nernst law is

$$E = E° - \frac{0.0592}{n} \log_{10}Q$$

where E is the potential (voltage) at some concentration other than standard, E° is the standard potential, 0.0592 is a constant (it equals 2.303 RT/F when T = 298 K), n is the number of electrons transferred, and Q is the reaction quotient. Q is simply the concentration of each product to the power of the coefficient of each reactant to the appropriate power multiplied together.

For the redox titration reaction in this experiment,

$$Fe^{2+} + Ce^{4+} \rightleftharpoons Fe^{3+} + Ce^{3+}$$

The Nernst law is

$$E = E° - \frac{0.0592}{n} \log_{10}Q \quad \frac{[Fe^{3+}][Ce^{3+}]}{[Fe^{2+}][Ce^{4+}]}$$

and E° = 1.44 V − 0.77 V for the reaction. As the titration proceeds and the concentrations of Fe^{2+}, Ce^{4+}, Fe^{3+}, and Ce^{4+} change, the E for the solution is measured (sensed) by the pencil lead (graphite) electrode in reference to the copper wire electrode. The reference electrode potential stays constant during the titration.

Section E. Part 2. The Fe^{2+} – $Cr_2O_7^{2-}$ Titration

Experimental Steps:

A potentiometric redox titration of Fe^{2+} with $Cr_2O_7^{2-}$ (dichromate ion) may be carried out in a manner exactly analogous to the Ce^{4+} – Fe^{2+} titration described in Part 1.

1. Serially titrate 1 drop of 0.06 M Fe^{2+} with 0.002 M $K_2Cr_2O_7$ solution. A suitable indicator is diphenylamine sulfonate. The redox reaction is

$$Cr_2O_7^{2-} + 6Fe^{2+} + 14H^+ \rightleftharpoons 2Cr^{3+} + 6Fe^{3+} + 7H_2O$$

The H^+ is already in the titration solutions, which are made to be 1 M in H_2SO_4.

3. Plot the potentiometric titration curve. Write the Nernst law for this titration.

$$Fe^{3+} + e^- \rightleftharpoons Fe^{2+} \qquad E° = 0.77 \text{ volt}$$

$$Cr_2O_7^{2-} + 14H^+ + 6e^- \rightleftharpoons 2Cr^{3+} + 7H_2O \qquad E° = 1.33 \text{ volt}$$

Section F. The Lead-Acid Automobile Battery

Goals: *(1) To build a single-cell, lead-acid battery. (2) To charge it to produce the electrode surface conditions necessary for the system to act as a battery.*

Experimental Steps:

1. Half-fill a well of a 24-well tray with 1 M H_2SO_4 solution.
2. Place 2 small strips of lead metal ($Pb_{(s)}$) into the solution in the well so that they *do not touch.*
3. Clip electrical leads to the 9 V battery and clip an electrical lead to each strip of lead metal.

   NOTE: There will be an immediate vigorous reaction as the H_2SO_4 is electrolyzed. *Do not allow the lead strips to touch!*

   - What will happen if they do touch? ?
4. Unclip the battery and examine the strips of lead.

   NOTE: One of the strips is a clean $Pb_{(s)}$ surface; however, the other has a brownish-colored layer of $PbO_{2(s)}$ on the surface. The initial charging reaction involves the electrolysis of dilute sulfuric acid,

$$2H_2O_{(l)} \rightleftharpoons 2H_{2(g)} + O_{2(g)}$$

   and the dioxygen oxidizes the $Pb_{(s)}$ surface to $PbO_{2(s)}$. Once the electrode surfaces are established, then the overall discharge-charge reaction is

$$Pb_{(s)} + PbO_{2(s)} + 2H_2SO_{4(aq)} \underset{\text{charge}}{\overset{\text{discharge}}{\rightleftharpoons}} 2PbSO_{4(s)} + 2H_2O_{(l)}$$

Chapter 16

Acid Deposition

"Science says the first word on everything and the last word on nothing."

Victor Hugo

Introduction

Acid deposition is the transfer of acidic rain, snow, gases, and aerosols from the atmosphere to terrestrial and aquatic environments. Acid deposition may occur by two different types of mechanisms — wet and dry — both of which are closely related to the chemical form of the deposition. *Wet deposition* denotes acidic substances and their precursors that are dissolved in rain, snow, dew, fog, frost, or hail. *Dry deposition* is the transfer of gaseous and particulate matter to natural surfaces by gravitational settling, turbulence, and vegetation uptake. Acid deposition, commonly called *acid rain*, is often defined as precipitation that has a pH below that of water saturated with atmospheric carbon dioxide, which has a pH of 5.6. This definition is somewhat misleading because it blurs the distinction between pH and acidity (see later), and it presumes that there exists a "normal" pH of unpolluted precipitation.

Acidic deposition is characterized by the presence of the strong acids nitric acid (HNO_3) and sulfuric acid (H_2SO_4) — and occasionally hydrochloric acid (HCl) — that mostly arise as a consequence of human interference in natural biospheric cycles. These acids and their derivatives are the end products of complex gas and aqueous phase chemistry that starts with the oxidation of carbon, nitrogen, and sulfur in the combustion of coal, oil, and gas (fossil fuels) and in smelting operations. The key atmospheric pollutants formed initially in these oxidations are nitric oxide (NO) and sulfur dioxide (SO_2). The environmental impact of acid deposition on natural and man-made systems occurs mostly in the northern hemisphere where most of the fossil fuels are burned. The enormous consumption of these fuels in many regions of the northern hemisphere is oxidation on such a large scale that it rivals natural oxidation, such as photosynthesis and respiration.

Of course, acidic deposition can arise from natural causes. Volcanoes, for example, can emit very large amounts of SO_2 and HCl into the atmosphere. However, these events generally lead to well-documented, local (mesoscale) acid deposition and can be differentiated from man-made deposition. It is perhaps worth pointing out that the atmosphere is very sensitive to anthropogenic pollution because it is a much smaller reservoir than the lithosphere or the hydrosphere. Atmospheric alterations also occur very rapidly compared with those changes that occur in the other reservoirs. Acid deposition is one of the inevitable environmental consequences of the production of energy by the combustion of fossil fuels.

Fossil fuels consist mostly of carbon and hydrogen together with smaller amounts of sulfur and nitrogen. These fuels, when burned in air, produce useful energy from the oxidation of carbon and hydrogen to carbon dioxide and water. At the high temperatures produced in these combustion reactions, sulfur and nitrogen compounds are also oxidized by oxygen to sulfur dioxide (SO_2) and nitric oxide (NO). The amount of SO_2 and NO produced in this manner will depend on the amount of sulfur and nitrogen in the fuel. Sulfur compounds behave differently from nitrogen compounds in combustion and atmospheric processes. Some fossil fuels (e.g., coal) contain much larger amounts of sulfur compounds than other fuels (e.g., gasoline). For these reasons it is useful to discuss the chemistry of sulfur and nitrogen compounds separately.

NITROGEN COMPOUNDS AND ACID DEPOSITION

Fuel nitrogen is not the only source of nitric oxide in high-temperature combustion processes. Reactions between dinitrogen and dioxygen from air can in fact be a dominant source of nitric oxide:

$$O + N_2 \rightleftharpoons NO + N$$

$$O_2 + N \rightleftharpoons NO + O$$

The rate of formation of NO in these reactions is highly dependant on the air/fuel ratio in the combustion and post-combustion reactions, on the flame temperature, and on the length of time of combustion.

After combustion, the exhaust gases rapidly cool, and a small amount (5–10%) of nitric oxide is oxidized by dioxygen to nitrogen dioxide (NO_2):

$$2NO + O_2 \rightleftharpoons 2NO_2$$

These emissions (NO plus NO_2) are commonly referred to as NO_x. The leading source of NO_x emissions is the high-temperature combustion of coal to generate steam in large power plants as well as in other large industrial boilers. In the early 1980s world coal consumption was estimated to be about 4 billion tons, which generates about 23 million tons of NO_x. Typically, coal used in electricity generation produces 9 kg of NO_x per ton of coal. Emissions of NO_x from the above stationary sources are rivalled by those from mobile sources, i.e., motor vehicles. In the United States, NO_x emissions from new motor vehicles have been regulated since 1968. Gasoline-fueled passenger cars remain the largest mobile NO_x source, with diesel-fueled trucks and buses a distant second. Emissions of NO_x by vehicles can vary drastically depending on the type, size, age, compression ratio, spark timing, and exhaust gas recirculation of the vehicle. The amounts of NO_x can range from below 1 g NO_x per km for a new passenger car to over 20 g per km for an old diesel truck.

The anthropogenic NO_x emissions seem to be rather large compared to natural emissions (e.g., from forest fires) and are increasing as the global consumption of fossil fuels and the number of cars increase. The geographic distribution of NO_x emissions reflects the distribution of large power plants and for transportation sources, the population density. In the United States, source mapping shows that the highest NO_x emissions occur from large power plants in Indiana, Ohio, and Pennsylvania. Figure 16.1 is a grid map of NO_x from mobile sources that illustrates the point about population density. It is quite apparent that the map mirrors the distribution of large cities on the east and south coasts.

One last word about sources of NO_x. The discussion so far has centered on outdoors emissions. It has been shown that indoor concentrations of NO_x, particularly in homes with gas ranges, can often exceed those outdoors!

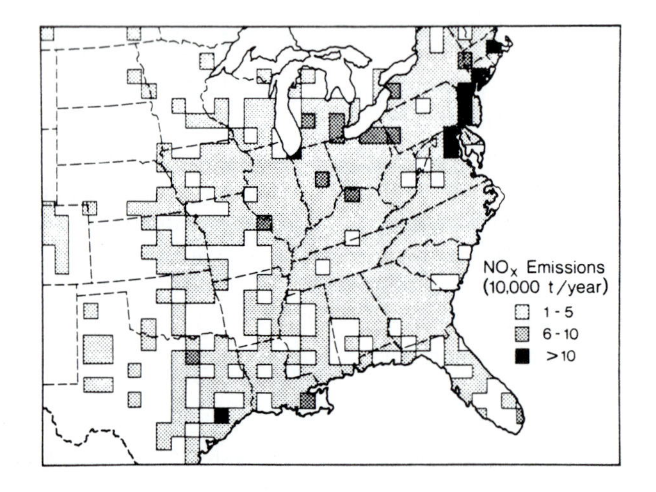

Atmospheric NO_x concentrations can vary tremendously, largely depending on such factors as the closeness and intensity of the source, the prevailing weather conditions, and the relative rates of removal by chemical reactions. Measurement of NO_x in places far from urban and industrial areas show concentrations of 0.2–1.5 $\mu g\ m^{-3}$, whereas in densely populated areas, values can be as high as 2 $mg\ m^{-3}$. The United States Environmental Protection Agency (EPA) has established National Ambient Air Quality Standards for six pollutants. The standard for NO_2 is an annual average of 100 $\mu g\ m^{-3}$. Nitrogen oxides are relatively reactive compounds and, therefore, have fairly short residence times.

The atmospheric chemistry of NO_x is extraordinarily complex and has been well researched, particularly in the air pollution classic: Los Angeles smog. In polluted, dry, sunny air, NO is oxidized to NO_2 and eventually to nitric acid (HNO_3) in a series of photochemical reactions that involve the hydroxyl radical (OH), hydrocarbons, carbon monoxide, and aldehydes. If water is present (as clouds, rain, or fog), the formation of nitric acid occurs by

$$2\,NO_2 + H_2O \rightleftharpoons NO_3^- + H^+ + HNO_2$$

The final production of acid (HNO_3 and HNO_2) can take place through two types of reaction: (1) the homogeneous, gas phase, photochemical sequence, and (2) via the heterogeneous, aqueous sequence.

SULFUR COMPOUNDS

Without doubt, the combustion of coal represents the largest contribution to sulfur emissions. Coal can have a sulfur content ranging from 1–8%, with the sulfur in both organic and inorganic form (pyrite, marcasite, and hydrated iron(II) sulfate). In the United States, power plants consume more than 75% of American coal, and in order to meet the EPA's emission standards, the coal is selected to have a sulfur content of about 2%. Combustion of this type of coal in high-temperature processes directly oxidizes the sulfur to give about 40 kg of sulfur dioxide (SO_2) per ton of coal. A small amount of sulfur dioxide is oxidized to sulfur trioxide (SO_3) in this process. These oxidation reactions are fundamentally different from those of the nitrogen compounds discussed in the previous section. Here, the sulfur comes from the fuel (coal), whereas the nitrogen comes largely from the *air* used in the combustion process. About 75% of SO_2 emissions in the United States are released east of the Mississippi. The grid map shown in Figure 16.2 emphasizes the dominant SO_2 sources as being power plants in the Ohio Valley.

Atmospheric SO_2 concentrations can vary tremendously and depend on the same factors discussed for NO_x. Concentrations range from about 0.1 $\mu g\ m^{-3}$ in unpolluted air to values greater than 1 $mg\ m^{-3}$ in heavily polluted locations. In London's famous smog of December 4–9, 1952, the highest observed value reached almost 4 $mg\ m^{-3}$ (the author was there!). The EPA National Ambient Air Quality Standard for SO_2 is shown in Table 16.1.

Primary	annual arithmetic mean	80 $\mu g\ m^{-3}$
Primary	24-hour*	365 $\mu g\ m^{-3}$
Secondary	3-hour*	1300 $\mu g\ m^{-3}$

*Not to be exceeded more than once per year

Table 16.1 National Ambient Air Quality Standard for SO_2

Sulfur dioxide is oxidized rather slowly to sulfur trioxide by photochemical gas phase oxidation (e.g., by OH), although recent research has shown that fast oxidation in water drops (e.g., rain and clouds) may be a very important route. Water drops in the atmosphere contain small concentrations of hydrogen peroxide (H_2O_2), metal ions, and dissolved oxygen. These dissolved substances can oxidize the highly soluble SO_2 in the following reactions:

$$H_2O_2 + SO_2 \rightarrow SO_3 + H_2O \rightarrow H_2SO_4$$

$$2\,SO_2 + O_2 \rightarrow 2\,SO_3 \text{ (metal ion catalyst)}$$

Once SO_3 is formed, reaction with water is instantaneous, forming sulfuric acid. By analogy with NO_x, the combination of atmospheric SO_2 and SO_3 is called SO_x.

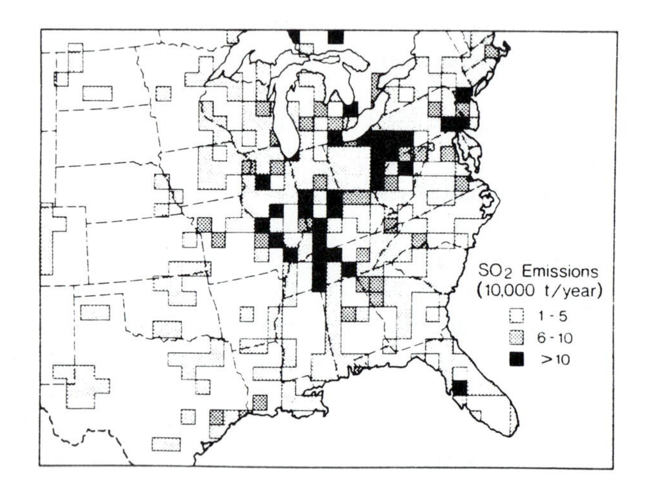

Figure 16.2 Annual Point Source Emissions of SO_2

TRANSPORT AND SUBSEQUENT REACTIONS

The transformation of NO and SO_2 into the various substances that are eventually removed from the atmosphere by precipitation and deposition takes many hours, often days. Residence times depend on many factors, such as the intensity, location, and type of source; temperature inversions; wind patterns; the distribution and volume of clouds; precipitation and sunshine frequency; and the topography of land and water.

Many further chemical reactions can occur while the pollutants are being transported by the wind. One major process is the formation of aerosols by reaction of acids with atmospheric bases. These bases are primarily ammonia (from fertilizers and bacterial reactions) and dust particles blown into the air by wind. The products are small particles called *aerosols* (radius < 0.1–$10\ \mu m$) that have highly variable composition, but often contain NH_4^+, Na^+, SO_4^{2+}, and NO_3^- ions. The long residence times for almost all of these substances (aerosols included) enable the transport of air pollutants to occur over distances of hundreds of kilometers from the source. In the United States regional-scale meteorological models indicate that the acid deposition problem in the eastern region and southeast Canada is probably caused by power plants in the midwest burning high sulfur coal. The cause of acid damage to more than 20,000 of Sweden's 85,000 lakes appears to be acid deposition blown from England and other parts of western Europe. The issue of the long-range transport of acids has even led to strained relations between the United States and Canada. Canadians argue that the United States produces most of the acid pollution — e.g., 84% of man-made emissions of SO_2 and 92% of NO_x. They also point out that Canadian lakes, soils, and forests are highly vulnerable to acid deposition damage and that the Canadian economy is highly dependent upon forest products and their export.

Of course, weather patterns hold the key to how long and in what form all of these pollutants remain in the atmosphere. Most of the pollutants discussed in earlier sections are polar species and are not only very soluble in water, but also can be strongly adsorbed on natural surfaces. Rain, snow, and fog are well known to scavenge most of these pollutants very efficiently and produce acid deposition.

THE EFFECTS

The crux of the controversy about acid deposition lies in assessing the environmental damage caused by acids. In the United States, the battle lines are drawn between the environmentalists on one side and industry and politicians from the major coal-producing states on the other. The controversy is exacerbated by the nature of atmospheric pollution. Long-range transport means that the environmental effects may occur a long way from the source of pollution. Establishing causal connections in these circumstances is extremely difficult, if not impossible.

The effects of polluted air on human health, ranging from *in vitro* studies of animal tissue exposed to pollutants to experimental inhalation of gases and aerosols by volunteers, have been extensively investigated since the 1950s. The toxicology of sulfur compounds of acid deposition (SO_2, H_2SO_4 aerosols, etc.) is now well established. Scientific opinion is that the *concentration* of these compounds currently encountered in the cities of industrialized western nations are too low to cause changes in lung function among healthy adults. One should add the caveat that chronic exposure — i.e., higher concentrations for longer periods — of a more susceptible population group should be viewed with great concern. In contrast to the many epidemiological studies of SO_2 effects, community studies on NO_x and nitrates are scarce. The fatal concentration of NO_2 is almost 300 mg m^{-3}, which is several thousand times the levels found in highly polluted urban air. At lower concentrations NO_2 can be detected by smell at 0.23 mg m^{-3} and by changes in the adaptation of the eye to darkness at 0.14–0.50 mg m^{-3}. Again, it is worth repeating that concentration, duration, and sensitivity are the important factors in determining human health effects.

Certain environmental impacts of acid deposition have now been well established. One of the most obvious degradative effects is the weathering and corrosion of paint, steel, and building materials. In Europe ancient statues, facades, and indeed entire buildings (e.g., the Parthenon) have deteriorated more in the last 50 years than in previous centuries or millenia. Acid deposition compounds react directly and irreversibly with the stone surface. Recent research has shown that a combination of sulfur compounds and soot damages marble ($CaCO_3$) by reacting to form gypsum crystals ($CaSO_4 \bullet 2H_2O$) that penetrate and push apart the granules of the porous marble. The result is crumbling and spalling (breaking off) of the stone. There appears to be no solution to this problem other than to remove the object to a controlled atmosphere in a museum or control the acid deposition! Given the inevitability of these degradative reactions, it is interesting to speculate about the fate of one of the world's most beautiful buildings — the Taj Mahal in India. This incredible marble and iron structure is situated just 30 km downwind from a new refinery in Agra. The prevailing winds will transport about 30 tons of SO_2 per day toward the structure for six months of the year. Both the marble surface and iron structural framework will be attacked by the acid deposition.

The effect of acid deposition on ecosystems, such as lakes, soils, and forests, has been cause for concern, and much research is now being carried out to assess the damage. The potential seriousness of the damage to lakes and rivers originally emerged from studies carried out in Scandinavia beginning in the late 1960s. Of Sweden's 85,000 lakes, about 20,000 now have pH values less than 4.5, and about 4,000 have no fish at all. In southern Norway, thousands of lakes have severely reduced or no fish stock. In North America hundreds of lakes in the Ontario and Adirondack region are showing severe acid stress. The chemistry of these degradative processes is highly dependent on the individual lake properties. The dominant factor is the lake alkalinity that provides a buffering capacity by neutralizing acid deposition. The buffers in lakes are calcium and magnesium bicarbonate ($Ca(HCO_3)_2$ and $Mg(HCO_3)_2$) and occasionally organic acids, which enter the lake from the watershed. Lakes that lie over nonbuffering bedrock (e.g., most of the Canadian Shield lakes — see Figure 16.3) are highly susceptible to damage from acid deposition.

Low pH values are not the only problem in acidified lakes. As the acidity of a lake increases, normally insoluble aluminum and heavy metal compounds in rocks and sediments start to dissolve, as demonstrated in the equation that follows.

$$Al(OH)_3 + 3H^+ \rightleftharpoons Al^{3+} + 3H_2O$$

An even more complex problem than lake acidification is that of soil acidification. The types of changes in soils that receive acid deposition involve loss of base cations Ca^{2+} and Mg^{2+}, an increase in SO_4^{2-} exchange, and increases in the mobilization of aluminum ions. One of the unresolved and controversial issues is the problem of assessing the potential soil sensitivity to acid deposition. A recent study argues that contrary to much published information, the total area of highly sensitive soils in the eastern region of the United States is relatively small.

Correlated with the problem of soil acidification is the effect of acid deposition on plants. Healthy plants have many defenses against such toxic air pollutants as acid deposition. However, studies have shown that some plants are much more sensitive than others, particularly to acid deposition in gaseous form (e.g., SO_2). Severe tree damage in large areas of forest in the northeastern region of the United States, in the Black Forest in Germany, and in forests in Czechoslovakia has been attributed to acid deposition. There have been numerous suggestions as to the actual mechanism of tree damage:

- Atmospheric NO_x is involved in photochemical reactions that produce ozone (O_3), and it is ozone that causes the damage.
- Acid deposition leaches Mg^{2+} from leaves and needles and mobilizes soluble aluminum ions in the soil.

Figure 16.3. Regions Containing Lakes Sensitive to Acid Precipitation

- A syndrome explanation theorizes that a *combination* of acids, ozone, hydrocarbons, heavy metal ions, and drought is the culprit.
- A recent hypothesis suggests that overfertilization of the forests by nitrates and sulfates from acid deposition eventually causes trees to shed leaves and die.

The length and diversity of "explanations" in this list epitomizes the real difficulty of establishing cause and effect in ecosystem damage. One is tempted to wonder whether nature has any baseline, "normal" or predictable behavior, that may be used to discern "abnormal" consequences! In a recent report by the Environmental Defense Fund (EDF) entitled "Polluted Coastal Waters: The Role of Acid Rain," the acid deposition debate is focused on marine ecosystems rather than lakes and forests. EDF calculates that the nitrate input to Chesapeake Bay from acid deposition has a significant nutritive effect on aquatic plants such as algae. The conclusion is that the bay's ecosystem becomes unbalanced and marine life is adversely affected due to pollutants acting as fertilizers. In reading the literature, we are repeatedly reminded that the use and demand for electricity and cars is increasing and so will acid deposition. Changes in ecosystems will occur whether the mechanisms are bad (toxic) or good (nutritive).

SOME CONCLUSIONS

Acid deposition is a direct consequence of the way we generate energy in modern civilization and is only one of many strong links between energy production and environmental change. James Lovelock, in his book *Gaia*, has put it most succinctly, "We as a species, aided by industries at our command have now significantly altered some of the major chemical cycles of the planet. We have increased the carbon cycle by 20%, the nitrogen cycle by 50%, and the sulfur cycle by over 100%. As our numbers and our use of fossil fuels increase, those perturbations will grow likewise."

The environmental consequences are beginning to be seen in the western America where this book was researched and written. During the past 30 years, the west's population and emissions of NO_x and SO_x have *doubled*. New power plants, built close to the coal fields, and the development of Colorado and Wyoming's abundant oil and natural gas resources all lead to a potential for drastic increases in acid deposition. Will the lakes of the Rocky Mountains be subject to the same atmospheric acid-base titration as Swedish lakes? Public awareness of the problems associated with large-scale acid deposition has occurred only recently and has led to pressures to try to find immediate solutions. We live in an energy-intensive society, and the solutions are going to require a long time to implement, are going to be very expensive, and are going to depend on close international cooperation. It is hoped that this series of experiments will enable you to explore the chemistry of acid deposition and perhaps stimulate you to embrace those twin virtues of conservation and environmental awareness. It is wise to remember that in environmental issues we are all part of the problem and that science does not provide solutions — people with knowledge, vision, and commitment do!

Background Chemistry

The Background Chemistry section for the Acid Deposition experiments is in a format different from other chapters in this book. The usual general discussion of the background chemistry of the experiments is replaced by a series of specific discussions, each pertinent to one of the individual experimental sections A–K. This approach is adopted because the experiments on Acid Deposition are of an interdisciplinary nature and are, therefore, rather complex. The experiments involve ideas and techniques from diverse fields, such as meteorology, cloud physics, environmental science, ecology, botany, geology, and rocketry, as well as chemistry. The aim is to concentrate on the major concepts that are necessary for your understanding of each experiment. The specific laboratory experiment section to which a particular discussion refers appears in parentheses after each title.

There are eleven experiments, together with any you may design or want to do. The first concerns the design of an acidity probe, the next four allow you to explore the synthesis, transport, and environmental reactions of NO_x. There are then four experiments on the synthesis, transport and acid deposition properties of SO_x. These are followed by an experiment in which the chemistry of an $NO_x - SO_x$ system in humid air is examined. The final experiment (optional) is about local hydrochloric acid deposition from solid fuel rocket booster engines.

DESIGN AND CHARACTERIZATION OF A PROBE SYSTEM FOR ACIDITY AND ALKALINITY (SECTION A)

The accurate measurement of pH and acidity of rain and other forms of precipitation, at different locations and over historical time, is essential in assessing the magnitude and impact of acid deposition. Perhaps surprisingly this measurement is not an easy one. Critical factors in the analysis are the design of the collection system for wet or dry deposition, the volume of sample collected, the type of container, the delay time before analysis, and the method of analysis. One example of the difficulties was the use of soft-glass storage containers for rainwater samples (in use before 1970). This practice led to considerable errors because of the neutralization of acid in the sample by bases leached from the glass surface. Carefully cleaned plastic containers are now the *standard*! Measurement of the pH of very dilute solutions with conventional glass membrane electrode systems is difficult. A number of scientists have suggested that pH measurement might be better done with colorimetric indicators and spectrophotometric techniques, especially in acid deposition samples.

It is important to remember that pure, unpolluted rain (distilled water) equilibrated with atmospheric CO_2 has a pH of 5.6 (at 15 °C) due to

$$CO_{2\,(g)} \leftrightharpoons CO_{2\,(aq)}$$
$$CO_{2\,(aq)} + H_2O \leftrightharpoons H_2CO_3$$
$$H_2CO_3 \leftrightharpoons H^+ + HCO_3^-$$
$$HCO_3^- \leftrightharpoons H^+ + CO_3^{2-}$$

A simple acid-base equilibrium calculation shows that the *acidity* — i.e., the total titratable acid (H_2CO_3 in this instance) — is 1.8×10^{-5} M. This is a very dilute solution indeed!

A CHEMICAL REACTION SOURCE FOR NITRIC OXIDE AND EXPLORATION OF THE ATMOSPHERIC TRANSPORT AND REACTIONS OF NO_x WITH RAINDROPS (SECTION B)

The chemical synthesis of $NO_{(g)}$ in this experiment produces concentrations similar to that found in tailpipe exhaust. Nitric oxide is not very soluble in water:

$$\text{For} \quad NO_{(g)} \leftrightharpoons NO_{(aq)} \qquad K_H = 1.9 \times 10^{-3} \text{ M atm}^{-1}$$

The direct oxidation of $NO_{(g)}$ to $NO_{2(g)}$ by $O_{2(g)}$ does occur in auto exhaust and in stack gases, but it has been shown to be too slow at normal atmospheric concentrations and is therefore not the main reaction for the production of NO_2. The reaction

$$2\,NO_{(g)} + O_{2(g)} \leftrightharpoons 2\,NO_{2(g)}$$

is rather unusual in that the rate law is overall third order:

$$\text{Rate} = \frac{d[NO_2]}{dt} = k[NO]^2[O_2]$$

for which the rate constant is 1.48×10^4 L^2 mol^{-2} s^{-1} at 25 °C. Notice that the rate is dependent on $[NO]^2$, hence the very slow rate at ambient [NO]. Oxidation of atmospheric $NO_{(g)}$ occurs photochemically by OH and in clouds by H_2O_2. Nitrogen dioxide is about five times more soluble in water than $NO_{(g)}$ and also reacts rapidly with water.

The transport of NO_x in the environmental chamber used in the experiment in Section B occurs mainly by diffusion, with some convection, until the gases interact with raindrops (actually half-raindrops!). Table 16.1 shows some typical sizes and properties of various types of water drops in the atmosphere. Both clouds and rain scavenge huge volumes of air over relatively large distances and are thus able to absorb gases and aerosols from a large region. The dynamics of the scavenging process in the chamber can be clearly seen and is an interesting combination of gas phase diffusion, chemical reaction, and diffusion through solution. The rate of the scavenging process is dependent on the surface area and volume of the water drops. The termination reactions show quite clearly the effect of aerosols on *atmospheric visibility*.

Type of Drop	Diameter µm	Diameter mm	Volume Relative to Cloud Drop	Terminal Velocity	Time to Fall 100 meters (cm s^{-1})
Cloud	10	0.01	1	0.3	5 days
Cloud	40	0.04	64 : 1	5.4	6 hours
Drizzle	1000	1	$10^6 : 1$	390	4 minutes
Rain	2000	2	$8 \times 10^6 : 1$	690	2.3 minutes
Rain	4000	4	$6.4 \times 10^7 : 1$	930	1.8 minutes

Table 16.1 Typical Sizes and Properties of Various Types of Water Drops in the Atmosphere

CLOUD FORMATION AND CLOUD SCAVENGING OF NO_x (SECTION C)

Clouds are vastly underrated! Go out and watch some today. Clouds consist of tiny water drops (sometimes very high clouds freeze into ice crystals). Clouds are formed when saturated humid air expands and/or cools, and water vapor condenses into liquid water drops. In the atmosphere condensation is catalyzed by tiny particles called *cloud condensation nuclei* (CCN). The polar surfaces of clay or salt particles act as very efficient CCN. In the experiment in Section C the humid air inside the dish, when cooled, will condense onto catalytic sites on the plastic, giving a two-dimensional cloud of tiny water drops. Cloud and fog drops are much smaller than raindrops, and this means that the total volume of liquid water in a cloud is very small — about 1×10^{-4} L m^{-3} air. Clouds and fogs can therefore be 10–50 times more acidic than rain. The total acidity of a small cloud is determined by acid-base titration in this experiment.

REDOX CHEMISTRY OF NO_x (SECTION D)

Nitrogen oxides, NO and NO_2, can take part in a vast array of redox reactions, some of which have very important environmental implications. Nitrogen dioxide is a fairly powerful oxidizing agent and, in fact, has been used extensively (in the solid form N_2O_4) as a rocket fuel oxidizer. In the experiments in Section D, you will be investigating several redox reactions of NO_x using a starch/KI oxidant probe. The reaction of NO_2 with water is an extraordinarily complicated disproportionation reaction:

$$2NO_2 + H_2O \rightleftharpoons HNO_2 + NO_3^-$$

The oxidation of I^- to I_2 by NO_2 is probably a complex surface reaction. The formation of I_2 can be followed very dramatically in the presence of starch by the production of the intense blue-black I_5^-/starch complex. The modern method of NO analysis involves chemiluminescence determination by reaction of NO with ozone. This sensitive, accurate method can also be used for analysis of NO_2 by reducing it to NO.

THE SUSCEPTIBILITY OF LAKES TO ACID DEPOSITION MODEL STUDIES (SECTION E)

Simulation models are important working tools for elucidating the extent to which atmospheric acid deposition contributes to the acidification of lakes and streams. Several models were developed in Scandinavia in the early 1980s, such as the Birkenes model. The models are usually based on the hydrological and chemical features of one specific lake catchment area. In the experiment in Section E, you can explore an unusual dry deposition model that involves an acid-base titration of lake bases by gas phase NO_x. Of course, the gross simplification here is that the acidic input only comes from dry deposited NO_x, nitrates, and aerosols. However, the model requires only a microlake and could be extended to incorporate a watershed.

SOURCE, TRANSPORT, AND DEPOSITION REACTIONS OF SULFUR DIOXIDE (SO_2) (SECTION F)

The chemical synthesis of $SO_{2(g)}$ in the Section F experiment produces concentrations that simulate those found in plumes exhausted from a large coal-fired power plant. Sulfur dioxide is quite soluble in water:

$$\text{For} \quad SO_{2\,(g)} \leftrightarrows SO_{2\,(aq)} \qquad K_H = 1.3 \text{ (at 25 °C)}$$

and for this reason is very efficiently removed from air by clouds and rain. The chemical form of dissolved SO_2 depends highly on pH, as shown in Figure 16.4.

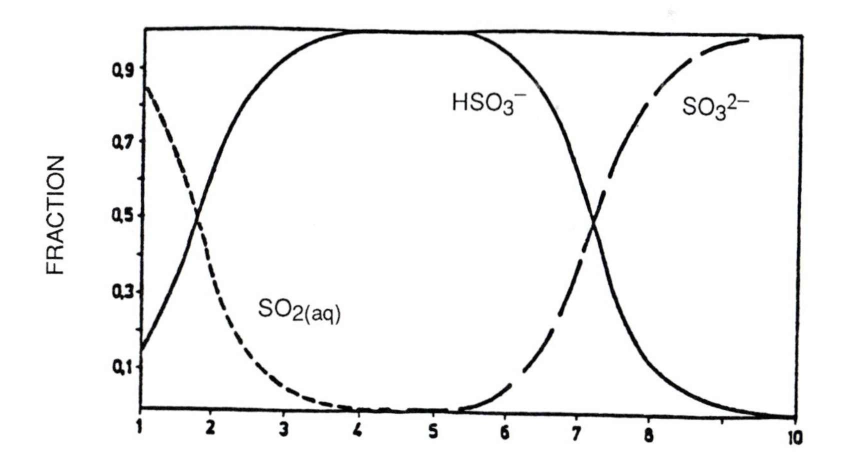

Figure 16.4 Fraction of Sulfur Species in Equilibrium at 25 °C as a Function of Aqueous Solution pH

HETEROGENEOUS OXIDATION OF SO_2 TO SULFUR TRIOXIDE (SO_3) AND SULFURIC ACID (H_2SO_4) BY HYDROGEN PEROXIDE (H_2O_2) IN RAINDROPS AND CLOUD DROPS (SECTION G)

Recent research has presented *field* evidence for the oxidation of dissolved SO_2 in cloud drops by hydrogen peroxide (H_2O_2). The experiments involved releasing SO_2 into a cloud over the Pennine Hills of England and then monitoring downwind cloud and air composition with a mobile laboratory. (See additional Reading, reference 14.) Hydrogen peroxide oxidation may be of major significance in determining the pattern of acid deposition over a region. The oxidation product is sulfuric acid, which may be detected by its reactions with barium chloride ($BaCl_2$):

$$H_2SO_4 + Ba\,Cl_2 \rightarrow BaSO_{4(s)} + 2HCl$$

The fine precipitate of $BaSO_4$ has been used in turbidimetric methods to quantitatively measure atmospheric sulfate concentration. Atmospheric H_2O_2 is produced by photochemical reactions involving OH, OH_2, and H_2O. Hydrogen peroxide is also the basis for a new method of removing SO_2 from exhaust gases from coal- or oil-fired power plants. According to the West German developer Degussa, the costs are very low and the method worked very well in a one-year test. In this section, small-scale techniques are used to investigate the heterogeneous oxidation of SO_2 by H_2O_2.

THE EFFECT OF ACID DEPOSITION ON NATURALLY OCCURRING MINERALS AND ON CONSTRUCTION MATERIALS (SECTION H)

Research into the chemistry of accelerated decay of stone and other structural materials by acid deposition is really just beginning. The processes appear to be extraordinarily complicated; however, some general features are now clear (at least for $CaCO_3$). Airborne SO_2 reacts with wet stone to form gypsum ($CaSO_4 \bullet 2H_2O$). The resulting mixture of calcite and gypsum builds up a surface layer that progressively consumes the stone. Occasionally the crust breaks away, exposing a fresh surface on which the decay process starts again. High concentrations of black soot particles (from burned fuel oil) are consistently found inside the crust. These soot particles have been found to catalytically oxidize SO_2 in the presence of moisture:

$$2SO_2 + 2H_2O + O_2 \rightarrow 2H_2SO_4$$

Sulfuric acid now reacts with calcite crystals:

$$CaCO_{3(s)} + H_2SO_4 \rightarrow CaSO_{4(s)} + CO_2 + H_2O$$

The product, gypsum, is 100 times more soluble in water than calcite, and by repeated dissolution and evaporation, a network of water-conducting channels is formed in the stone. These channels now expose fresh stone to soot and SO_2, accelerating the formation of crust. In Section H, a series of rather dramatic experiments is presented to investigate the effect of acid deposition on naturally occurring minerals. Very small

small amounts (<100 mg) of minerals and materials are exposed to rain and an acidic environment.

THE EFFECT OF SO_2 ON PLANTS (SECTION I)

Plants control the exchange of gases with the atmosphere by regulating stomatal openings present in leaf and flower epidermis. Leaf and flower protection against general assault from the atmosphere and predators is provided by a surface layer of wax called the epicuticular wax. The effect of acid deposition on plants depends on the chemical form of the deposition — gas, solution, aerosol, etc., and/or any structural damage that might be present in the epidermis. Changes in soil chemistry as a result of acid deposition will, of course, also have a strong influence on plants via the root system. Experiments have also shown that acidic fog and rain can leach essential cations such as Mg^{2+} and K^+, from leaves, leaving plant tissue more susceptible to attack by insects. In Section H, small-scale techniques allow you to explore the effect of air pollutants such as SO_2 on plants.

PUTTING IT ALL TOGETHER: ACID DEPOSITION FROM NO_x PLUS SO_x (SECTION J)

Studies on acid deposition caused by combinations of atmospheric pollutants are really just beginning. The outcome in systems like this will depend highly on the relative rates of the individual processes. The strong interaction between chemical, meteorological, and substrate processes will make this type of research intriguing and deliciously difficult. Again, the convenience and flexibility of small-scale techniques allow you to examine the kinetics and topochemistry of NO_x/SO_x interactions.

SPACE SHUTTLE LAUNCHES: AN EXAMPLE OF SEVERE LOCAL HYDROCHLORIC ACID DEPOSITION (SECTION K)

The resumption, in 1989, of the United States space shuttle launches at Cape Canaveral in Florida, has led to a renewed potential for severe, local hydrochloric acid deposition. The space shuttle has two solid rocket boosters (SRBs) that each contain 1.1 million pounds of solid rocket propellant. Combustion of the fuel (aluminum powder) with an oxidizer (ammonium perchlorate) produces, in less than two minutes, about 240 tons of hydrochloric acid. About 60 tons of HCl is released close enough to the ground to cause damage to the launch pad and to areas close to the launch site. Within 24 hours of a shuttle launch, highly acidic conditions have been shown to occur in clouds, rainstorms, lakes, and the ocean.

In spite of extensive investigations into the acid deposition, NASA has not been able to eliminate the problem completely. Direct erosion of the concrete launch pad during the first few seconds of the rocket burn is reduced by releasing a huge deluge of water directly under the exhaust plume. However, once the exhaust cloud billows out to a large volume, any subsequent acid deposition is then completely at the mercy of the prevailing weather conditions. The suggestion has been made that it might be feasible to neutralize the low-level exhaust cloud by releasing a volatile base, e.g., ammonia, from an aircraft flown through the cloud. Unfortunately, the resulting formation of an ammonium chloride aerosol might be a worse problem than the original hydrochloric acid deposition. In Section K, you can explore some interesting spatial characteristics of an aerosol formed in the gas-phase neutralization of hydrochloric acid plume by ammonia.

ADDITIONAL READING

1. Smil, V., *Carbon-Nitrogen-Sulfur: Human Interference in Grand Biospheric Cycles,* Plenum Press, New York, 1985.

   A wonderful book, not just on acid deposition but on every aspect of the carbon, nitrogen, and sulfur cycles. Has all the major references!

2. Johnson, R.W. (ed.), et al., *The Chemistry of Acid Rain: Sources and Atmospheric Processes,* ACS Symposium Series, no. 349. American Chemical Society, Washington, D.C., 1987.

   In-depth papers developed from a symposium sponsored at the 191st meeting of ACS, New York, 1986.

3. Canter, L.W., *Acid Rain and Dry Deposition,* Lewis Publishers, Inc., Chelsea, Mich., 1986.

   An excellent summary of information published up to the mid-1980s.

4. Durham, J.L. (ed.), *Chemistry of Particles, Fogs, and Rain,* Butterworth Publishers, Boston, 1984.

   Excellent research papers on aerosols, fogs, and rain.

5. Calvert, J. G., *SO_2, NO, and NO_2 Oxidation Mechanisms: Atmospheric Considerations,* Butterworth Publishers, Boston, 1984.

   In-depth research papers on oxidation mechanisms.

6. Committee on Monitoring and Assessment of Trends in Acid Deposition, NRC, *Acid Deposition: Long-Term Trend,* National Academy Press, Washington, D.C., 1986.

   One report in a long series. Gives the "official" viewpoint.

7. Zajicek, O.T., "Why Isn't My Rain as Acidic as Yours?," *J. Chem. Ed., 62,* 1985, p. 158.

   A short, to-the-point reminder of the difference between pH and acidity.

8. Pruppacher, H.R., and J. D. Klett, *Microphysics of Clouds and Precipitation,* D. Reidel Publishing Co., Boston, 1980.

   The book on clouds.

9. Udall, J.R., and T. Moore, (photograph), Finis Mitchell, "Lord of the Winds," *Audubon,* 1986, p. 73.

   A beautifully written story of a man, a mountain range, a few million trout, and acid rain.

10. Andersson, F., and B. Olsson (eds.), *Lake Gårdsjön: An Acid Forest Lake and Its Catchment,* Ecological Bulletins, Stockholm, Sweden, 1985.

    The complete story of the effect of acid deposition on a Swedish lake and watershed.

11. Lodge, J.P. (ed.), *The Smoake of London: Two Prophecies,* Evelyn, J., *Fumifugium* (1772), and Barr, R., *The Doom of London* (1877), Maxwell Reprint Company, Elmsford, N.Y., 1969.

Two quite incredible prophecies about coal and the effects of its burning. Well worth reading — not much has changed!

12. Schaefer, V.J., and J.A. Dan, *A Field Guide to the Atmosphere,* Houghton Mifflin Co., Boston, 1981.

All you want to know about the atmosphere — in one book.

13. Seinfeld, J. H., *Lectures in Atmospheric Chemistry,* American Institute of Chemical Engineers, New York, 1980.

A clear, concise, and brilliant exposition of atmospheric chemistry from the chemical engineer's point of view.

14. Gervat, G.P., et al., *Nature* 333, 1988, p. 241.

British study. The first direct experiment on oxidation of SO_2 by H_2O_2 in real clouds.

15. Mohnen, V.A., *Scientific American* 259, 1988, p. 33.

16. Wedzicha, B.L., *Chemistry of Sulphur Dioxide in Foods,* Elsevier Applied Science Publishers, New York, 1984.

Technical, but encyclopedic.

Pre-Laboratory Quiz

1. Why does unpolluted rain have a pH of about 5.6?

2. Why is the atmosphere very sensitive to anthropogenic pollution?

3. Acid deposition is primarily caused by the oxidation of what substances?

4. Which gaseous air pollutants are the precursors to acid deposition?

5. How much NO_x per km does a new car exhaust to the atmosphere?

6. Give a chemical reaction for a raindrop process by which sulfur dioxide is oxidized to sulfuric acid.

7. What is an aerosol?

8. What are the natural buffers that are present in lakes that can neutralize acid deposition?

9. Give one suggested mechanism for tree damage by acid deposition.

10. In North America acid deposition appears to be a more serious environmental problem in the northeastern United States and southeastern Canada than elsewhere. What factors do you think are responsible for this regional imbalance?

Laboratory Experiments

Flowchart of the Experiments

Section A.	Design and Characterization of a Probe System for Acidity and Alkalinity
Section B.	A Chemical Reaction Source for Nitric Oxide and Exploration of the Atmospheric Transport and Reactions of NO_x with Raindrops
Section C.	Cloud Formation and Cloud Scavenging of NO_x
Section D.	Redox Chemistry of NO_x
Section E.	The Susceptibility of Lakes to Acid Deposition — Model Studies
Section F.	Source, Transport, and Deposition Reactions of Sulfur Dioxide (SO_2)
Section G.	Heterogeneous Oxidation of SO_2 to Sulfur Trioxide (SO_3) and Sulfuric Acid (H_2SO_4) by Hydrogen Peroxide (H_2O_2) in Raindrops and Cloud Drops
Section H.	The Effect of Acid Deposition on Naturally Occurring Minerals and on Construction Materials
Section I.	The Effect of SO_2 on Plants
Section J.	Putting It All Together: Acid Deposition from NO_x plus SO_x
Section K.	Space Shuttle Launches: An Example of Severe Local Hydrochloric Acid Deposition

Requires two three-hour class periods to complete

CAUTION: In these experiments you will be using a wide variety of chemicals and synthesizing air pollutants. Even though the amounts and concentrations are very small, *please be careful. Remember* — acids are corrosive; sulfites, when ingested, can cause allergic reactions; NO_x, SO_x, and I_2 are all toxic and should be treated appropriately before disposal. If you get any of these solutions on your skin, wash well with cold water and report to your Instructor!

Before You Begin:

These experiments will be carried out in a polystyrene environmental chamber (a petri dish). You will have the opportunity to observe many unusual chemical processes and reactions. Please take care of your plastic apparatus. Use microtowels (TP) to clean and dry your dish. Water spots and scratches will make observations more difficult to interpret. Use the dish on top of a plastic reaction surface that has backgrounds suitable for this experiment. Most of the reagents will either be at your desk or at Reagent Central. Use correct transfer technique. Think and act conservation and environmental awareness.

Section A. Design and Characterization of a Probe System for Acidity and Alkalinity

Goal:

To recognize the various dynamic color changes that occur when bromocresol green solution is used to characterize a variety of acids, bases, salts, and acid-base reactions.

Discussion:

Colorimetric acid-base indicators can be used to probe both the pH and acidity of environmental atmospheric samples (see Background Chemistry section). Once the data base (discussed above in the goal) has been established, bromocresol green may then be used to explore both qualitatively and quantitatively acid deposition phenomena. If you know that you are color deficient, check with your instructor, who will provide you with an alternate indicator system.

Experimental Steps:

1. Use a clean, dry petri dish against a white background for most of these experiments.
2. Establish the color of the indicator in acidic and basic solution by using diluted sulfuric acid and diluted ammonia. Dilute 1 drop of 2 M H_2SO_4 by dropwise dilution with distilled H_2O to about 0.2 M H_2SO_4.
3. Similarly, dilute 2 M NH_3 to about 0.2 M NH_3.
4. Add 1 drop of 0.2 M H_2SO_4 to 1 drop of indicator.
   - Record the color.

   Color comparison of small volume samples is remarkably easy if the various samples are close together and observed simultaneously.
5. Place 1 drop indicator plus 1 drop H_2O next to the acid/indicator solution.
6. Add 1 drop of 0.2 M NH_3 to 1 drop of indicator.
   - Record all the color changes.

- Do you think that the order of addition would affect any of these color changes? Why (not)?
- Does the color change occur rapidly? Why?

Now let's look at the effect of salts on the indicator. In your reagent tray there are $NaHCO_3$, KNO_2, and Na_2SO_3 solutions. (They will be used extensively later on in the module.) Continue to use the same petri dish (don't clean it yet).

7. Probe the acid-base nature of these solutions as you did in the above experiments.
   - Explain any color changes you observe with appropriate chemical reactions — e.g., H_2SO_4 is diprotic acid and in water

$$H_2SO_4 \longrightarrow H^+ + HSO_4^-$$

$$HSO_4^- \rightleftharpoons H^+ + SO_4^{2-}$$

8. Place 1 drop of 2 M NH_3 close to any yellow solution.
   - Watch carefully and describe what you see.
   - What can be the only explanation for this phenomenon?

The indicator system that you have been working with is a 0.03% wt solution of bromocresol green ($C_{21}H_{14}O_5Br_4S$, F. Wt. 698) that has a pK_a of 4.6 for the reaction

$$HIn \rightleftharpoons H^+ + In^-$$

- What color are the 2 species HIn and In^-?
- Calculate the transition range of this indicator.

Remember (from your Acid-Base Equilibria laboratory, Chapter 14) that for a person with normal color vision,

a) For color in acid solution, $\frac{[In^-]}{[HIn]} \approx \frac{1}{10}$

b) For color in basic solution, $\frac{[In^-]}{[HIn]} \approx \frac{10}{1}$

c) For color at $pH = pK_a = 4.6$, $\frac{[In^-]}{[HIn]} \approx 1$

9. Flush the solutions in the petri dish into the waste container. Then wash it (at the sink) with soap and water. Rinse with distilled H_2O and dab dry with microtowel.
   - Design (and carry out) an experiment to prove that H_2SO_4 is a nonvolatile acid.
   - Design and carry out an experiment to show that acid-base reactions initiated through the gas phase are strongly dependent on surface area.

HINT: Think small drops and/or microtowels!

- Make a brief list of what you consider to be the important criteria in selecting an indicator probe for acidity in environmental atmospheric samples.

Section B. A Chemical Reaction Source for Nitric Oxide and Exploration of the Atmospheric Transport and Reactions of NO_x with Raindrops

Goals: *(1) To synthesize NO_x's. (2) To use the probe system to explore some of the factors that play a part in the formation of nitrous and nitric acid in raindrops.*

Discussion: Since it is rather difficult to fit an automobile (even a VW bug!) into the environmental chamber, you can synthesize nitric oxide (NO) by a simple aqueous chemical reaction. NO is not very soluble in aqueous solutions and will be slowly released as a gas, which is then partially oxidized to nitrogen dioxide (NO_2). The gaseous mixture is called NO_x. Transport of NO_x then occurs by diffusion and convection through the chamber atmosphere until the pollutants interact with water drops (rain, fog, cloud). In this experiment you will be incorporating the acidity probe in various size raindrops and exploring the many factors that play a part in the formation of nitrous and nitric acids in acid rain.

NOTE: For this experiment the acidity probe (bromocresol green) has already been dissolved in pure rainwater. This will enable you to study some of the dynamics of the interaction of NO_x with rain on a real time basis. This solution is called rainwater throughout Section B. However, in Sections C through H the same solution (acidity probe) is used, but is referred to as acidity probe rather than as rainwater.

Experimental Steps:

1. Place a clean dry petri dish onto the circular reaction grid and line it up.
2. Drop single drops of the rainwater into the dish in the positions shown in the diagram below. This arrangement will enable you to understand how the NO_x gases move out spatially from the source (which will be in the center).

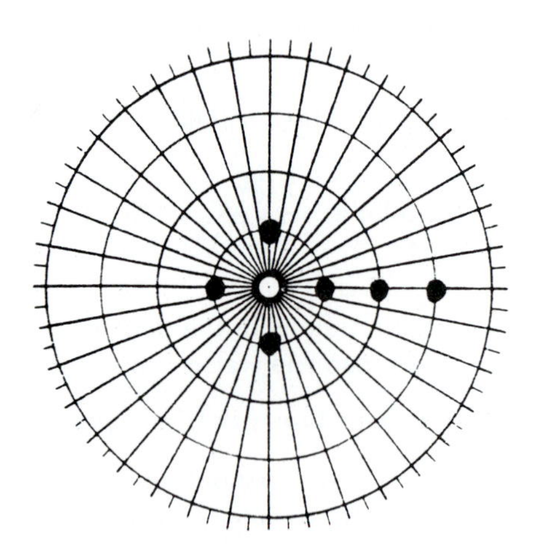

3. Now produce a series of much smaller raindrops on several radii of the grid by putting 2 drops of rainwater onto a plastic sheet and then dipping a piece of cutoff thin stem into it. (A small amount of liquid will adhere to the tip.) Now transfer it to the dish by simply touching it down in the appropriate position. These steps are shown below. Now you are ready to carry out the chemical reaction to produce NO. Make sure your notebook is close by because there will be lots of observations to record.

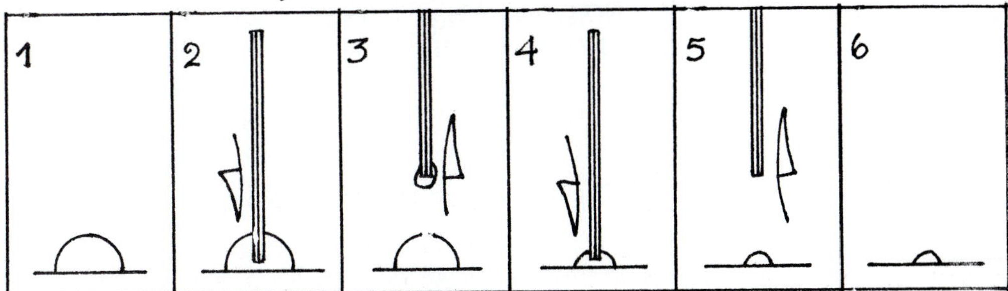

4. As pictured below, drop 1 drop of 0.5 M KNO_2 onto the center of the dish. Hold the top at an angle over the dish and add 2 drops of 2 M H_2SO_4 to the KNO_2. Immediately place the top on the dish.

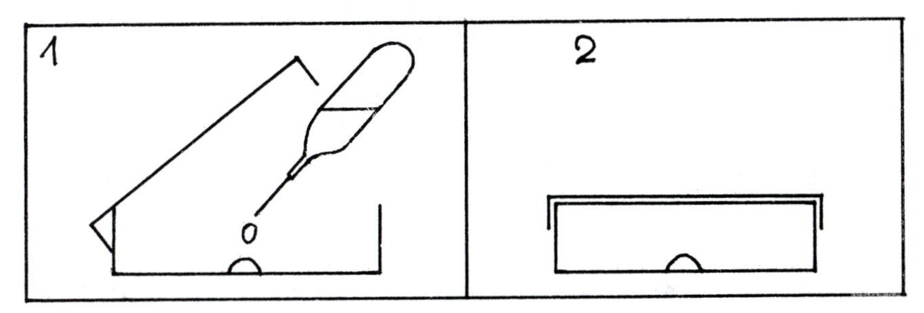

- Watch carefully and record what happens. You might even want to time (approximately) how long the various processes take.
- Draw rough diagrams of the processes that you see happening.

Now you can explore the effect of naturally-occurring bases — e.g., NH_3 — on an NO_x air pollution episode.

5. Terminate the air pollution by adding 2 drops of 2 M NH_3 to the dish by lifting the lid up slightly, at an angle, dropping the NH_3 into the dish, and then replacing the lid.

   - Watch carefully, first against a white background and, after 2 or 3 minutes, move the petri dish to a black background. Record your observations.

Now let's see if we can interpret the complex sequence of events in these air pollution phenomena. Nitric oxide is produced in the reaction

$$3\,NO_2^- + 2\,H^+ \rightarrow 2NO(g) + H_2O + NO_3^-$$

As soon as the $NO_{(g)}$ leaves the solution, some of it is slowly oxidized to nitrogen dioxide (NO_2):

$$2\,NO_{(g)} + O_{2(g)} \rightarrow 2\,NO_{2(g)}$$

Both $NO_{(g)}$ and $NO_{2(g)}$ diffuse throughout the humid air and dissolve in the rain drops. NO_2 reacts with water to give nitrous acid HNO_2 and nitric acid HNO_3:

$$2\,NO_2 + H_2O \rightarrow HNO_2 + NO_3^- + H^+$$

- Did you see an "eye" shape in the large raindrops?

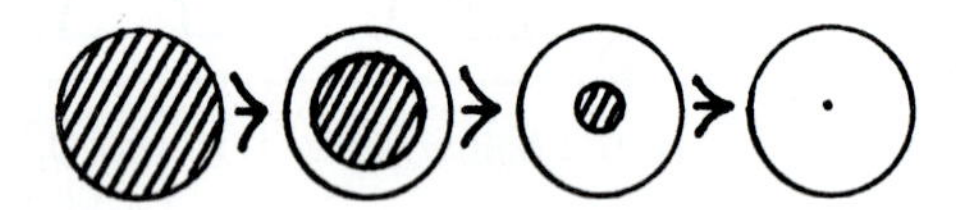

- Explain this phenomena. Compare the "eye" phenomenon with what happened in the small raindrops.
- Why did the raindrops closest to the source turn yellow more quickly than the drops farther away?

In the termination sequence the 2 M NH_3 solution gives off $NH_{3(g)}$, which diffuses through the humid air and then takes part in *two major chemical reactions*:

a) $NH_{3(g)}$ is very soluble in water, and as soon as it reaches a raindrop, it dissolves and reacts with water:

$$NH_3 + H_2O \leftrightarrows NH_{3\,(aq)}$$

dissolution

$$NH_3 + H_2O \leftrightarrows NH_4^+ + OH^-$$

reaction

Since NH_3 is a base, it will neutralize the acid in the raindrop.

- Why do you see no "eye" in this reaction sequence?

b) The second reaction is a gas phase reaction between $NO_{2(g)}$ and $NH_{3(g)}$, probably catalyzed by $H_2O(g)$ to give an *aerosol*. The tiny white particles can be seen against a black background. The exact chemical formula of this aerosol is not known.

- Why does the aerosol appear close to the source of $NO_{(g)}$?

6. Finish the experiment by using a wash bottle to flood the dish with water. Wash with water at the sink. Rinse once with distilled water and dry with a microtowel.

Section C. Cloud Formation and Cloud Scavenging of NO_x

Goals:

(1) To explore techniques for the formation of two-dimensional clouds. (2) To investigate the rapid and efficient scavenging of soluble air pollutants by high surface area clouds.

Discussion:

A polystyrene petri dish can be used to produce two-dimensional clouds consisting of large numbers of extremely small water drops. These clouds form when cold water is used to cool the plastic surface. Water vapor inside the dish condenses onto the surface, and individual tiny drops are formed. The clouds can be seen easily when viewed against a black background because of the light that is scattered from the tiny drops. In this experiment you can show that the huge surface area of a cloud produces very efficient scavenging (collection) of atmospheric pollutants, such as NO_x.

Experimental Steps:

1. Use 2 clean, dry petri dishes and tops for this experiment. In one dish set up to produce $NO_{(g)}$, but *don't* add the 2 M H_2SO_4 yet! Drop in single drops of acidity probe so that you can monitor the NO_x transport. Place the lid on the dish.
2. Make an ice bath by filling a coffee cup with 2/3 ice, 1/3 water.
3. Use a microburet to transfer ice cold water to form a pool (about 2 cm diameter) on the outer surface of the lid. Slide the dish over a black background so that you can observe cloud formation. Wait until there is a substantial cloud (2–5 min) on the underside of the lid.

   NOTE: If a cloud does not form, add some ice chips to the pool and wait.
4. Now gently lift the lid to keep the pool in place and place 2 drops 2 M H_2SO_4 onto the nitrite solution in order to generate NO_x. Replace the lid and wait for about 2 minutes for the cloud to scavenge the NO_x. Keep the pool cold.
5. Remove the pool by sucking it up with the microburet.
6. Remove the lid and place *another* lid onto the dish.
7. Place the lid, cloud up, onto black background.
8. Add 2 drops water to the middle of the cloud, then tilt and tap the lid gently to move the water around in order to collect all the cloud.
9. Add 1 drop of acidity probe, stir, and titrate the collected cloud dropwise with 0.001 M $NaHCO_3$ solution until a blue end point is reached. Keep stirring during the titration.

   - Was the cloud acidic? Give the chemical reaction that made the cloud acidic.

   Let's assume that the actual total volume of a 2 cm diameter cloud is about 1×10^{-7} L and that the 0.001 M $NaHCO_3$ drops are about 20×10^{-6} L each.

   - Calculate an approximate molarity of the cloud acidity.
   - Design an experiment to prove that the acid does not come from the polystyrene plastic of the lid.

Very low clouds are called fog — and there is now a lot of evidence that scavenging of mesoscale (local) pollution by fog can produce extremely acidic conditions!

10. Clean the petri dishes in the usual manner.

Section D. Redox Chemistry of NO_x

Goal: *To investigate the redox properties of NO_x.*

Discussion: Nitrogen-containing air pollutants undergo an almost bewildering series of redox reactions. NO and NO_2 are particularly important in this respect and are called *oxidants*. This chemical property has been found very useful in developing several sensitive methods of analysis for NO_2. In this experiment you can explore the use of a starch/KI solution as a sensitive test for NO_2.

Experimental Steps:

1. Use a clean, dry petri dish.
2. Design and carry out an experiment using single drops of starch/KI solution as test points for NO_2, which will enable you to describe the transport of NO_2 through the atmosphere in the dish.

   The color reaction is very interesting. The $NO_{(g)}$ reacts in air to give $NO_{2(g)}$, which diffuses to the surface of a starch/KI drop. NO_2 oxidizes I^- to I_2 and is reduced to nitrite ion (NO_2^-):

$$2\,NO_2 + 2I^- \rightarrow 2\,NO_2^- + I_2$$

   The I_2 can now react with excess I^- (from the KI) to give an I_5^- ion (pentaiodide ion) that just fits down the hole in the center of the amylose (starch) helix, as in the following structure.

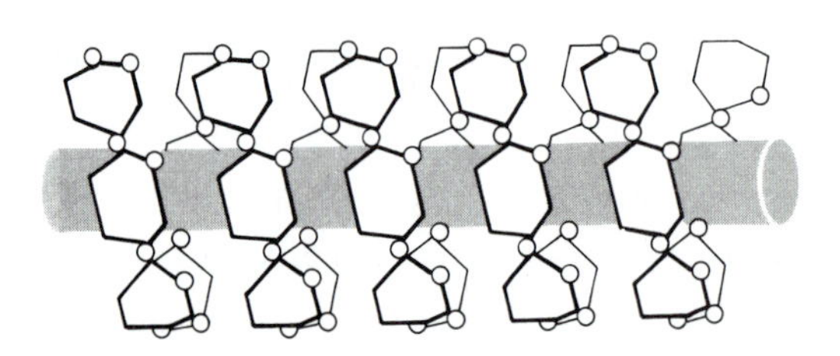

3. Terminate your experiment by adding 2 drops of NH_3 to the dish.
4. Now add 2 drops of 0.5 M Na_2SO_3 (sodium sulfite) and flood with water. Replace the lid and swirl.

The sulfite will reduce the I_5^- back to I^- and will be oxidized to sulfate. The reason for doing this is that iodine in the form of I_5^- is a hazardous substance. Reducing I_5^- to I^- produces a colorless, nontoxic solution.

5. Clean the petri dish in the usual manner.

Section E. The Susceptibility of Lakes to Acid Deposition — Model Studies

Goal:

To qualitatively and quantitatively examine and model the effect of dry deposition of NO_x on a variety of natural and synthetic microlakes.

Discussion:

The susceptibility of lakes to acid deposition depends on many factors. One of the most important is the concentration of natural buffers in the lake water. These buffers are mostly calcium and magnesium bicarbonates ($Ca(HCO_3)_2$ and $Mg(HCO_3)_2$). In this experiment you can model the effect of *dry deposition* of NO_x on microlakes. The model is that of a titration of lake bicarbonate by gas phase NO_x. The time required for complete titration is a semiquantitative measure of the ability of a lake to resist acid input from the atmosphere.

Experimental Steps:

1. Clean and dry 2 petri dishes and lids.

   Various lake samples are in plastic bottles at Reagent Central. You are provided with a microburet containing 0.01 M $NaHCO_3$ solution.

2. By appropriate dilution of 0.01 M $NaHCO_3$, prepare at least 10 drops each of 0.001 M, 0.0025 M, 0.004 M, and 0.005 M $NaHCO_3$. Use a 1 x 12 well strip for the dilution.

   - Record how you did this.

3. Place a dish on the circular grid. Place 1 drop distilled H_2O into the dish on the second circle out from the center.

4. Now go round the circle dropping 1 drop of 0.001 M, 0.0025, and 0.005 M $NaHCO_3$ and various lake and/or spring water samples at 40° *intervals.*

   NOTE: The $NaHCO_3$ solutions and lake and water samples should now all be in a circle, equidistant from the center of the dish.

5. Add 1 drop of acidity probe to each drop. Stir gently to mix while maintaining the circular shape of each sample.

   - Note the color of each sample in your lab book.

6. Drop 2 drops of 0.5 M KNO_2 into the dish at the center mark. Have the lid *ready* to be placed on the dish.

7. Add 1 drop 2 M H_2SO_4 to generate NO_x. Quickly place the lid on the dish. Start timing how long it takes for the "eye" of green to go completely acidic in each of the samples.

   - Estimate the approximate pH of the original samples.

- Plot a graph of time (seconds) versus concentration for the $NaHCO_3$ standards.
- Compare the lake samples and rate their susceptibility to dry acid deposition.
- Why were the samples placed in a circle around the NO_x source?
- The distilled H_2O control sample was also titrated by the NO_x. What is the base in this sample?
- What assumptions have to be made in this experiment in order to make the correlation between time and lake susceptibility?
- What are some of the other factors that determine lake susceptibility to acid deposition?

8. Terminate the experiment by adding 2 drops of NH_3 to the petri dish and cleaning the dish in the usual manner.

Section F. Source, Transport, and Deposition Reactions of Sulfur Dioxide (SO_2)

Goal:

To synthesize SO_2 and investigate the dissolution reactions of this air pollutant with raindrops and with ammonia.

Discussion:

By far the two most important precursors to acid deposition are NO and SO_2. Sulfur dioxide comes mainly from *stationary* combustion sources, such as power plants. In this section you can use a chemical reaction source of $SO_{2(g)}$ to explore the transport and sink reactions in raindrops, etc. Gas phase neutralization of SO_2 (and SO_3, later) by ammonia gives ammonium salt aerosols.

Experimental Steps:

1. Clean and dry a petri dish.
2. Design and carry out an experiment to examine the source, transport, and sink reactions of SO_2. Use Section B as a design model for your experiment.
3. To generate $SO_{2(g)}$ simply add 2 drops 2 M H_2SO_4 to 1 drop of 0.5 M Na_2SO_3 (sodium sulfite).
   - Record your observations in your notebook.
4. Terminate the reactions by adding 2 drops NH_3 to the dish.
   - Watch carefully and record the characteristics of any aerosols that may be formed.
   - Give the chemical reaction for $SO_{2(g)}$ generation.
   - What acid is produced when SO_2 dissolves in raindrops?
   - Do you find any major differences between source, transport, and sink reactions of SO_2 compared with NO_x?
   - Assuming the $SO_{2(g)}$ is released instantaneously (which it is not!) in the source reaction, calculate the number of μg of $SO_{2(g)}$ produced in the source reaction.

- Now calculate the concentration of $SO_{2(g)}$ in the dish in µg m^{-3}.

5. Clean the apparatus in the usual manner.

Section G. Heterogeneous Oxidation of SO_2 to Sulfur Trioxide (SO_3) and Sulfuric Acid (H_2SO_4) by Hydrogen Peroxide (H_2O_2) in Raindrops and Cloud Drops

Goal: *To investigate the heterogeneous in-cloud oxidation of SO_2 to H_2SO_4 by hydrogen peroxide.*

Discussion: Recently, it was shown that naturally occurring H_2O_2 in cloud water drops is one of the major routes by which SO_2 is oxidized to SO_3. Water then immediately converts SO_3 into the strong acid H_2SO_4. H_2SO_4 dissociates into HSO_4^- and SO_4^{2-} ions that can be selectively detected by means of a barium chloride probe reaction.

Experimental Steps:

1. Place a clean, dry petri dish on the circular grid. Equidistant from the center and at 90° to each other, prepare the following:

   a) 1 drop H_2O + 1 drop 3% H_2O_2 (This is a raindrop + hydrogen peroxide.)

   b) 1 drop H_2O + 1 drop 0.5 M $BaCl_2$ (This is a raindrop + sulfate ion detector.)

   c) 1 drop H_2O_2 + 1 drop 0.5 M $BaCl_2$ (This is a raindrop + hydrogen peroxide + sulfate ion detector.)

   d) 1 drop of acidity probe (This is a raindrop + acidity probe.)

2. Put 1 drop 3% H_2O_2 on the underside of the lid. Turn it over without sliding the drop and place aside for a moment.

3. In the center of the dish, drop 2 drops 0.5 M Na_2SO_3. Pick up the lid and have it ready and add 2 drops 2 M H_2SO_4 to the Na_2SO_3 to produce $SO_{2(g)}$. Put the lid on quickly.

4. Slide dish over the plastic surface until it is over a black background and watch carefully for about 3 minutes.

   - Record what you see.

   - Interpret the observations you have made on each solution. Give chemical reactions wherever possible, e.g.,

$$H_2O_2 + SO_2 \rightarrow SO_3 + H_2O$$

   - What do you think the white "halo" is that surrounds the raindrop on the lid?

Section H. The Effect of Acid Deposition on Naturally Occurring Minerals and on Construction Materials

Goal: *To examine the effect of SO_x and acid deposition on naturally occurring minerals and on construction materials.*

Discussion:

Acid deposition can react with basic minerals (in rocks and soils) and with construction materials, such as concrete, marble, metals, etc. The combined effect of acids and water is to slowly dissolve these solids, producing increased solution concentrations of dissolved cations and leaving eroded solids. The increased cation concentration can be more damaging in the environment than the direct effect of acids, particularly if the cation is Al^{3+}. In this experiment you can examine the effects of SO_x and acidic rain on very small amounts of solids.

Experimental Steps:

1. Place a clean dry petri dish on the circular grid. At various points (at least 40° apart) on the second circle out from center, place (using tweezers or straw spatula) a very small piece (~ 1–2 mm) of the following: (a) concrete, (b) marble, (c) marble dust, (d) alumina, (e) quartz, (f) road dust (damp), and (g) magnesium metal.

   CAUTION: Try not to scratch the dish!

2. Drop 2 or 3 drops of acidity probe (bromocresol green) onto each sample so that the sample is covered by rain.
   - Record the colors!
3. Have the lid ready and drop 2 drops 0.5 M Na_2SO_3 onto the center point of the dish. Generate SO_2 by adding 2 drops 2 M H_2SO_4. Place lid on dish immediately.
4. Watch for about 3 minutes.
   - Record what you see.
5. Remove the SO_2 source with a cotton swab and replace lid. Place the swab in waste cup containing water.
6. Wait and watch.
   - Record what happens.
   - By referring to the initial color changes (Step 2), can you tell which samples give basic reactions?
   - What is the chemical composition (approx.) of samples (a) through (e)?
   - What effect does mineral particle "size" have on the reaction with acid rain?
   - Look at the metal sample through your lens. What is the gas being produced?
7. Clean the dish. Try not to scratch the dish and lid.

Section I. The Effect of SO_2 on Plants

Goal:

To investigate the effect of both dry and wet deposition of SO_x on plants.

Discussion:

Many plants suffer SO_2 induced necrosis (death) when subject to acute SO_2 doses. Other plants show relatively high resistance, presumably due to natural defense mechanisms, such as resistant epicuticular waxes and fast stomatal closure. In this section you can investigate plant/SO_2 interactions by means of a very interesting *in vivo* color probe —

anthocyanins! These natural pigments, which are often present in epidermal cells of leaves and flowers, give dramatic color changes with SO_2. Cellular damage (anthropogenic or natural) can be shown to lead to rapid entry of gaseous air pollutants.

Experimental Steps:

1. Obtain 2 flower samples, a vegetable sample, and an anthocyanin solution sample from Reagent Central.
2. Place them appropriately into a clean, dry petri dish. Damage the epidermis of one of the flower samples by holding it up with tweezers and make several holes with a pin.
3. Drop 1 drop of rainwater onto one of the undamaged flowers.
   - What shape does the rainwater form?
4. Generate SO_2 in the usual manner.
5. Wait and watch.
   - Record what happens.

The chemical reaction that you see is due to the formation of a bisulfite addition compound between the anthocyanin and HSO_3^- ion formed by $SO_{2(g)}$ dissolving in cell fluid. For the formation of bisulfite ion, the sequence is:

$$SO_{2\,(g)} \leftrightharpoons SO_{2\,(aq)}$$

$$SO_{2\,(aq)} + H_2O \leftrightharpoons H^+ + HSO_3^-$$

Anthocyanin then reacts with bisulfite ion:

HO, OH, OH, +, OR, OR + HSO_3^- ⇌ HO, SO_3^-, OH, OH, OR, OR + H^+

Red to blue
(anthocyanin)

Colorless
(bisulfite addition compound)

You can see from the above scheme that this reaction can be reversed by adding strong acid to the colorless compound.

6. Try it. Add 1 drop 2 M H_2SO_4 to the colorless part of the plant.
   - Compare the effect of gaseous SO_2 versus the acidic raindrop on anthocyanin in the undamaged flower.

Interestingly, these anthocyanin reactions are of great importance in food, beverage, and wine production. Sulfur dioxide is used extensively as a preservative in fruit juices and in some wines. Obviously, color retention is an absolute necessity in these products and the concentration of SO_2 used must be carefully controlled.

7. Terminate the experiment in the usual manner.

Section J. Putting It All Together: Acid Deposition from NO_x plus SO_x

Goal: *To explore the complex gas and solution phase interactions of NO_x and SO_x.*

Discussion: One of the main reasons for studying air pollution chemistry in environmental chambers in the laboratory is the incredible complexity of the chemistry, biology, meteorology, etc., and the enormous difficulty of studying "real world" phenomena, which often occur over distances of hundreds of kilometers. In this system there is the fascinating prospect of gas phase reactions between $NO_{2(g)}$ and $SO_{2(g)}$ with subsequent aerosol formation. In this section, you can get a feel for some of the "real world" complexity by studying the combined NO_x/SO_x system.

Experimental Steps:

1. Place a clean, dry dish on the grid. At the positions indicated, drop the following solutions:

■ 1 drop 0.5 M KNO_2

▲ 2 drops 0.5 M Na_2SO_3

● 1 drop starch/KI

○ 1 drop 0.5 M $BaCl_2$

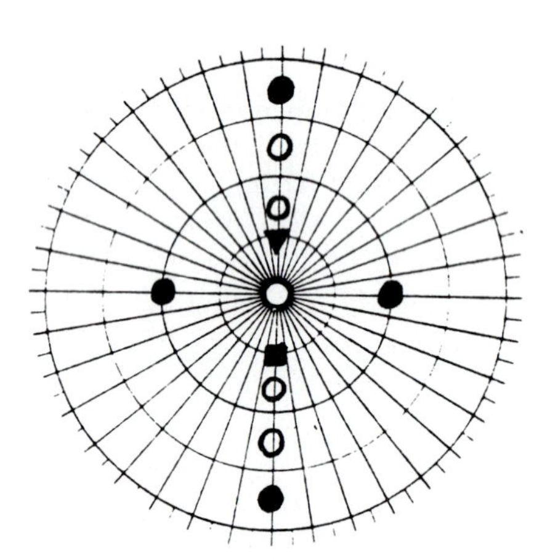

2. Generate SO_x and NO_x by adding 2 drops 2 M H_2SO_4 *first* to Na_2SO_3 and then to KNO_2. Replace the lid quickly.

Now before you record what you see, let's try an interesting experiment. Here's a fact — $NO_{2(g)}$ can oxidize $SO_{2(g)}$ to $SO_{3(g)}$ in the gas phase:

$$NO_{2(g)} + SO_{2(g)} \rightarrow NO_{(g)} + SO_{3(g)}$$

- Now, from the chemistry you have learned in this module and the above fact, can you predict what will happen to the various probe drops?
- In your notes describe the following:

  a) What each $BaCl_2$ drop will appear to look like

  b) The appearance of each starch/KI drop

  c) Where you think that aerosols will appear

  NOTE: Remember, there is no H_2O_2 in the $BaCl_2$. Any oxidation of $SO_{2(g)}$ is being done by $NO_{2(g)}$!

3. Now look at the solutions in the dish.
   - Record your observations in your lab book.
   - Are your predictions correct? If not, perhaps it would be worthwhile to try to interpret what you see.
4. Terminate and wash in the usual manner.

Section K. Space Shuttle Launches: An Example of Severe Local Hydrochloric Acid Deposition

Goal: *To investigate an example of severe local and possibly stratospheric pollution by hydrochloric acid from space shuttle launches. This experiment is optional.*

Discussion: We have seen that the major acidic components in acid deposition are usually nitric and sulfuric acids. There are, however, some very unusual examples of severe local acid deposition that are principally due to *hydrochloric acid.* The space shuttle affords us one example. The space shuttle has two solid rocket boosters that operate for the first two minutes of flight to provide the additional thrust needed for the shuttle to escape the gravitational pull of the earth. Each booster contains 1.1 million pounds of solid rocket propellant. The propellant looks like rubber and has the following composition:

- 70% ammonium perchlorate (NH_4ClO_4)
- 16% aluminum powder (Al)
- 12% polybutadiene acrylic acid acrylo nitrite (complex polymeric binder)
- 2% epoxy curing agent
- ~ 0.17% iron oxide powder (catalyst)

The incredible amount of energy comes from the oxidation of the aluminum to alumina (Al_2O_3) by NH_4ClO_4. This chemical reaction produces (in less than two minutes) about 240 tons of hydrochloric acid, 26 tons of chlorine, 7 tons of nitric oxide, and 304 tons of alumina. About 60 tons of HCl is released close enough to the ground to cause severe environmental damage in areas within about 20 miles of the launch site (Cape Canaveral, Florida). The concrete launch pad itself suffers severe erosion due to the high concentration of hydrochloric acid produced by wash out of the exhaust.

Numerous acid rain episodes have been reported after shuttle launches. In-cloud measurements made by plane as well as by ground sampling have shown this rain to be strongly acidic (>2 M HCl, pH < 1). NASA has explored the possibility of neutralizing the low-level exhaust cloud with bases (e.g., ammonia). In this extra-credit section, you can carry out a rather bizarre experiment involving the gas phase neutralization of $HCl_{(g)}$ with $NH_{3(g)}$.

Experimental Steps:

1. Take the lid off a clean, dry dish. Turn it over and drop 1 drop 6 M HCl onto the center. Drop 1 drop 6 M NH_3 onto the center of the dish bottom. You can use the grid to find the center.
2. Now you can demonstrate great skill and technique! Turn the lid over while keeping the drop of HCl in the center.
3. Place the lid on the dish and slide the dish over a black background.
   - Examine the phenomenon with your hand lens. Draw a picture.
   - Can you explain what you see? Give a chemical reaction. Do you think water is involved in this reaction?
   - Why would NASA be concerned about inadvertent weather modification of the Florida peninsula resulting from neutralization of shuttle exhaust clouds?

Chapter 17

Semimicro Qualitative Analysis and Metal Ions

The Control of Equilibria to Achieve Desired Separations and Analysis

Introduction

The technological society in which we live is driven by the tremendous consumption of fossil fuels and the mobilization of heavy metals. Heavy metal (unlike the musical connotation) is a name used to describe those metals in the periodic chart that occur after scandium (atomic number 21). Modern industrial societies have enormous requirements for many heavy metals and alloys, particularly iron, copper, chromium, nickel, and manganese. The end products are all around us: buildings, bridges, automobiles, ships, storage tanks (and other tanks), wiring, trains, airplanes, guns, and the list goes on.

All these metals come from rocks that must be reduced from the ionic state. The enormous scale on which this reduction is done requires the digging up, moving, sorting, and roasting (smelting) of millions of tons of minerals. Mining and industrial operations on this scale generate many jobs (and coincidentally, gaping holes in the earth, soil erosion, and severe air and water pollution) and the metals our industrial society demands. A prototypical example of heavy metal water pollution is discussed later in the Introduction.

Inevitably, the production of heavy metals redistributes and broadcasts these same heavy metals far and wide throughout the global biosphere. What was once deep and isolated in the earth's crust is now everywhere exposed simultaneously to dioxygen and water. The combination of oxidation, dissolution, erosion, and transport reactions in abandoned shaft mines, strip mines, and tailings dumps produces acid and heavy metal ion runoff into streams, rivers, and oceans. Heavy metal pollution continues to plague the metallurgical processing that is carried out in furnaces, smelters, and electrolytic cells. Metal forming, machining, plating, and finishing processes can all contribute to significant loading of heavy metals in the environment. Ultimately the free metals disintegrate as they are slowly attacked by dioxygen and water. As a result, the thermodynamically unstable metals electrochemically corrode back to the natural, ionic form. Predictably, this global cycle of ion to metal to ion strongly impacts on plants, animals, and, yes, eventually on humans, often with bizarre consequences.

However, several heavy metal ions are essential to life — e.g., iron, zinc, copper, chromium, and molybdenum — and, of course, are normally ingested in roughly the amounts required in the diet. Most of these ions are necessary for the proper functioning of many types of enzymes. Unsurprisingly, dietary deficiencies lead to impaired functions and, eventually, death. In the United States dietary deficiencies of iron and zinc are especially common, mainly because of poor diet and poor cooking practices. The nutritional requirements for heavy metal ions fall into a relatively narrow range. Too little in the diet results in deficiency symptoms, and too much produces toxic effects. It is interesting to note that nutrition studies in large animals (cattle, sheep, and pigs) have shown that dietary supplements of iron and zinc can lead to decreases in disease and dramatic increases in meat production. Severe zinc deficiencies in humans have been reported to cause dwarfism; with dietary supplements of zinc, "cures" have been reported in some cases. For more details see Chapter 22.

Ingestion of the wrong type of metal ion can cause illness, trauma, mental retardation, and death. Medical diagnosis of metal ion poisoning is extraordinarily difficult. Unfortunately, there have been many proven instances of widespread heavy metal ion poisoning in modern industrial societies, including our own. Perhaps the classic case is the mercury poisoning that occurred in the town of Minamata, Kyushu Island, Japan. At the end of 1953, a strange disease began to afflict the people of this region and by 1956 had assumed epidemic proportions. The symptoms were uncontrollable shaking, salivating, staggering walk, mental retardation, grotesque body deformation — and death was the eventual consequence. Scientists and doctors tested for various communicable diseases, and eventually (three years later), a link was established between fish caught in Minamata Bay and the disease. Many heavy metals — e.g., manganese, thallium, copper, lead, and arsenic — were investigated and ruled out. In 1959 the causative agent was found to be mercury compounds present in the fish. The problem was then to find the source of the toxic compounds. The town had one large factory, the Chisso Company, that manufactured chemicals for making plastics and had been discharging waste water into the bay for many years. For ten years, however, the company frustrated every outside attempt to obtain samples. Eventually, the waste was found to contain mercury compounds that had then accumulated in the bay. The company steadfastly refused to admit any liability. Only after six years of courageous protest by the victims was the matter resolved in a landmark decision in the Japanese Supreme Court (1974). Hence, we have the prototypical example of *water* pollution due to a heavy metal.

The Minamata scandal shook other countries into action, including Sweden, Canada, and the United States. A national research program was carried out in the United States to investigate the problem of lead in the environment and the public health. After about *ten* years of research and testing, the conclusion was that lead exhausted from automobiles was a serious health problem. Consequently, the lead antiknock compounds in gasoline are *slowly* being phased out in order to reduce the environmental loading of these heavy metal pollutants.

Industrial societies would be crippled and economies would collapse if the use of heavy metals were to be abruptly discontinued. It is important, however, to study the implications and consequences of the vast mobilization and redistribution of heavy metals. Only when there is sufficient information about, and interest in, these complex environmental problems will public health disasters be avoided. Any research involving a study of the environmental and health aspects of heavy metals must be based on the following criteria:

- A knowledge of the chemistry of the metals, particularly of the chemical forms that exist in natural aquatic and soil habitats, and their chemical toxicology
- Methods and techniques for the separation and analysis of complex mixtures of the heavy metals (often when present in small concentrations)

One of the main objectives of this laboratory module is to learn metal ion chemistry and, therefore, obtain a good perspective on some of the environmental consequences of heavy metal ion pollution. First, however, it will be helpful to look at a very brief survey of some of the principal sources, uses, and health considerations of the heavy metals that you will be studying. The metals are silver, lead, mercury, copper, cadmium, iron, chromium, manganese, zinc, and nickel.

SOURCES, USES, AND EFFECTS OF HEAVY METALS

Silver

(Symbol Ag, atomic number 47, second transition period)

Used widely in photography (in the form of film emulsions), coinage (prior to 1965 in the United States), and in solders. Also still used in antiseptics. One of the most common forms is silver nitrate, $AgNO_3$. Minimum lethal dose (MLD) for $AgNO_3$ is approximately 2 grams/150 pound man. Relatively nontoxic but cumulative. Symptoms of poisoning include severe gastro-intestinal irritaiton, nausea, vomiting, yellow burns on mouth and skin. Permanent blue-black pigmentation of skin (argyria) may occur after long exposure.

Lead

(Symbol Pb, atomic number 82, group IV A)

Uses include paints, ceramic glazes, auto batteries, china ware, alloys, solder, foil wrapper (e.g., around tops of wine bottles), insecticides, vulcanized rubber, shot bullets, gasoline (as an antiknock additive). The MLD for soluble lead salts is approximately 5–10 g/150 pound man; in lead tetraethyl (antiknock agent in gasoline), 100 mg/150 pound man. Cumulative poisoning occurs when absorption exceeds elimination. Children are especially susceptible and show signs of convulsions (encephalitis). Symptoms include central nervous system damage, mental retardation, peripheral neuritis, partial paralysis, collapse, and coma. Eventually results in death.

Mercury

(Symbol Hg, atomic number 80, third transition period)

Uses include insecticides and antifungal agents (e.g., processing of seeds and as preservatives in cosmetics), and in industry such as slimicides (in the paper pulp industry), in plastics, lamps, and amalgams. Many industrial uses are in the production of organic chemicals. The MLD vary widely, depending on the compound, for example, mercuric chloride ($HgCl_2$) is approximately 0.5 g/150 pound man. Some are extremely poisonous in very, very small amounts, such as methyl and dimethyl mercury. Symptoms include rapid weak pulse, pallor, cold moist skin, slow breathing, severe acidemia, leukocytosis, shock, collapse, and eventually lead to death. Central nervous system disorders and symptoms include tremors (shakes), partial paralysis, loss of memory, easily upset (the Mad Hatter!)

Copper

(Symbol Cu, atomic number 29, first transition period)

A large variety of industrial uses include alloys, e.g., brass, wire, plating of electronic parts, insecticides, fungicides, algaecides, and some paints. The MLD, in copper sulfate ($CuSO_4$) for instance, is approximately 15 g/150 pound man. The metal is relatively nontoxic and cumulative. Symptoms include kidney irritation, thirst, irregular pulse and respiration, cold sweat, delirium, convulsions, exhaustion, pain in chest, and coma.

Cadmium

(Symbol Cd, atomic number 48, second transition period)

Examples of uses are alloys, plating (for screws, etc.), welding, batteries, and paint pigments. The MLD is approximately 1 g/150 pound man. Symptoms include headache, rapid pulse, liver damage, convulsions, extreme pain in joints ("itai-itai" disease), pulmonary edema, shock, collapse, and eventually results in death.

Iron

(Symbol Fe, atomic number 26, first transition period)

Iron is mainly used in the production of steel and other alloys. Iron is one of man's dietary requirements and is relatively nontoxic.

Chromium

(Symbol Cr, atomic number 26, first transition period)

Major uses include alloys (stainless steel), plating, magnets, tanning, and dyeing pigments (especially for ceramics). Not much is known about its toxicity.

Manganese

(Symbol Mn, atomic number 25, first transition period)

Manganese is mainly used in the production of alloys. Manganese ion (Mn^{2+}) is necessary for the human diet. Not much is known about its toxicity.

Zinc

(Symbol Zn, atomic number 30, first transition period)

Uses include alloys (brass), plating (galvanizing), batteries, and solder flux. Zinc ion is a human dietary requirement. The MLD in Zinc Sulfate ($ZnSO_4$), for example, is approximately 15 g/150 pound man. Relatively nontoxic in small amounts. Symptoms include abdominal pain, leukocytosis, fever, and a metallic taste in the mouth.

Nickel

(Symbol Ni, atomic number 28, first transition period)

Nickel is mainly used in the production of alloys (stainless steel), batteries, and plating. The MLD depends on the compound — some, such as nickel carbonyl ($NiCO_4$), are extremely toxic. The metal itself is relatively nontoxic.

Background Chemistry

In this module you are going to be doing semimicro qualitative analysis (SMQA) of samples for heavy metal ions. *Semimicro* refers to the relative scale of sample sizes, volumes, and equipment size. *Qualitative* refers to an analysis in which the presence or absence of certain chemical species is determined, but not the amount or concentration. Generally, the major steps involved in a qualitative chemical analysis are

1. Dissolution of the sample in some type of solvent
2. Separation of the various chemical species present
3. Identification of a separated species on the basis of cumulative chemical differences

Samples for analysis, particularly environmental ones, may contain all the heavy metal ions that can be analyzed. It is impossible to carry out specific tests for each metal ion when the ions are in a complex mixture. The reason is that the chemistries of many of the ions are very similar, and they interfere with each other in the identification reactions. The interference problem can be overcome by taking advantage of the differences that do exist and effecting a *group separation*. In this module you will be able to analyze a very complex mixture of up to 11 metal ions by first separating them into three groups:

- Group I: containing Ag^+ (silver ion), Pb^{2+} (lead ion), and $Hg_2{}^{2+}$ (mercury (I) ion)
- Group II: containing Pb^{2+} (lead ion), Cu^{2+} (copper (II) ion), Cd^{2+} (cadmium ion), and Hg^{2+} (mercury (II) ion)
- Group III: containing Fe^{3+} (iron(III) ion), Cr^{3+} (chromium (III) ion), Mn^{2+} (manganese (II) ion), Zn^{2+} (zinc ion), and Ni^{2+} (nickel ion)

Separation of metal ions into these three groups is obtained by adding a group reagent that will precipitate a sparingly soluble compound of the ions in that group. Once a solid precipitate is obtained, it is then physically separated from the remaining solution (containing other metal ions) by centrifugation, and the supernatant liquid is removed by means of a pipet. The mixture of solids from each group precipitation is then further separated by utilizing chemical differences between the group species. Finally, when individual species have been separated, confirmatory tests are done to show unequivocally that certain ions are present or absent.

It must be emphasized that throughout the entire analysis scheme, subtle differences in the metal ion chemistry have to be exploited in order to achieve *clean* separations. The solution conditions must be carefully controlled in order to optimize the extent of the chemical equilibria because the proof of the presence of a particular metal ion in a sample rests on all the chemical evidence accumulated from the beginning of group separation, through all the intermediate reactions, and to the final confirming reaction. Each metal ion must show a distinct chemical profile in order to be identified. All the types of equilibria that occur in aqueous solution — i.e., acid-base, complexation, redox, and precipitation — must be manipulated to produce a good analysis. An analysis of a complex sample demands an integration of a variety of principles, technique, timing, and flow. The chemistry is going to come fast and furiously and you must be prepared to focus and think through what you are doing (and why) at each stage of the scheme. If you don't, the result will be mindless mixing followed by an incorrect analysis.

The following sections of Background Chemistry give a concise description of the chemistry pertinent to each of the group separations. Please read through it, but don't try to memorize any of it. The chemical principles will slowly make sense as you apply them during the actual analysis.

In the first laboratory period, your instructor will give a demonstration of the various semimicro techniques. You may then start an SMQA on a sample solution that contains all of the 11 cations described earlier.

NOTE: In order to keep the following discussion as short as possible, chemical formulas are used rather than chemical names. A useful way to brush up on your chemical nomenclature is to try to read in the names as you go through the text. Formulas and names for many of the compounds you will encounter in the Laboratory Experiments section of SMQA are listed at the end of Background Chemistry. Please refer to this list if you are having nomenclatural amnesia.

GROUP I (Ag^+, Pb^{2+} Hg_2^{2+}) CHEMISTRY

The group reagent is HCl. The addition of a small excess of HCl to a solution containing Ag^+, Pb^{2+}, and Hg_2^{2+} will result in the formation of three sparingly soluble precipitates:

$$Ag^+ + Cl^- \rightleftharpoons AgCl_{(s)}$$

$$Pb^{2+} + 2Cl^- \rightleftharpoons PbCl_{2(s)}$$

$$Hg_2^{2+} + 2Cl^- \rightleftharpoons Hg_2Cl_{2(s)}$$

Of course, the group I reagent must completely precipitate group I ions, but not group II or III ions. A small excess of HCl accomplishes this precipitation; however, a large excess would cause problems due to the formation of soluble chlorocomplexes of the precipitates, e.g.,

$$AgCl_{(s)} + Cl^- \rightleftharpoons AgCl_2^-$$

The solid is separated from the supernatant liquid (which is reserved for group II and III separations). Use is now made of the fact that $PbCl_{2(s)}$ is reasonably soluble in hot water, whereas $AgCl_{(s)}$ and $Hg_2Cl_{2\,(s)}$ are not:

$$PbCl_{2(s)} \overset{\text{heat}}{\rightleftharpoons} Pb^{2+} + 2\,Cl^- \text{ (in solution)}$$

Pb^{2+} is then confirmed by adding a K_2CrO_4 solution:

$$Pb^{2+} + CrO_4^{2-} \rightleftharpoons PbCrO_{4\,(s)}$$

and a precipitate of $PbCrO_{4(s)}$ is formed.

The residue from the hot H_2O treatment is a mixture of two solids: $AgCl_{(s)}$ and $Hg_2Cl_{2(s)}$. Concentrated NH_3 is added, and Ag^+ is complexed to form a soluble complex ion:

$$AgCl_{(s)} + 2NH_3 \rightleftharpoons Ag(NH_3)_2^+ + Cl^-$$

The residue that remains after treatment with NH_3 is a mixture of $Hg_{(s)}$, $Hg_2O_{(s)}$, and $HgNH_2Cl_{(s)}$ in basic solution. The solution containing soluble $Ag(NH_3)_{(s)}^+$ is made acidic with HNO_3, and the dissolution process is reversed:

$$Ag(NH_3)_2^+ + Cl^- + H^+ \rightleftharpoons AgCl_{(s)} + 2NH_4^+$$

The precipitation of $AgCl_{(s)}$ confirms the presence of Ag^+ in the original sample. The residue, described earlier, confirms the presence of Hg_2^{2+}.

GROUP II (Hg^{2+}, Pb^{2+}, Cu^{2+}, Cd^{2+}) CHEMISTRY

The basis for separation of this group (and the Pb^{2+} that came through from group I) is the precipitation of the metal ions as sparingly soluble sulfides from *acid* solution. The group reagent is H_2S, which is produced *in situ* by

$$CH_3CSNH_2 + H_2O \xrightarrow{\text{heat}} CH_3CONH_2 + H_2S$$

This rather unusual way of producing the group reagent is done for safety reasons. H_2S is an extremely toxic, evil-smelling gas. However, if it is prepared as above, very little escapes from solution, and enough is generated to produce the group precipitates. The pH of the aqueous solution at the start of this group is critical to the separation. The $[H^+]$ controls the formation of S^{2-} through the equilibria of H_2S acting as a weak diprotic acid:

$$H_2S \rightleftharpoons H^+ + HS^-$$

$$HS^- \rightleftharpoons H^+ + S^{2-}$$

The solution is acidic, mainly due to H^+ from the strong acid HNO_3. A series of dilutions with H_2O ensures that the final pH is about 0.5. At this low pH, the $[S^{2-}]$ is very low, and only those metal ions that produce sparingly soluble sulfides with a very small K_{sp} will precipitate. The group III ions with relatively large K_{sp} values will not precipitate under these conditions. The real problem in this group is Cd^{2+}, which has an intermediate K_{sp} value. Dilution of the solution decreases the $[H^+]$ (and, hence, increases the pH) enough to give a $[S^{2-}]$ sufficient to precipitate Cd^{2+}, but not Zn^{2+}. The precipitation reactions are

$$Hg^{2+} + S^{2-} \rightleftharpoons HgS_{(s)}$$

$$Pb^{2+} + S^{2-} \rightleftharpoons PbS_{(s)}$$

$$Cu^{2+} + S^{2-} \rightleftharpoons CuS_{(s)}$$

$$Cd^{2+} + S^{2-} \rightleftharpoons CdS_{(s)}$$

The group II precipitation separates the ions from group III ions, which remain in solution. Group II sulfides must now be redissolved in order to separate the individual ions. A relatively high concentration of HNO_3 (6 M) drives the precipitation equilibria back towards soluble ions by pushing the H_2S equilibria towards free H_2S, except for $HgS_{(s)}$, which has a very small K_{sp} (3×10^{-52}).

The residue of $HgS_{(s)}$ is dissolved only by the combined effect of very high $[H^+]$ and an oxidizing environment. A mixture of 12 M HCl and 16 M HNO_3 does the job:

$$3HgS_{(s)} + 2HNO_3 + 6HCl \rightleftharpoons 3HgCl_{2(s)} + 4H_2O + 2NO_{(g)} + 3S_{(s)}$$

The $HgCl_{2(s)}$ formed dissociates to give Hg^{2+}, and this is reduced with $SnCl_2$ in a redox reaction:

$$2Hg^{2+} + Sn^{2+} \rightleftharpoons Hg_2{}^{2+} + Sn^{4+}$$

The $Hg_2{}^{2+}$ is reacted with excess Cl^- to give

$$Hg_2{}^{2+} + 2Cl^- \rightleftharpoons Hg_2Cl_{2(s)}$$

which confirms the presence of Hg^{2+} in the original sample.

The solution containing the other group II ions is treated with a high concentration of $(NH_4)_2SO_4$, yielding a precipitate of $PbSO_{4(s)}$

$$Pb^{2+} + SO_4{}^{2-} \rightleftharpoons PbSO_{4\,(s)}$$

The other ions, Cu^{2+} and Cd^{2+}, stay in solution. The $PbSO_{4(s)}$ is dissolved by an excess of acetate:

$$PbSO_{4\,(s)} + C_2H_3O_2{}^- \rightleftharpoons Pb(C_2H_3O_2)^+ + SO_4{}^{2-}$$

The addition of K_2CrO_4 will again confirm Pb^{2+} by a precipitate of $PbCr_2O_4$.

A high concentration of NH_3 is added to the solution that contains Cu^{2+} and Cd^{2+}, and both are converted to ammine complex ions:

$$Cu^{2+} + 4NH_3 \rightleftharpoons Cu(NH_3)_4^{2+}$$

$$Cd^{2+} + 4NH_3 \rightleftharpoons Cd(NH_3)_4^{2+}$$

The resulting solution is divided into two parts. One part is made acidic and $K_4Fe(CN)_6$ is added:

$$2Cu(NH_3)_4^{2+} + Fe(CN)_6^{4-} \rightleftharpoons Cu_2Fe(CN)_{6\,(s)} + 8NH_3$$

To test for Cd^{2+} in the presence of Cu^{2+}, the $Cu(NH_3)_4^{2+}$ is first reduced by $Na_2S_2O_4$ to $Cu_{(s)}$, which is removed by centrifugation. The Cd^{2+} is detected by providing more S^{2-} to produce a confirmatory precipitate of $CdS_{(s)}$.

GROUP III (Fe^{3+}, Cr^{3+}, Ni^{2+}, Zn^{2+}, Mn^{2+}) CHEMISTRY

In order to analyze for the ions in group III, any H_2S left from group II must be removed. Evaporation and oxidation accomplish this removal. Any Fe^{2+} is oxidized to Fe^{3+} by H_2O_2:

$$2H^+ + H_2O_2 + 2Fe^{2+} \rightarrow Fe^{3+} + 2H_2O$$

The solution is adjusted to a pH of 10 with an NH_3/NH_4Cl buffer. At this pH, Fe^{3+} and Cr^{3+} precipitate as the hydroxides,

$$Fe^{3+} + 3OH^- \rightleftharpoons Fe(OH)_{3\,(s)}$$

$$Cr^{3+} + 3OH^- \rightleftharpoons Cr(OH)_{3\,(s)}$$

whereas Ni^{2+} and Zn^{2+} are held in solution as soluble complex ions:

$$Ni^{2+} + 6NH_3 \rightleftharpoons Ni(NH_3)_6^{2+}$$

$$Zn^{2+} + 4NH_3 \rightleftharpoons Zn(NH_3)_4^{2+}$$

The Mn^{2+} does not form a complex ion and does not precipitate because the $[OH^-]$ is not high enough.

The solid mixture of $Fe(OH)_{3(s)}$ and $Cr(OH)_{3(s)}$ is oxidized with H_2O_2 in basic solution, and $Cr(OH)_{3(s)}$ is oxidized to CrO_4^{2-}, which is confirmed with $Pb(C_2H_3O_2)_2$. The remaining $Fe(OH)_{3(s)}$ is dissolved in acid,

$$Fe(OH)_{3\,(s)} + 3H^+ \rightleftharpoons Fe^{3+} + 3H_2O$$

and confirmed by the addition of NH_4SCN:

$$Fe^{3+} + SCN^- \rightleftharpoons FeSCN^{2+}$$

The solution containing the soluble complex ions $Ni(NH_3)_6{}^{2+}$, $Zn(NH_3)_4{}^{2+}$, and Mn^{2+} is treated with H_2S when

$$Ni(NH_3)_6{}^{2+} + S^{2-} \rightleftharpoons NiS_{(s)} + 6NH_3$$

Precipitates of $ZnS_{(s)}$ and $MnS_{(s)}$ form in a similar manner. The $ZnS_{(s)}$ and $MnS_{(s)}$ are dissolved in 6 M HCl.

The $NiS_{(s)}$ is dissolved in strong acid at high concentration,

$$NiS_{(s)} + 2H^+ \rightleftharpoons Ni^{2+} + H_2S_{(g)}$$

and, in basic solution, Ni^{2+} with dimethylglyoxime gives the confirmatory precipitate of nickel dimethylglyoximate:

$$Ni^{2+} + 2\,CH_3-\underset{\underset{HO}{\|}\,N}{C}-\underset{\underset{OH}{\|}\,N}{C}-CH_3 \rightleftharpoons Ni(CH_3C(=NO)C(=NOH)CH_3)_2 + 2H^+$$

The remaining liquid contains Mn^{2+} and Zn^{2+}. A high $[OH^-]$ is added, and Mn^{2+} precipitates:

$$Mn^{2+} + 2\,OH^- \rightleftharpoons Mn\,(OH)_{2\,(s)}$$

The $Mn(OH)_{2(s)}$ is redissolved in acid and oxidized by $NaBiO_3$ to permanganate ion ($MnO_4{}^-$), which confirms Mn^{2+}:

$$2Mn^{2+} + 5HBiO_3 + 9H^+ \rightleftharpoons 5Bi^{3+} + 7H_2O + 2MnO_4{}^-$$

The remaining solution contains $Zn(OH)_4{}^{2-}$. A complexing agent, dithizone, is added, and the resulting complex confirms the presence of Zn^{2+}:

$$Zn^{2+} + 2DH \rightleftharpoons Zn(D)_2 + 2H^+$$

NOTE: D is the dithizone molecule.

A COMMENT ABOUT WRITING CHEMICAL REACTIONS

In the preceding discussions of the chemistry of each group, chemical reactions have been written as net ionic reactions in the direction in which they are carried out experimentally in an analysis. An example is the addition of the group I reagent HCl to a solution containing Ag^+, Pb^{2+}, and Hg_2^{2+} ions in order to precipitate the sparingly soluble chlorides:

$$Ag^+ + Cl^- \rightleftharpoons AgCl_{(s)}$$

$$Pb^{2+} + 2Cl^- \rightleftharpoons PbCl_{2\,(s)}$$

$$Hg_2^{2+} + 2Cl^- \rightleftharpoons Hg_2Cl_{2\,(s)}$$

These precipitation reactions are very fast, and when equilibrium is achieved, the equilibrium constants are the reciprocals of the solubility product constants (K_{sp}). Normally, these heterogeneous equilibria are written as solubility (dissolution) equilibria:

$$AgCl_{(s)} \rightleftharpoons Ag^+ + Cl^-$$

$$PbCl_{2(s)} \rightleftharpoons Pb^{2+} + 2Cl^-$$

$$Hg_2Cl_{2(s)} \rightleftharpoons Hg_2^{2+} + 2Cl^-$$

The solubility product constants and expressions are

- For $AgCl_{(s)}$, $K_{sp} = 1.8 \times 10^{-10} = [Ag^+][Cl^-]$
- For $PbCl_{2(s)}$, $K_{sp} = 1.7 \times 10^{-15} = [Pb^{2+}][Cl^-]^2$
- For $Hg_2Cl_{2(s)}$, $K_{sp} = 1.2 \times 10^{-18} = [Hg_2^{2+}][Cl^-]^2$

NOTE: Be aware that the equilibrium constants for precipitation reactions are very big, whereas the equilibrium constants for the dissolution reactions of sparingly soluble solids are very small. In a chemical analysis involving the separation of ions by precipitation, complete removal of the precipitated ions is necessary in order to proceed through the scheme.

CHEMICAL FORMULAS AND NAMES

The qualitative analysis of samples for metal ions involves an enormous variety of chemical elements, compounds, and reactions. Keeping track of all the chemical names for these substances is not an easy task. Although you may be familiar with many of the simple acids, bases, and salts, some of the coordination compounds (complexes) and organic compounds may be new to you. Tables 17.1–17.3 provide a list of the chemical names of the major substances you will encounter in group I, II, and III metal-ion analysis.

Table 17.1	*Group I*
HCl	hydrochloric acid
$AgCl$	silver chloride
$PbCl_2$	lead chloride
Hg	mercury
Hg_2O	mercury (I) oxide
Hg_2Cl_2	mercury (I) chloride
K_2CrO_4	potassium chromate
$Ag(NH_3)_2^+$	diamminesilver (I) ion
HNO_3	nitric acid
NH_3	ammonia

Table 17.2	*Group II*
H_2S	hydrogen sulfide
CH_3CSNH_2	thioacetamide
HgS	mercury (II) sulfide
PbS	lead sulfide
CuS	copper (II) sulfide
CdS	cadmium sulfide
$SnCl_2$	tin (II) chloride
$(NH_4)_2SO_4$	ammonium sulfate
$C_2H_4O_2$	acetic acid
$NH_4C_2H_3O_2$	ammonium acetate
$C_2H_3O_2^-$	acetate ion
$Cu(NH_3)_4^{2+}$	tetraamminecopper (II) ion
$Cd(NH_3)_4^{2+}$	tetraamminecadmium ion
$K_4Fe(CN)_6$	potassium ferrocyanide
$Na_2S_2O_4$	sodium dithionite

Table 17.3	*Group III*
H_2O_2	hydrogen peroxide
NH_4Cl	ammonium chloride
$Fe(OH)_3$	iron (III) hydroxide
$Cr(OH)_3$	chromium (III) hydroxide
$Ni(NH_3)_6^{2+}$	hexaamminenickel (II) ion
$Zn(NH_3)_4^{2+}$	tetraamminezinc (II) ion
$Pb(C_2H_3O_2)_2$	lead acetate
NH_4SCN	ammonium thiocyanate
$FeSCN^{2+}$	thiocyanato iron (II) ion
NiS	nickel sulfide
ZnS	zinc (II) sulfide
MnS	manganese (II) sulfide
$Mn(OH)_2$	manganese (II) hydroxide
$NaBiO_3$	sodium bismuthate
$NaOH$	sodium hydroxide
MnO_4^-	permanganate ion
$Zn(OH)_4^{2-}$	tetrahydroxozincate (II) ion
$C_4H_8N_2O_2$	dimethylglyoxime

Pre-Laboratory Quiz

1. Name 3 heavy metal ions that are regarded as being essential to life.

2. What heavy metal pollution was responsible for "Minamata" disease?

3. Give 2 uses for chromium metal.

4. What are the 3 major steps in any qualitative chemical analysis?

5. Give the formula for the cations of group I.

6. What is the group reagent for group I separation?

7. In group II separation, the $[S^{2-}]$ is kept very low. How is this done?

8. Give the formula and name for a soluble complex ion.

9. Write out the equilibrium reaction for the dissolution of lead chloride.

10. What is the chemical name for MnO_4^-?

Laboratory Experiments

Flowchart of the Experiments

Section A. Group I: The Separation and Detection of Ag^+, Pb^{2+}, and Hg_2^{2+}

Section B. Group II: The Separation and Detection of Hg^{2+}, Pb^{2+}, Cu^{2+}, and Cd^{2+}

Section C. Group III: The Separation and Detection of Fe^{3+}, Cr^{3+}, Ni^{2+}, Zn^{2+}, and Mn^{2+}

Section D. Case History and Analysis of an Unknown Sample

Requires three three-hour class periods to complete

CAUTION: Semimicro qualitative analysis of aqueous samples for 11 cations requires the use of many corrosive and toxic chemicals. Wear your safety goggles (over your eyes!). Use good technique and think about what you are doing. If you get any chemicals on your hands, etc., wash well with cold water and inform your instructor as quickly as you can.

TECHNIQUES IN SEMIMICRO QUALITATIVE ANALYSIS (SMQA)

Semimicro qualitative analysis involves analyzing samples that weigh only about 1–10 mg or liquid samples consisting of a few drops. The range of solution volumes in group separations is 2 drops to about 3 mL. You have been using a variety of small-scale techniques throughout this chemistry laboratory course, and you are now familiar with how much care is necessary when working with small samples. However, the process of analyzing a sample for 11 metal ions requires even more care. Please remember that surface effects are important. These effects often make mixing and transferring solutions somewhat difficult. It is also very easy to contaminate solutions when you are working on a small scale. A clean work area and clean apparatus are essential. A brief description follows of all the techniques that you will be performing, and your instructor will demonstrate them for you.

Measuring and Transferring Solutions. Solution volumes are measured in drops and, occasionally, in milliliters. There are about 20–25 drops (from droppers, pipets, etc.) per milliliter. It is imperative that transfer pipets and droppers do not touch the sides of test tubes or dip into solutions unless they are clean. Do not contaminate reagent droppers with your analysis solutions.

Mixing Solutions. Because of the small diameter of the test tubes used in SMQA, solutions are rather difficult to mix. However, there are several ways to accomplish well-mixed homogeneous solutions or uniform suspensions. If the solution fills no more than half the test tube, grasp the top of the tube with the forefinger and thumb of one hand and flick the bottom of the tube with the forefinger of the other hand. *Wear your safety goggles.* (Your neighbor's flicking may be more enthusiastic than safe!) Another simple mixing technique is to suck up a portion of the solution into a pipet and then expel it back out. Repeat this action several times (gently), and the solution will be well mixed. Solutions may also be stirred with a small glass rod. This method is particularly effective for dislodging and resuspending precipitates. However, you must take care not to push the rod through the bottom of the rather fragile tube.

Heating Solutions. Solutions in small test tubes should *never* be heated directly in any type of flame because they bump and shoot solution out. Always heat solutions in test tubes in a boiling water bath. The test tube can be held vertically in a hot water bath by 2 test tube clamps resting on the edge of the bath. Your instructor will demonstrate. Solutions in porcelain casseroles may be heated directly over a small flame. Keep the casserole moving to stop the liquid from spitting.

Evaporating Solutions. Casseroles are very convenient containers for evaporating a few milliliters of solution. Make sure that the handle has a cork insulator and carry out the evaporation in a fume hood or in a very well ventilated area. Withdraw the casserole from the flame *before* complete dryness is reached and allow the heat retained in the casserole to complete the evaporation.

Centrifugation. Centrifugation is a quick, simple way in which to separate the usually more dense solids from less dense liquids. Centrifugation involves much less loss of material than filtration when separations are done on a small scale. Always make sure that the centrifuge is well placed on the bench. Balance the centrifuge head by placing a tube (filled to the same level with water) symmetrically across from where you are placing your tube. It is generally not necessary to centrifuge for more than 5–10 seconds at top speed. Do not try to slow the centrifuge head down too quickly. Remember that a loose test tube ejected from a centrifuge is a lethal projectile!

Separation of Supernatant From a Solid. This operation is carried out with a pipet and a steady hand. Squeeze the bulb before inserting it into the solution. Try drawing the solution off in several portions rather than all at once. Don't forget to use clean pipets. Clean the pipet after *each* use and then you won't inadvertently contaminate your solution.

Washing Precipitates. It is sometimes necessary to *completely* remove a reagent solution (supernatant) from a precipitate. Removal of the supernatant from a solid usually leaves the solid wet with the solution. To remove this residual solution, add wash liquid (often water) and break up the solid with a stirring rod. Centrifuge and remove the supernatant wash liquid with a pipet. Two washes are usually sufficient to ensure complete removal.

Transferring Precipitates. This operation must be done when the volume of the solution is so large that a precipitation had to be carried out in 2 or 3 tubes. The problem is how to collect the precipitates into a single tube. Add water to one of the tubes containing a precipitate and squirt the mixture in and out of the pipet to resuspend the solid. Suck the suspension up and transfer it to a clean, empty tube. Transfer the other precipitate portions in the same way. Centrifuge and remove the supernatant wash liquid. You now have all the precipitates in one place.
REMEMBER: When in doubt about technique, ask your instructor. Of course, there are other ways of doing semimicro qualitative analysis, but discuss your ideas with the experienced person before trying them out!

KEEPING YOUR LABORATORY RECORD OF SMQA

You can see from the rather complicated discussion of the background chemistry and techniques that SMQA involves observing and recording lots of experimental data. You must be well organized and have a clear idea of where you are going. The experimental steps are not listed one after the other as they have been thus far. Each of the groups

(I, II, and III) is arranged in the form of a flowchart with the instructions superimposed. This is done to clarify each step of the process.

As you go along through the analysis, you will see many types of chemical phenomena — e.g., color changes, precipitation of solids, evolution of gases, changes of texture, and so on. Record your observations, as concisely as you can, on the template sheets provided at the end of the chapter. The templates are exact replicas of the flowchart for each group except that they are empty — you fill them in. It is suggested that you tear out the templates and use them alongside the group flowchart. Try to write net ionic reactions that correlate with the observations you are making. Most of the reactions are given in the Background Chemistry section, and it should be fairly easy to match them with what you see. Write a short conclusion at the end of each group experiment. Briefly discuss tests that you found to be better or perhaps more difficult to perform than others. In the final section, Section D, of this module, you will analyze an unknown heavy metal ion sample. The points you highlight in the conclusions of the groups I–III analyses of your known sample will help immensely. Write your laboratory record while you are in the laboratory. A firsthand, real time, accurate description of phenomena is the first and perhaps most important task of doing science!

CAUTION: Wear your safety goggles!

NOTE: To assist you in working through the flowcharts, the procedure for the *confirmation of a metal ion* is in a box with double outline (so that you are cued to check the product).

Section A. Group I: The Separation and Detection of Ag^+, Pb^{2+}, and Hg_2^{2+}

You will be analyzing a known solution that contains *all* the 11 heavy metal ions: Ag^+, Pb^{2+}, Hg_2^{2+}, Hg^{2+}, Cu^{2+}, Cd^{2+}, Fe^{3+}, Cr^{3+}, Mn^{2+}, Zn^{2+}, and Ni^{2+}. In Section A, you will be separating and detecting group I ions.

Add 10 drops of known solution to a clean test tube. Add 1 drop of 6 M HCl, stir, and centrifuge. Add 1 more drop of 6 M HCl to check for completeness of precipitation. Centrifuge. Separate solution from precipitate.

PRECIPITATE: $AgCl$, $PbCl_2$, Hg_2Cl_2 (all white)
Wash the precipitate with a few drops of very dilute HCl (10 drops H_2O + 1 drop 6 M HCl).
Centrifuge and add the wash to the solution to be saved for groups II and III.

SOLUTION: SAVE for group II and group III.

Wash the PRECIPITATE 3 times with 1 mL portions of boiling H_2O. Quickly centrifuge each time (in order not to let the H_2O cool down). Transfer hot solutions to a single clean test tube.

RESIDUE: $AgCl$, Hg_2Cl_2
Wash 4 times with 1 mL portions of boiling H_2O. Centrifuge each time and discard the hot washings.

SOLUTION: $PbCl_2$ (dissolved)
Add 1 drop K_2CrO_4. A yellow precipitate of $PbCrO_4$ *confirms* the presence of Pb^{2+}.

Add 5 drops of 15 M NH_3 to the *residue*. Stir thoroughly. Centrifuge and separate residue and solution.

RESIDUE: Hg, Hg_2O, and $HgNH_2Cl$

A residue that is grey to black, at this point, *confirms* Hg_2^{2+}.

SOLUTION: $Ag(NH_3)_2^+$

Add 6 M HNO_3 drop by drop, while stirring, until the solution is acidic. A precipitate of AgCl *confirms* the presence of Ag^+.

Section B. Group II: The Separation and Detection of Hg^{2+}, Pb^{2+}, Cu^{2+}, and Cd^{2+}

NOTE: The solution for group II is the SOLUTION that you *saved* from group I separation. The solution contains Hg^{2+}, Pb^{2+}, Cu^{2+}, and Cd^{2+}, plus group III ions.

Pour the SOLUTION into a clean casserole. Wash the test tube with 3 drops of H_2O and add washings to the casserole. Evaporate to a paste (slightly damp). Dissolve the residue in 4 drops of 6 M HNO_3 and 1 mL H_2O. You will get a precipitate and a solution. Transfer the solution to a test tube.

Add 2 mL of 5% thioacetamide to the SOLUTION. Heat on a boiling water bath for 2 minutes. Transfer half of the resulting mixture to a clean test tube. To *each* test tube add 10 drops of thioacetamide and 10 drops H_2O. Heat both tubes in a hot water bath 5 minutes. Centrifuge. Transfer the supernatant solutions from both test tubes into a third test tube. Transfer the precipitates from both test tubes into a fourth test tube.

PRECIPITATE: PbS, HgS, CuS (all black), CdS (yellow)
Wash the precipitate once with 2 mL of H_2O.
Discard washing.

SOLUTION: SAVE for group III analysis.

Add 4 drops 6 M HNO_3 to the PRECIPITATE and heat for 2–3 minutes on bath. Centrifuge and transfer liquid to a clean test tube. Add 4 more drops of 6 M HNO_3 to the residue and repeat the heating. Centrifuge and transfer the liquid to the test tube containing previous solution (combine solutions).

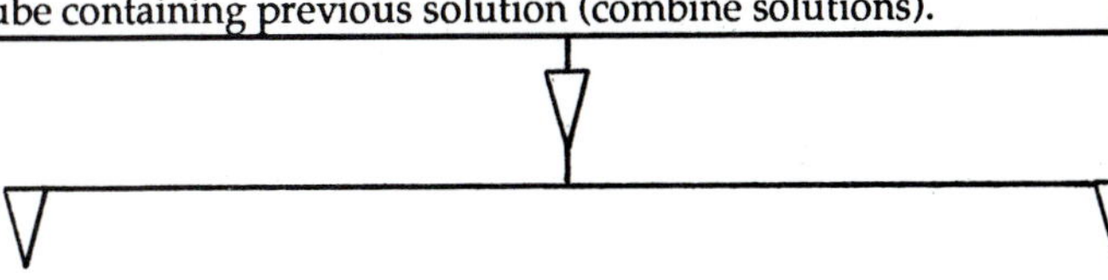

RESIDUE: HgS, $Hg(NO_3)_2$ (black to gray)
Dissolve in 1 drop 16 M HNO_3 and 3 drops 12 M HCl, heating if necessary. Add 1 mL H_2O, stir, and boil. Remove, and add 2 drops $SnCl_2$. A precipitate of Hg_2Cl_2 and/or Hg indicates the presence of Hg^{2+}.

SOLUTION: (in solution) Pb^{2+}, Cu^{2+}, Cd^{2+}

Add solid $(NH_4)_2SO_4$ to the SOLUTION until a slight excess of solid is present. Stir and allow to stand for 2 minutes. Centrifuge. Separate precipitate and solution.

PRECIPITATE: $PbSO_4$ (white)
Wash the precipitate 3 times with hot H_2O and discard washings. Dissolve precipitate in 3 drops of $NH_4C_2H_4O_2$. Add 1 drop of K_2CrO_4. A yellow precipitate of $PbCrO_4$ confirms Pb^{2+}.

SOLUTION: Cu^{2+} and Cd^{2+}
Add 15 M NH_3 dropwise until solution is strongly alkaline. Stir. Divide the solution into 2 parts. *Follow with Part 1 below, then do Part 2.*

PART 1

Make the solution acid with 6 M $C_2H_4O_2$. Add 2 drops of $K_4Fe(CN)_6$. A red precipitate of $Cu_2Fe(CN)_6$ *confirms* the presence of Cu^{2+}. In the absence of Cu^{2+}, a white precipitate of $Cd_2Fe(CN)_6$ may appear.

PART 2

If Cu^{2+} was not present in Part 1, start at the *bottom* paragraph below.

If Cu^{2+} was found to be present in Part 1: Heat the solution to boiling. Remove from heat, add a spatula full (~ 0.2 g) of $Na_2S_2O_4$ (sodium dithionite). Allow to stand a few minutes. Centrifuge and transfer clear, colorless liquid to a clean test tube. (Cu^{2+} has now been removed.)

If Cu^{2+} was not present in Part 1: To the clear, colorless solution, add 10 drops of thioacetamide and heat in boiling H_2O bath just until yellow precipitate appears. Yellow precipitate confirms Cd^{2+}.

Section C. Group III: The Separation and Detection of Fe^{3+}, Cr^{3+}, Ni^{2+}, Zn^{2+}, and Mn^{2+}

The solution from group II separation is poured into a casserole and evaporated *just* to dryness. Add 2 drops of 6 M HCl and 20 drops of 3% H_2O_2 and again evaporate *just* to dryness. Take up residue in 2 drops of 6 M HCl and 10 drops of water. Stir and transfer solution to test tube.

↓

Add 6 drops of 6 M NH_4Cl and sufficient 15 M NH_3 to make the solution *definitely* basic. Mix well and quickly centrifuge. Transfer supernatant solution to a clean test tube. Test for completeness of precipitation by adding 1 drop of 6 M NH_3 to solution. Reserve precipitate.

↓

PRECIPITATE: $Fe(OH)_3$ (red brown) and $Cr(OH)_3$ (gray)

SOLUTION: Save for a later time (within this group).

↓ (from PRECIPITATE)

Add 4 drops of 6 M NaOH and 10 drops of 3% H_2O_2 to the PRECIPITATE and *stir vigorously*! When the gas bubbling stops, boil gently for 1 minute. Dilute with H_2O to 2 mL. Stir well, centrifuge, and separate.

↓

RESIDUE: $Fe(OH)_3$ (dark brown)
Wash the precipitate once with H_2O and discard the washing. Dissolve the residue in 6 M HCl and stir. Add 2 drops of NH_4SCN. A deep red color of $Fe(SCN)^{2+}$ confirms Fe^{3+}.

SOLUTION: CrO_4^{2-} (yellow)
Acidify with $C_2H_4O_2$ and add a few drops of $Pb(Ac)_2$ solution. A yellow precipitate indicates Cr^{3+}.

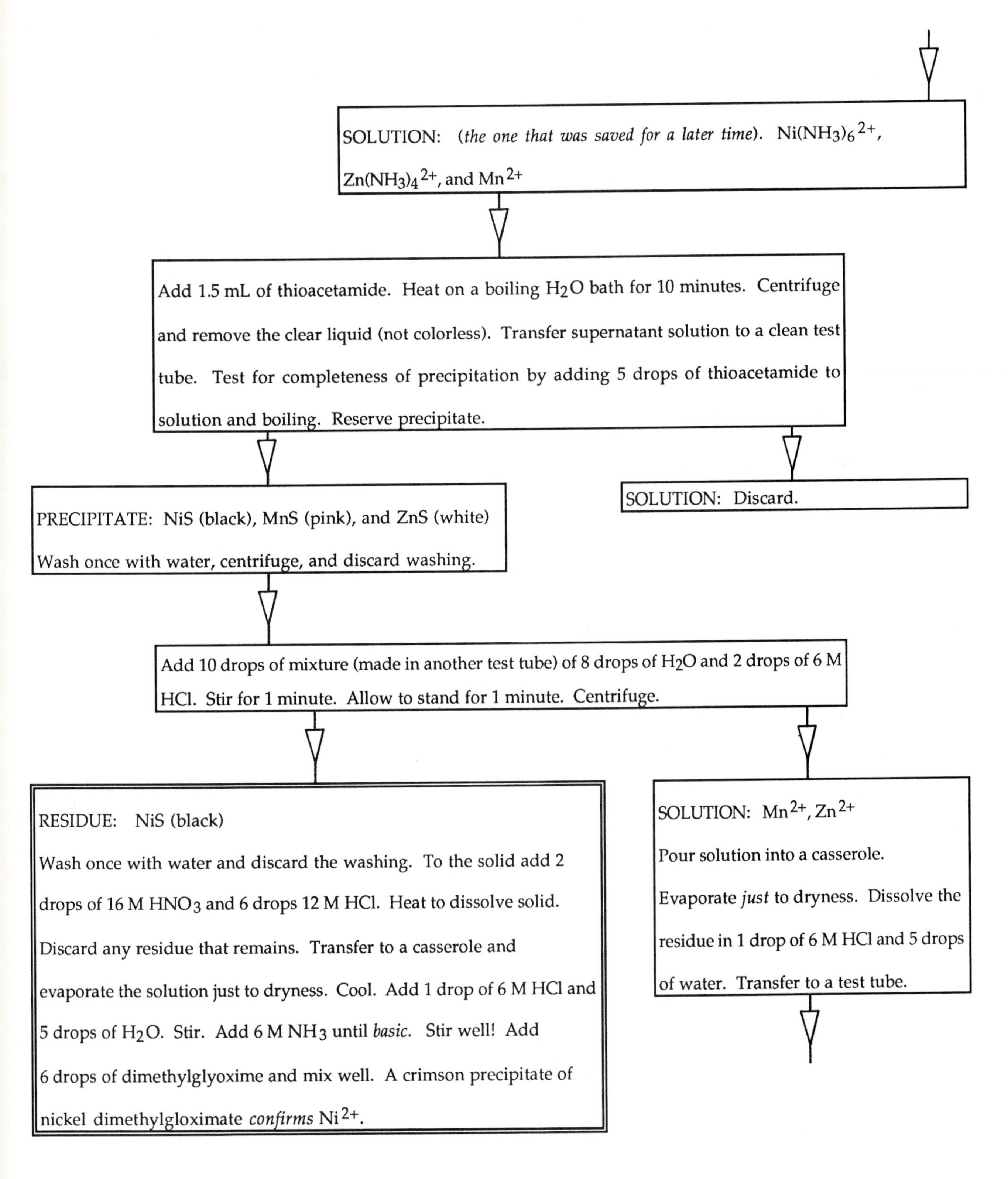
SOLUTION: (*the one that was saved for a later time*). $Ni(NH_3)_6^{2+}$, $Zn(NH_3)_4^{2+}$, and Mn^{2+}
Add 1.5 mL of thioacetamide. Heat on a boiling H_2O bath for 10 minutes. Centrifuge and remove the clear liquid (not colorless). Transfer supernatant solution to a clean test tube. Test for completeness of precipitation by adding 5 drops of thioacetamide to solution and boiling. Reserve precipitate.
PRECIPITATE: NiS (black), MnS (pink), and ZnS (white)
Wash once with water, centrifuge, and discard washing.
SOLUTION: Discard.
Add 10 drops of mixture (made in another test tube) of 8 drops of H_2O and 2 drops of 6 M HCl. Stir for 1 minute. Allow to stand for 1 minute. Centrifuge.
RESIDUE: NiS (black)
Wash once with water and discard the washing. To the solid add 2 drops of 16 M HNO_3 and 6 drops 12 M HCl. Heat to dissolve solid. Discard any residue that remains. Transfer to a casserole and evaporate the solution just to dryness. Cool. Add 1 drop of 6 M HCl and 5 drops of H_2O. Stir. Add 6 M NH_3 until *basic.* Stir well! Add 6 drops of dimethylglyoxime and mix well. A crimson precipitate of nickel dimethylgloximate *confirms* Ni^{2+}.
SOLUTION: Mn^{2+}, Zn^{2+}
Pour solution into a casserole.
Evaporate *just* to dryness. Dissolve the residue in 1 drop of 6 M HCl and 5 drops of water. Transfer to a test tube.

Add 6 drops of 6 M NaOH. *Mix well* and centrifuge. Separate.

RESIDUE: $Mn(OH)_2$ (black)

Wash once with 4 drops of H_2O. Dissolve the precipitate in 5 drops of 6 M HNO_3. Add a spatula end of $NaBiO_3$ so that you have about 3 mm of solid in tube. *Stir thoroughly.* Let stand for 1 minute. Centrifuge. Purple color in liquid supernatant *confirms* Mn^{2+}.

SOLUTION: $Zn(OH)_4^{2-}$

Apply 1 or 2 drops to dithizone paper. A dark rose spot confirms Zn^{2+}.

Section D. Case History and Analysis of an Unknown Sample

When you have finished analyzing the known sample containing 11 metal ions, your instructor will allow you to choose an unknown sample case history. The case history presents a brief discussion of a research problem in which a qualitative analysis was crucial to understanding and solving the problem. The case histories are real situations taken from recent scientific, technological, or forensic literature. Please choose one that interests you.

Each case history has a code number that matches an unknown sample. Obtain the unknown sample from your instructor. The sample is an aqueous solution and contains 3 of the 11 metal ions you can analyze for. Your task is to find out which metal ions are present in your sample. You also need to prove the absence of all other metal ions. When you have finished, obtain an unknown sample *report form*, fill it in, sign and date it, and give it to your instructor. Your instructor will tell you whether or not you were successful. Make sure that you write your observations as you carry out the analysis. Your grade for the experiment will depend on the quality of your observations and records, as well as on the results of the analysis.

Section A. Group I: The Separation and Detection of Ag^+, Pb^{2+}, and Hg_2^{2+}

You will be analyzing a known solution that contains *all* the 11 heavy metal ions: Ag^+, Pb^{2+}, Hg_2^{2+}, Hg^{2+}, Cu^{2+}, Cd^{2+}, Fe^{3+}, Cr^{3+}, Mn^{2+}, Zn^{2+}, and Ni^{2+}. In Section A, you will be separating and detecting group I ions.

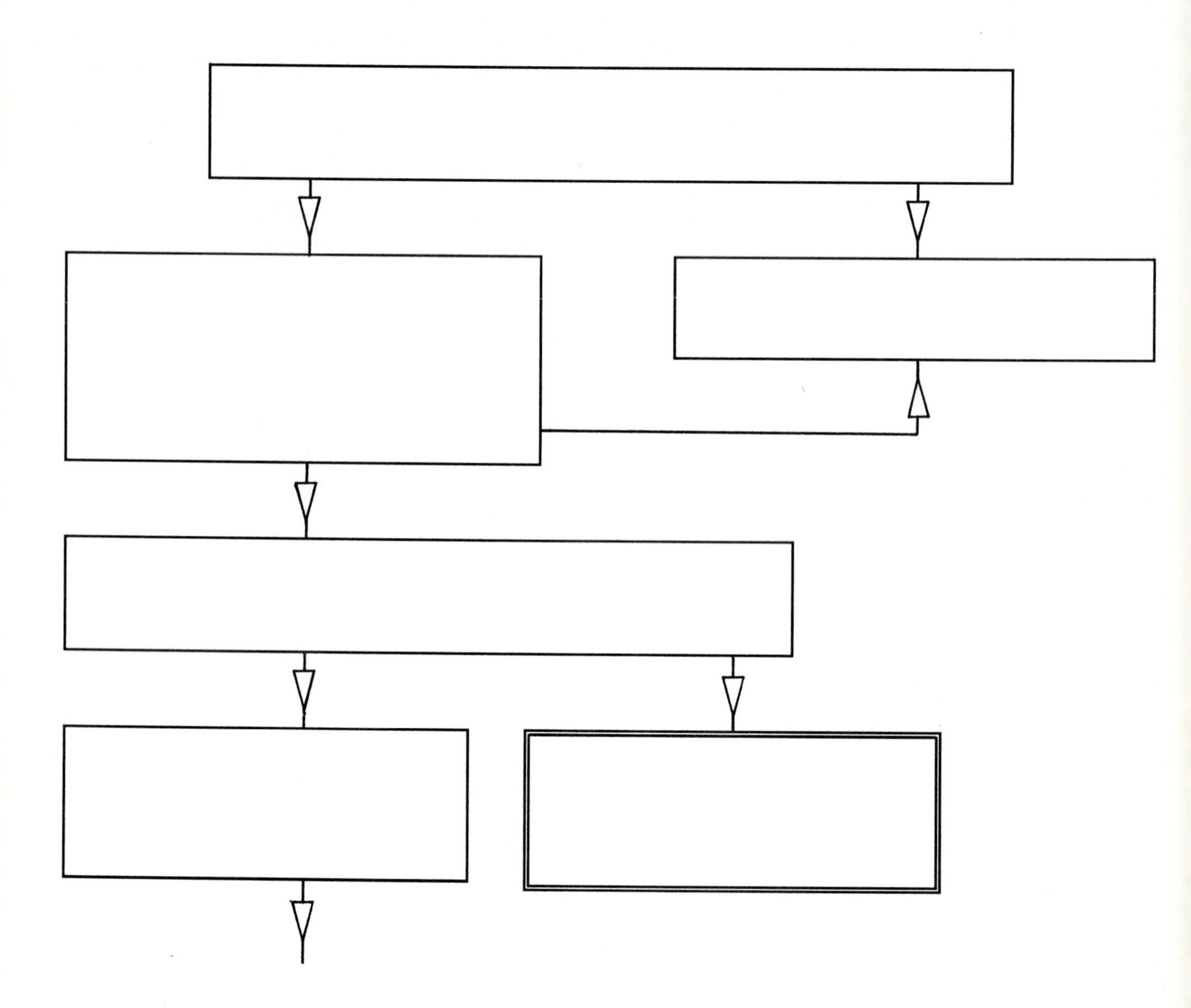

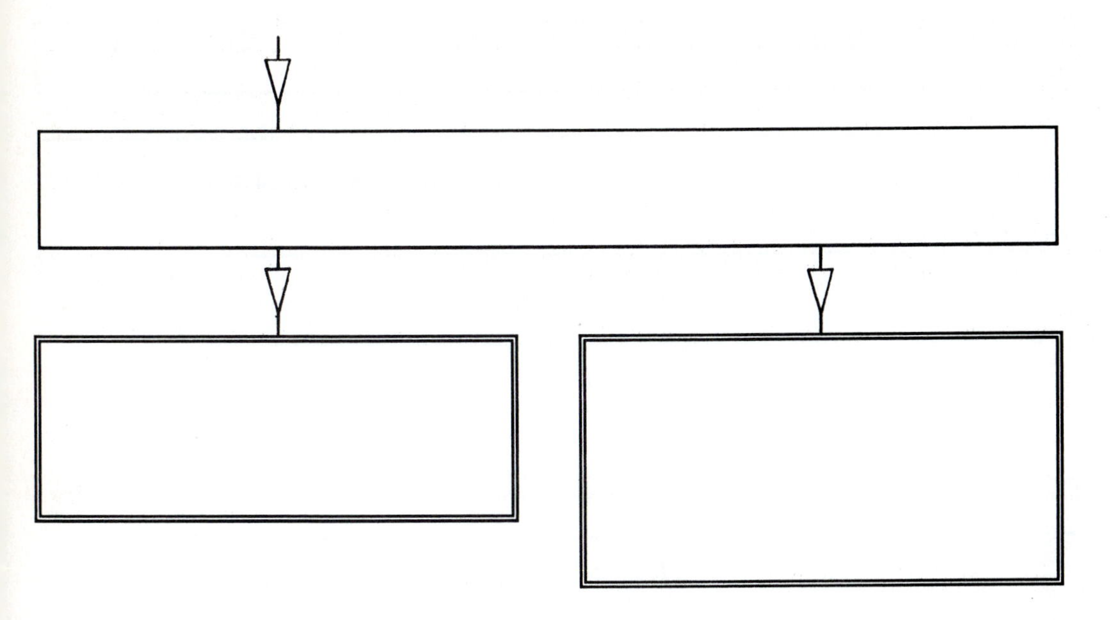

Section B. Group II: The Separation and Detection of Hg^{2+}, Pb^{2+}, Cu^{2+}, and Cd^{2+}

NOTE: The solution for group II is the SOLUTION that you *saved* from group I separation. The solution contains Hg^{2+}, Pb^{2+}, Cu^{2+}, and Cd^{2+}, plus group III ions.

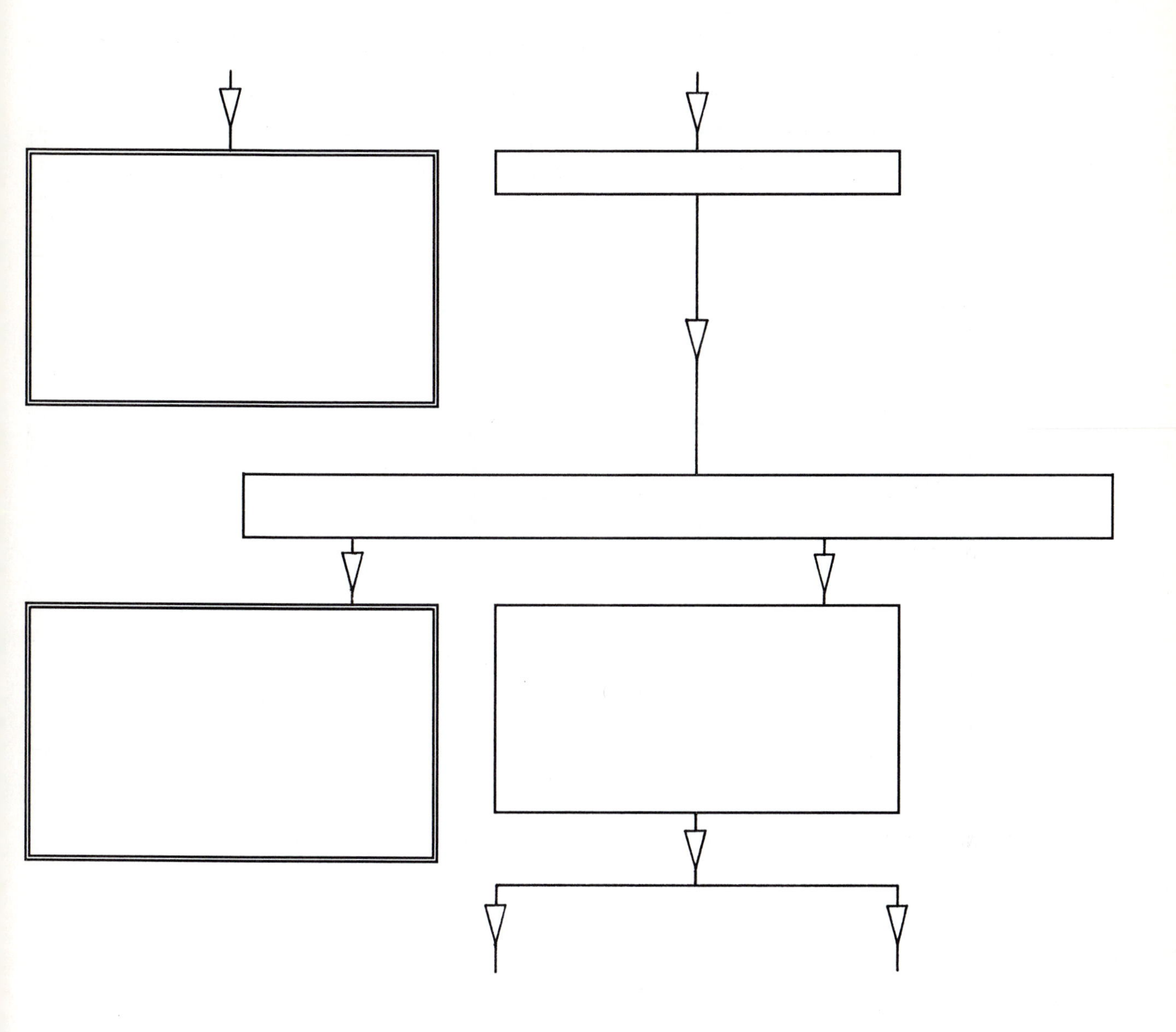

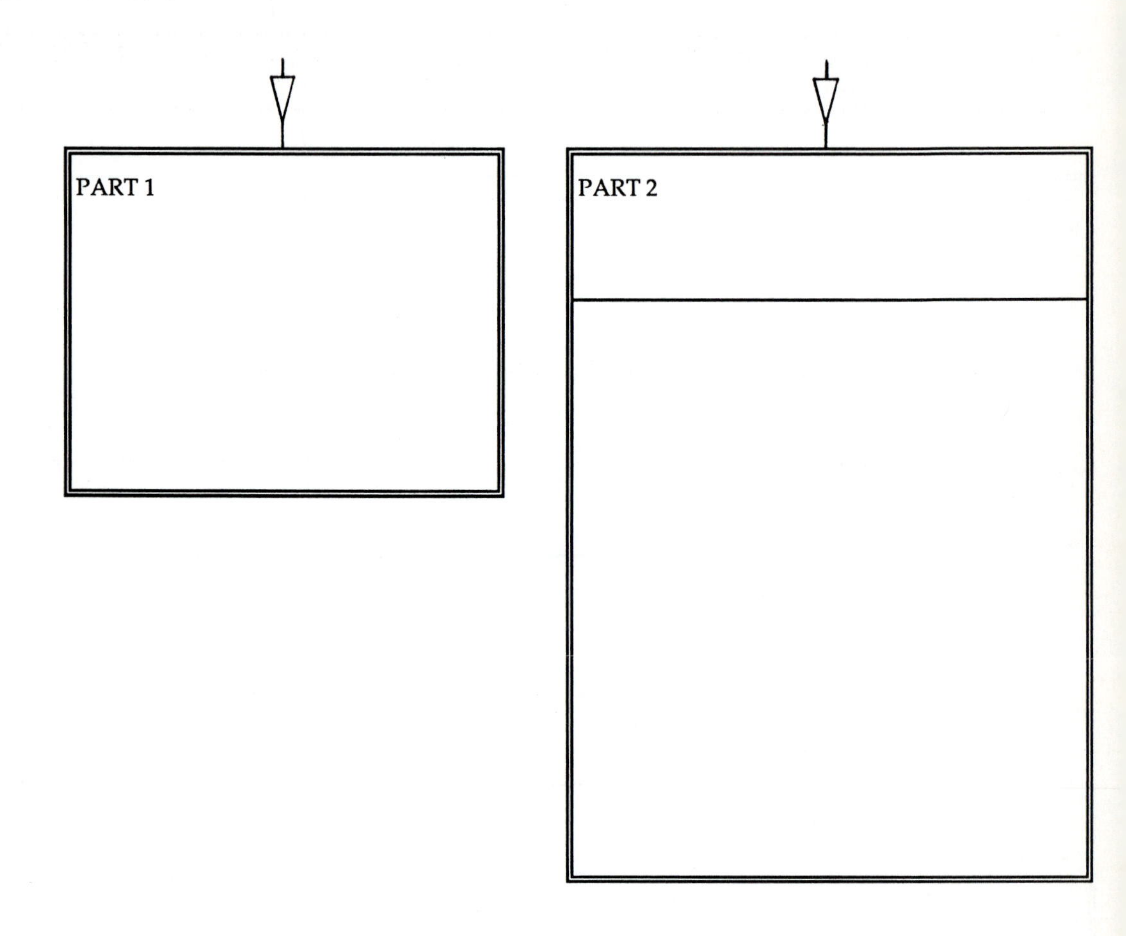
PART 1
PART 2

Section C. Group III: The Separation and Detection of Fe^{3+}, Cr^{3+}, Ni^{2+}, Zn^{2+}, and Mn^{2+}

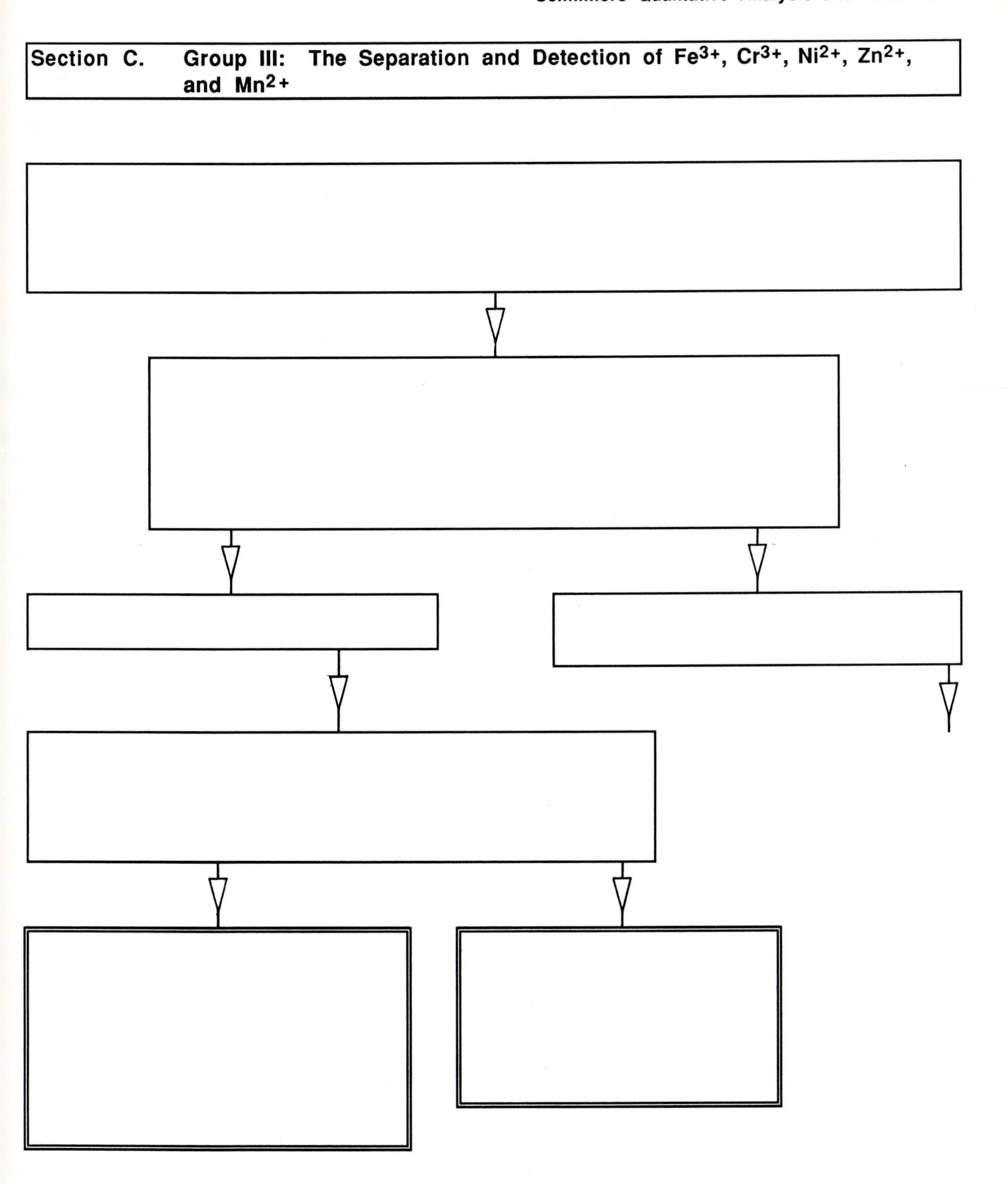

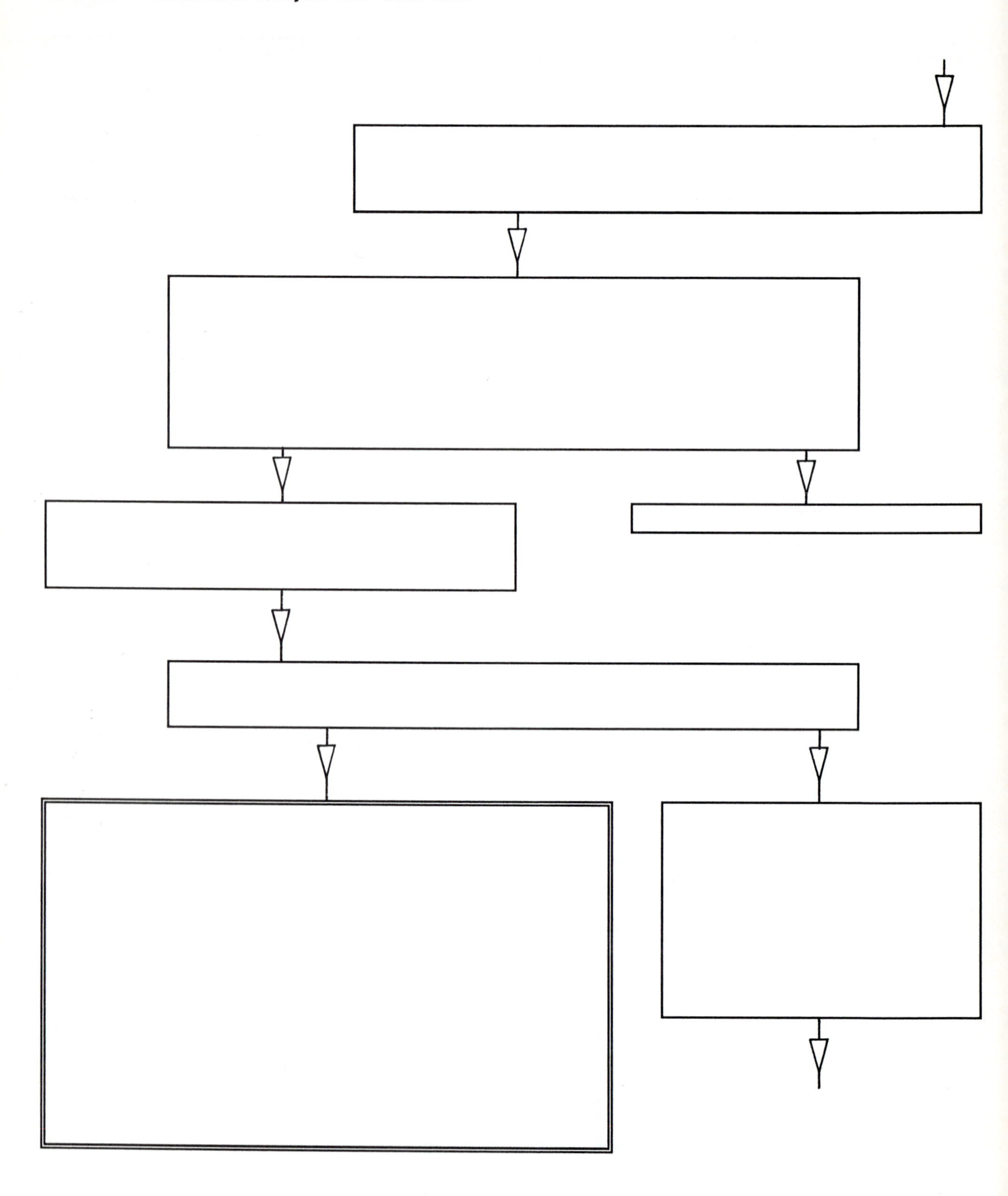

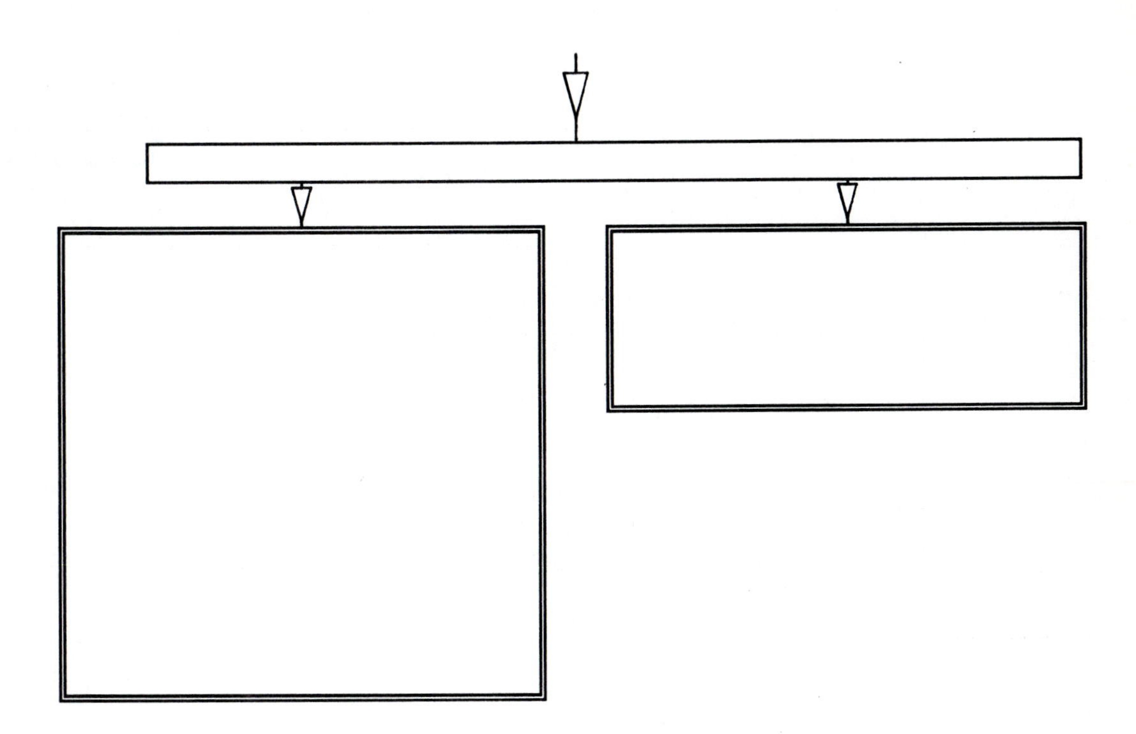

Chapter 18

Paper Chromatography and Liquid Chromatography

Introduction

It is perhaps true that great progress in science occurs after new inventions in technique occur. In the history of science, the battery and the vacuum pump are examples of such inventions. In modern science, the invention and application of two techniques, spectroscopy and chromatography, have revolutionized chemistry and biology. Chromatography is probably the most widely used and most powerful of all the techniques of chemical analysis. Chromatography was invented in the late nineteenth century by the Russian botanist Mikhail Tswett, who used it to separate naturally occurring chlorophylls. He extracted green plant material with organic solvents and allowed the extract to percolate through glass tubes full of powdered solids (e.g., sugar and calcium carbonate). A slightly more polar solvent was then used to wash the extract through the powder. Broad bands of color separated down the length of the column of powder. Tswett claimed that the different bands were different types of chlorophyl and other pigments and that chromatography was indeed the best way to investigate the chemistry of complex natural mixtures. As with many other inventions of true genius in the history of science, Mikhail Tswett's claims were dismissed as nonsense by the scientific establishment of the time, and the new technique was neglected for almost three decades. Chromatography was rediscovered, albeit in a different form called partition chromatography, by the English chemists Martin and Synge in the late 1930s. Martin's research group at the Wool Industries Research Association then developed the first microanalytical chromatography method, called paper chromatography, as a means to analyzing the structure of proteins. Martin and Synge received the Nobel Prize in chemistry in 1952 for the invention of partition chromatography.

The power of chromatography as a scientific tool lies in the fact that it is a simple, gentle, inexpensive, and general way of unmixing (separating) and analyzing complex mixtures of substances. Of course, once a mixture has been separated into pure components, it becomes a relatively straightforward matter to investigate the properties of each component. An interesting historical example, briefly mentioned earlier, is the application of partition chromatography to the study of proteins. The properties and behavior of fibers (e.g., wool) are very much determined by the particular chemical structure of the fiber. Martin *et al.* were trying to elucidate the structure of wool by breaking down the fiber with hot acid. Unfortunately, the resulting mixture of products from the destruction of fiber was so complicated that there was little progress. The problem was resolved by the invention of the partition chromatographic method. This method enabled the horrendous reaction mixtures to be easily separated into the individual amino acid and peptide products and eventually led to a complete structure of wool protein. It is interesting to note that partition chromatographic methods are now being applied extensively to solve gene sequencing problems in molecular biology.

Chromatography is now one of the major methods for chemical analysis and purification. Advances in technology, particularly in the miniaturization of columns and detectors, have resulted in the development of extraordinarily sensitive and quantitative chromatographic instruments (chromatographs). Micro gas chromatographs have been used in outer space probes, and micro liquid chromatographs have recently been explored as *in vivo* implantable analytical monitors. The objective in this series of chromatography experiments is to construct several small-scale chromatographic instruments (PC, LC, and GC) and to investigate the nature of chromatographic processes in several useful applications.

Background Chemistry

Chromatography is a method of separation in which the components to be separated are distributed between two phases, one of these being a porous substance or stationary phase, the other being a fluid that flows through the porous stationary phase. A small volume of the original sample containing the components to be separated is placed at the start of the porous stationary phase. It is important in most types of separation to try to place the sample in as small a volume of the stationary phase as possible or the separation becomes more difficult. The fluid, called the *mobile phase,* is allowed to flow through the porous bed, and as it does, the sample components begin to migrate through the bed. Each component will have a different affinity for the stationary phase and for the mobile phase. Components that have a higher affinity for the stationary phase will be slowed down relative to the other components. Components that tend to stay in the mobile phase will move farther along the stationary phase than the other components. The greater the fraction of time a component spends in the mobile phase, the farther it will move from the start. A component that remains only on the stationary phase will not move at all!

On the molecular scale, component molecules (or ions, atoms, etc.) do not simply move directly along the stationary phase in straight lines. Even though the component molecules are being pushed along in the general direction of the mobile phase, there is a tremendous amount of molecular jostling and bumping. Component molecules, therefore, diffuse in all directions as they migrate in the general direction in which the mobile phase is going. Thus, the chromatographic process is the sum of the billions and billions of molecular events involving diffusion to and from the surface of the stationary phase, random molecular bumping, and migration along the stationary phase in the general direction of the mobile phase. Chromatography may be described graphically in terms of the concentration profiles of the various components as they move along the stationary phase. A computer modelling of the chromatographic process for the separation of four components from a sample is shown in the computer graphic sequence in Figure 18.1.

The chromatographic sequence illustrated in Figure 18.1 represents the behavior of sample components in *elution chromatography.* Elution chromatography, by far the most commmonly used type of chromatography, is also one of the three major types of chromatography. The elution technique involves the chromatography of very small amounts of sample and results in concentration profile shapes similar to those shown in Figure 18.1.

The objective of chromatography is to obtain a complete separation of the individual components of a sample mixture. In chromatographic instruments separation is obtained by trying to

- Maximize the component migration *differences* of all components
- Minimize the component *spreading* that occurs during the chromatographic process

If you look at the computer model pictures, you can see that as the chromatography proceeds, the center (or peak) of the concentration profile of each component gets farther and farther apart, leading to a complete separation of the four components (as in the last picture of Figure 18.1). At the same time, all components spread out and occupy more and more of the space in the system. You can see that if the sample contained a large number of components, then overlapping might occur and a good separation might not be obtained.

In modern chromatography component migration differences are maximized by choosing the correct stationary phase and mobile phase based on an understanding of the chemical interactions involved. Sample components can interact with the stationary phase in a variety of ways — e.g., by partition, by adsorption, by ion exchange, by size, or by exclusion. A careful choice of the type of chemistry used in the system generally results in good separations. It must be emphasized that chromatography is now carried out on almost every conceivable type of chemical component, including proteins, gases, food dyes, etc., and on extraordinarily complex mixtures, such as oils, gasoline, urine, etc. Often, different stationary phases and different mobile phases are required in order to separate different groups of components from the same sample!

Much effort has been spent over the last 20 years to find ways of minimizing component spreading during chromatography. Narrow component bonds are particularly important if complex mixtures are to be resolved, i.e., well separated. The magnitude of the spreading process depends on several factors — e.g., the particle size of a solid stationary phase — and is usually measured in terms of the *efficiency* of the chromatographic system. The efficiency of a particular system is quantitatively expressed by the *number of theoretical plates* (N). N may be calculated from

$$N = 16\left(\frac{V_R}{W_b}\right)^2$$

where V_R, the *retention volume,* is defined as the volume of mobile phase required to carry a component through the chromatographic system. W_b is the width of the base of the component concentration profile (often called the *band width*). A better way of expressing the efficiency is to use the *height equivalent to a theoretical plate* (H), where

$$H = \frac{L}{N}$$

and where the *column* is the porous stationary phase and L is its length. Modern high-performance chromatography instruments can provide many thousands of plates per meter of column (i.e., $H < 0.1$ mm); consequently, the component bands are extraordinarily narrow. These high efficiencies have been achieved by technological developments in the manufacture of solid stationary phases and the tubes (columns) in which the phases are contained. In *high-performance liquid chromatography* (HPLC), the particles of stationary phase are often spherical, very small (diameters less than 5×10^{-6} m), and packed into the tube in a highly uniform manner. One unfortunate aspect of all this efficiency is that it has become very difficult to force the mobile

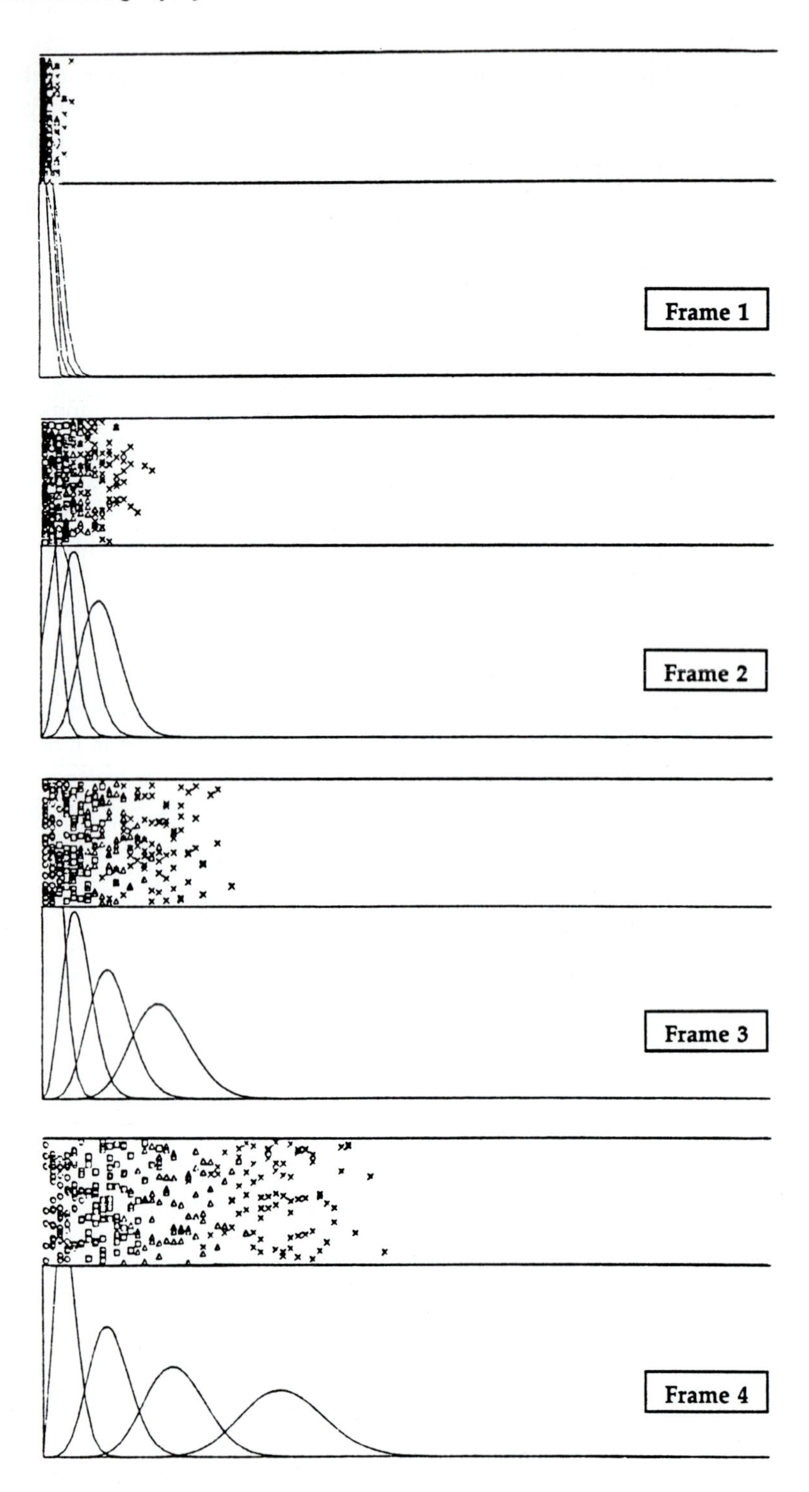
Frame 1
Frame 2
Frame 3
Frame 4

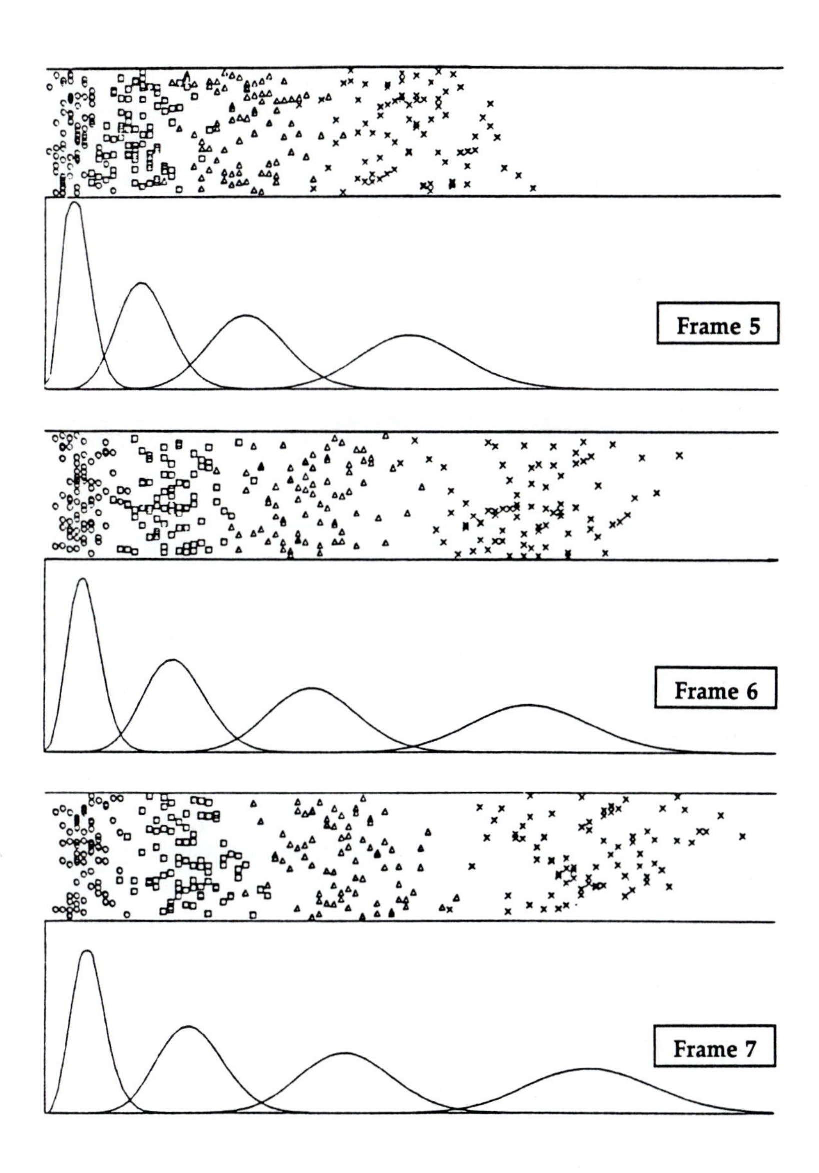

Figure 18.1 A Computer Model of the Chromatographic Process

Frame 1 A sample consisting of four components is injected into the chromatographic system. The mobile phase is allowed to flow and the various component molecules (represented by ❑, O, Δ, and X) begin to move along the column. The concentration profile of each component is shown directly below the picture of the column.

Frames 2–7 The components migrate along the column in the direction of the mobile phase. The component's migration differences increase until a complete separation (almost) of all four components is obtained (Frame 7). Notice that the molecular spreading increases as the components move farther along the column.

phase through the very small particles packed into a tube. Most modern HPLC instruments contain very high pressure pumps to force the mobile phase through the column at reasonable flow rates.

There are literally thousands of different chromatographic methods in use in a vast array of applications. The acronyms are legion — e.g., PC, IC, LC, GC, HPLC, HPTLRPC, GC/MS, GPC/EC, etc., etc. — and the techniques are used in all areas of science and technology. The ubiquity of chromatography makes it difficult to organize a simple classification of chromatographic systems. However, it is useful to classify systems on the basis of the types of phases and on the major distribution processes, shown in Table 18.1.

Nature of the Distribution Process	Mobile Phase	Stationary Phase	Type of Chromatography and Acronym
Partition	liquid	liquid	paper and thin layer on cellulose (PC, TLC) reverse phase liquid (HPRPLC)
Partition	gas	liquid	gas liquid (GLC)
Adsorption	liquid	solid	normal phase liquid, thin layer, and ion exchange (LC, TLC, IC)
Adsorption	gas	solid	gas solid (GSC)
Size sorting	liquid	gel	gel permeation (GPC)

Table 18.1 Types of Phases and Distribution Processes in Chromatography

In this laboratory text you will have the opportunity of constructing and investigating three major types of chromatographic systems: paper chromatography (PC), liquid chromatography (LC), and gas chromatography (GC). The sections that follow provide you with some background chemistry that is pertinent to paper and liquid chromatography. Gas chromatography is discussed in Chapter 19, although you should note that the introduction given earlier pertains to all elution chromatography, including GC.

PAPER CHROMATOGRAPHY

Paper chromatography (PC) is a technique developed in England in the 1940s by Martin, Synge, and Consden that uses paper for separating the complex mixtures of amino acids obtained from the breakdown of wool. The separation and subsequent identification of individual amino acids and peptides led to the elucidation of the structure of wool protein. Paper chromatography is the name that is used to describe chromatography carried out on a stationary phase consisiting of specially-prepared, porous paper. The sample, usually dissolved in a solvent, is directly applied to one end of a sheet of paper. The sheet is then placed in a large glass or plastic chromatography tank and mobile phase is allowed to contact the paper. The mobile phase is "pulled through" the paper by capillary action. PC has a somewhat limited use and is not used extensively now, although it is interesting to note that there have been several recent applications to the separation of complex enzyme mixtures with modified celluloses.

Paper is cellulose and has the structure shown in Figure 18.2. The hydroxyl

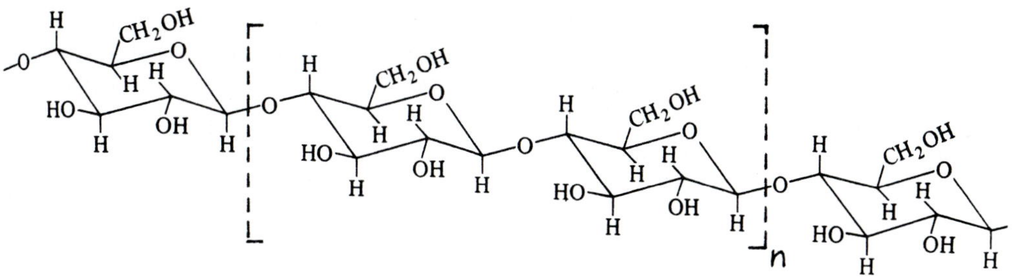

Figure 18.2 The Cellobiose Repeating Unit in Cellulose

groups (– OH) in the cellulose are responsible for the hydrogen bonding of water, which makes up about 6% of the weight of the paper. It is this water layer, along with more water that is sometimes adsorbed during the chromatographic process, that forms the stationary phase in PC. Chromatography paper is unsized and is carefully manufactured to produce a highly porous paper with relatively uniform fiber structure. If the edge of a sheet of chromatography paper is placed in a liquid, the liquid will be pulled through the paper by capillary action. The liquid moving through the pores in the paper constitutes a mobile phase, and chromatography can occur. The mobile phase may be pulled along, up, or down, depending on where the liquid is initially applied.

Let us now look at the factors that control the way in which components move in a paper chromatographic separation. Consider a single nonelectrolyte component applied to the paper at the start (see Figure 18.3).

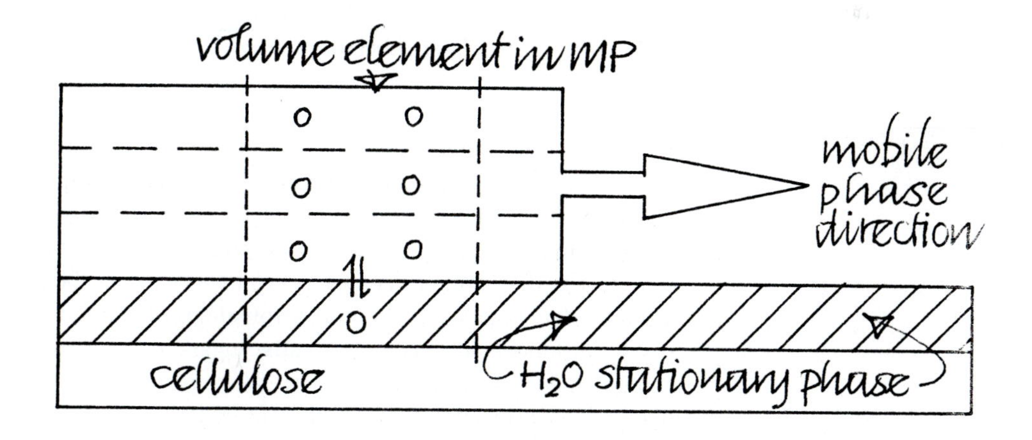

Figure 18.3 Distribution of a Single Nonelectrolyte Component in PC

The component molecules are distributed by *partition* between the aqueous stationary phase and the nonpolar mobile phase. The partition coefficient k is defined as the ratio of the concentration of the component in the mobile phase to that in the stationary phase. In the above diagram, k = 2 because 1 volume of the mobile phase contains 2 molecules, and an equal volume of stationary phase contains 1 molecule. In practice, however, the volumes of the two liquid phases are not equal. For many chromatography papers the mobile phase volume is 3 times the stationary phase volume, and the phase ratio r is

$$r = \frac{\text{volume of mobile phase}}{\text{volume of stationary phase}} = \frac{3}{1}$$

The *distribution ratio* is defined as kr — i.e., the number of molecules in the mobile phase per solute molecule in the stationary phase. In our example, kr = 6. In PC the *measure of retention* is the R_f. R_f is defined as the fraction of time spent by an "average" molecule in the mobile phase and is equal to the fraction of molecules present in the mobile phase. Since, in general, the total number of molecules is kr + 1 (kr molecules in the mobile phase and 1 in the stationary phase) then,

$$R_f = \frac{kr}{kr + 1}$$

Rearranging,

$$kr = \frac{R_f}{1 - R_f}$$

In an actual PC separation, the R_f value is defined experimentally by

$$R_f = \frac{\text{distance moved by the component}}{\text{distance moved by the mobile phase front}}$$

As you can see, the relative migration distance R_f depends on the partition coefficient k, the value of which depends on the chemical nature of the component and mobile phase. In a chromatographic experiment in which a sample mixture containing several components is applied to the paper at the start, the R_f value of each component will depend on the partition coefficient for each component in the system employed.

The discussion thus far has been restricted to the chromatography of a single *nonelectrolyte* component. Paper chromatography was really introduced to separate polar substances, such as amino acids, sugars, drugs, metabolites, etc. Most of these compounds are weak acids or bases. We must therefore blend together acid-base and distribution principles in order to explain how various components behave in a PC system.

The migration of a weak electrolyte in chromatography is very dependent on the state of the acid-base equilibria that can occur. Consider a weak organic base B (e.g., nicotine) that may accept a proton to give the conjugate acid BH^+:

$$B + H_2O \rightleftharpoons BH^+ + OH^-$$

$$BH^+ \rightleftharpoons B + H^+$$

In a paper chromatography system in which the mobile phase is a nonpolar solvent and the stationary phase is an aqueous buffer of known pH, we can picture the system as shown in Figure 18.4.

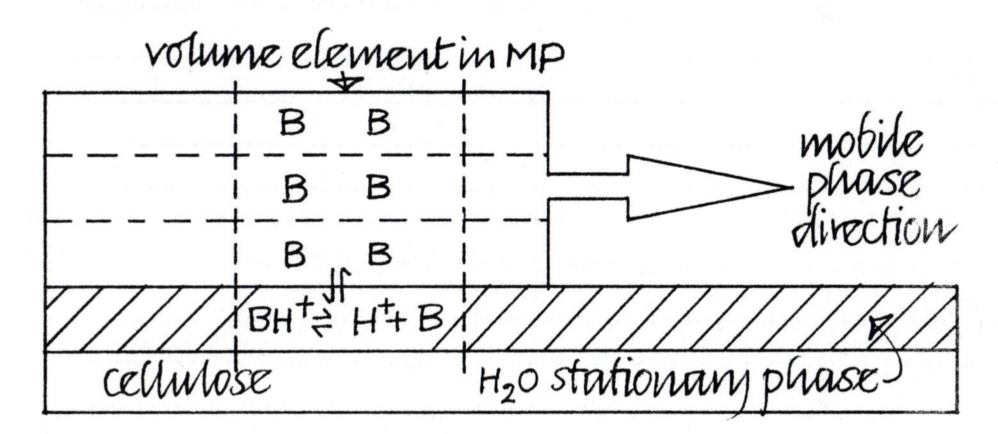

Figure 18.4 Distribution of a Weak Base in PC

The only species that will distribute into the nonpolar mobile phase is B. The charged species BH^+ remains in the aqueous stationary phase. The extraction coefficient D is equal to the ratio of the total concentration of component in the mobile phase to that in the stationary phase:

$$D = \frac{[Borg]}{[B] + [BH^+]}$$

NOTE: an absence of subscript refers to the aqueous phase.
Dividing by [B] and substituting,

$$k = \frac{[Borg]}{[B]} \quad \text{and} \quad \frac{[BH^+]}{[B]} = \frac{[H^+]}{K_a}$$

Then

$$D = \frac{k}{1 + \frac{[H^+]}{K_a}}$$

Now for weak electrolytes,

$$R_f = \frac{Dr}{Dr + 1}$$

Substituting the expression for D and rearranging gives

$$R_f = \frac{kr}{kr + 1 + \frac{[H^+]}{K_a}}$$

and

$$K_a = \frac{[H^+]}{\frac{kr}{R_f} - 1 - kr}$$

The R_f expression shows how the migration of a weak base depends on k, r, K_a, and on the $[H^+]$ of the aqueous stationary phase. Analogous expressions can be derived for weak acids, i.e.,

$$R_f = \frac{kr}{kr + 1 + \frac{K_a}{[H^+]}}$$

All these expressions (and many others for different types of equilibria) were originally derived by Soczewinski, Waksmundski, et al., and a group of chemists, biochemists, and pharmacologists working in Lublin, Poland. Graphs of R_f versus pH of the stationary phase in paper chromatography for various weak electrolytes are shown in the diagrams of Figure 18.5.

The most important results that can be obtained from the mathematical analysis of R_f versus pH data are the following:

- The chromatographic separation of weak acids and bases can be obtained simply by choosing an optimal pH of the stationary phase.
- Chromatography can be used to measure the K_a, K_b, and k values of any weak electrolyte.

The technique is fast, easy, inexpensive, and will work with extraordinarily small amounts of material.

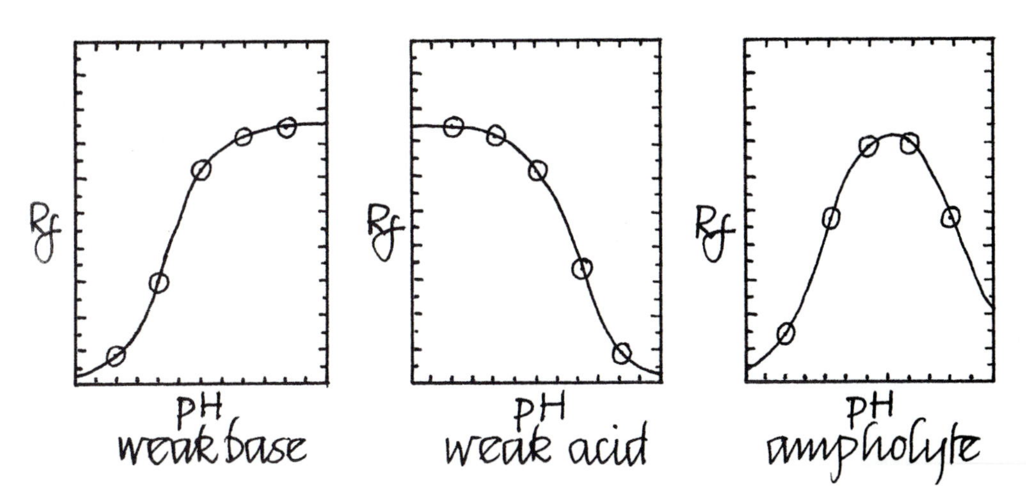

Figure 18.5 R_f Versus pH for Weak Electrolytes

LIQUID CHROMATOGRAPHY

Liquid chromatography (LC) is the name that is now universally used to describe chromatographic separations carried out in small tubular columns packed with small particles of stationary phase. As the name implies, the mobile phase is liquid and may be a pure solvent or a variable composition mixture of solvents. The LC apparatus may be a simple plastic tube with gravity flow of liquid or a very complex instrument with stainless steel tubes, high-pressure mobile phase pumps, and sophisticated, sensitive analytical detectors. Much of the chemical analysis that is done in clinical, industrial, and research laboratories is carried out on commercial high-performance liquid chromatography (HPLC) instruments costing thousands of dollars. A quantitative analysis is obtained by injecting a known volume of sample into the instrument and by quantitatively measuring the concentration of each separated component with a detector placed at the end of the column. HPLC is often performed in the reverse phase mode (HPRPLC). A chromatographic process in which the liquid mobile phase is more *polar* than the stationary phase is referred to as a *reverse phase process*. One of the most common HPRPLC stationary phases consists of 5 μ particles of silica gel that has been *derivatized* (chemically reacted) with octadecyl hydrocarbon groups, as shown in Figure 18.6.

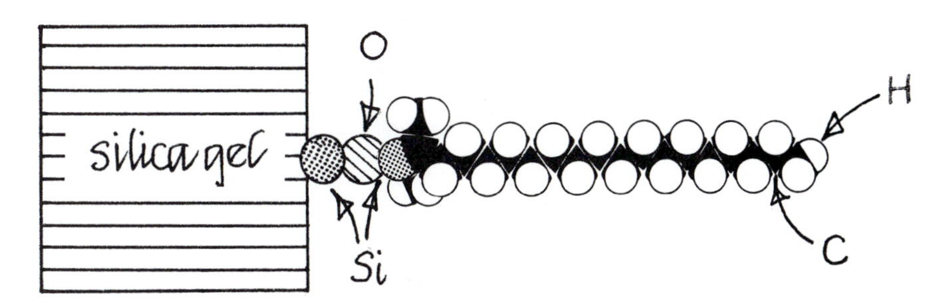

Figure 18.6 Octadecyl-Derivatized Silica Gel

The mobile phase is often aqueous buffer or a water/alcohol mixture. HPLC columns are extremely efficient, producing very narrow component bands. The efficiency is mostly due to the extremely small uniform particle size of the stationary phase and homogeneous packing in the column.

Pre-Laboratory Quiz

1. Who first used chromatography for the separation of natural products?

2. What type of chromatography was invented to help discover the structure of proteins?

3. Give a definition of chromatography.

4. In order to obtain good separations in chromatography, what factor must be maximized and what factor must be minimized?

5. What is one measure of the efficiency of a chromatographic system?

6. What is partition?

7. What does HPRPLC stand for?

8. In the partition chromatography of a nonelectrolyte, give the expression that relates R_f to partition coefficient and distribution ratio.

9. Draw a simple diagram showing how the R_f of a weak base varies as the pH of the stationary phase is changed.

10. Describe 1 reverse phase-stationary phase used in HPLC.

Laboratory Experiments

Flowchart of the Experiments

Section A.	Paper Chromatography of Dyes
Section B.	Moist Buffered Phase Chromatography of Nicotine
Section C.	Calculation of the K_b of Nicotine from Chromatographic Data
Section D.	Preparation of a Liquid Chromatography Column
Section E.	Investigations of Column Parameters and Processes
Section F.	Derivatization of the Silica Gel Stationary Phase
Section G.	Chromatography of Selected Synthetic Dyes
Section H.	LC of Beet Pigments

Requires one three-hour class period to complete

Section A. Paper Chromatography of Dyes

Goal: *To carry out a series of ascending paper chromatographic separations of inks and food dyes in a small-scale tank system.*

Discussion: You will also examine the effect of mobile phase composition on separation and use R_f parameters to characterize individual dye components in commercially available dye mixtures.

Before You Begin: Paper chromatography is a microanalytical technique. Extraordinarily small sample volumes are used (microliters), and the separation often yields very small amounts (often less than nanomoles) of product. It is particularly important to maintain good technique and to try to avoid contamination of the chromatography paper. Work on a clean piece of paper or paper towel when you are drawing lines and spotting samples.

Experimental Steps:

1. Obtain 2 pieces of chromatography paper from Reagent Central. On one piece, draw in the lines and points with a *pencil*, as shown below.

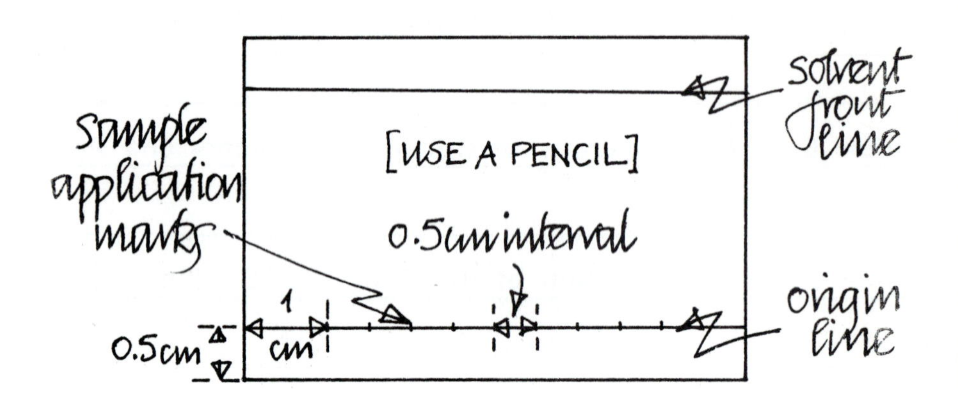

2. At your lab place will be a plastic chromatography tank and a solvent tray. Take the tray to Reagent Central and deliver 3 pipets of the mobile phase solvent, 2:1 1-propanol/water, to the tray.
3. Take the tray to your bench and place the tank over it.
4. Take a clean 1 x 12 well strip to Reagent Central and deliver 2 drops each of standard FD and C food dye solution to different wells. In the remaining wells deliver 2 drops of commercial food dye mixtures (you may select these from the variety available).
   - Note in your laboratory record the location of each dye.

5. Obtain a piece of cutoff thin stem (cut at an angle) as an applicator. Dip the applicator into the dye solution in the first well and then transfer the dye solution to the first application mark on the paper. Keep the applicator vertical.

   NOTE: Apply enough dye to see the color easily, but keep the spot *small*!

6. Spot the rest of the food dye samples, cleaning the applicator after you spot each dye.

   - Record where each dye is spotted.

7. Obtain the pens (which contain water-soluble inks) and touch each pen tip to a sample application mark.

   NOTE: Keep the pen vertical and try to make the spots small by touching and removing the tip quickly.

8. Roll the paper into a cylinder and hold it while you staple it as shown below. *Do not overlap* the 2 edges of the cylinder.

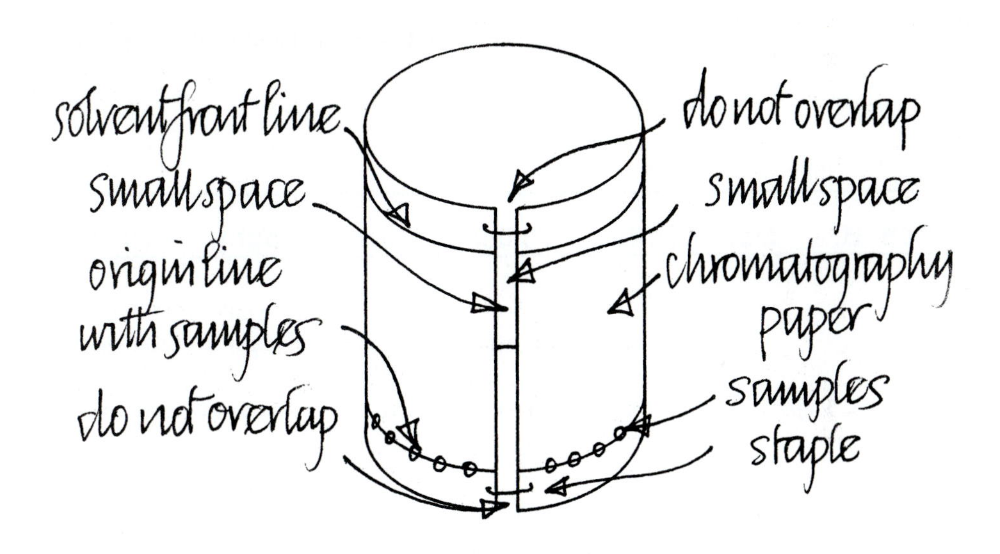

9. Lift the tank from the solvent tray and place the paper cylinder into the solvent as evenly and centrally as possible.

10. Place the tank over the cylinder.

11. Watch the mobile phase for a few moments so that you can see the mobile phase front as it climbs up the paper by capillary action.

    The chromatography will take about 20–25 minutes for the front to reach the solvent front line. From time to time check the progress of the front. Organize your time so that you do other things — e.g., keeping up your notes and preparing for the next section — while the chromatography is occurring.

12. When the front reaches the solvent front line, lift the tank and quickly remove the cylinder. Place it onto a clean paper towel. Replace the tank.

13. Pull the cylinder apart and hang it up to dry on the straw drying system (tray + straw + clothespin). The drying time is about 20 minutes.

14. While the paper is drying, start marking out the second piece of chromatography paper, as shown in Step 1 of Section B.

15. When the paper is dry (or when you are ready), place it flat on a towel on the table and locate the component zones or spots. This is done by drawing a circle around the leading edge of the oval color area.

    - Measure the distance that each component migrated from the original sample application mark.
    - Calculate the R_f value for each component, where

$$R_f = \frac{\text{component migration distance}}{\text{mobile phase migration distance}}$$

    - Use R_f values and color to identify the unknown food dyes in the commercial mixtures.

16. Attach the chromatogram to your notebook.

Section B. Moist Buffered Phase Chromatography of Nicotine

Goal: *To obtain an R_f-pH profile for a cigarette extract and for a pure nicotine standard by moist buffered phase paper chromatography.*

Experimental Steps:

1. Mark out a second piece of chromatography paper as shown.

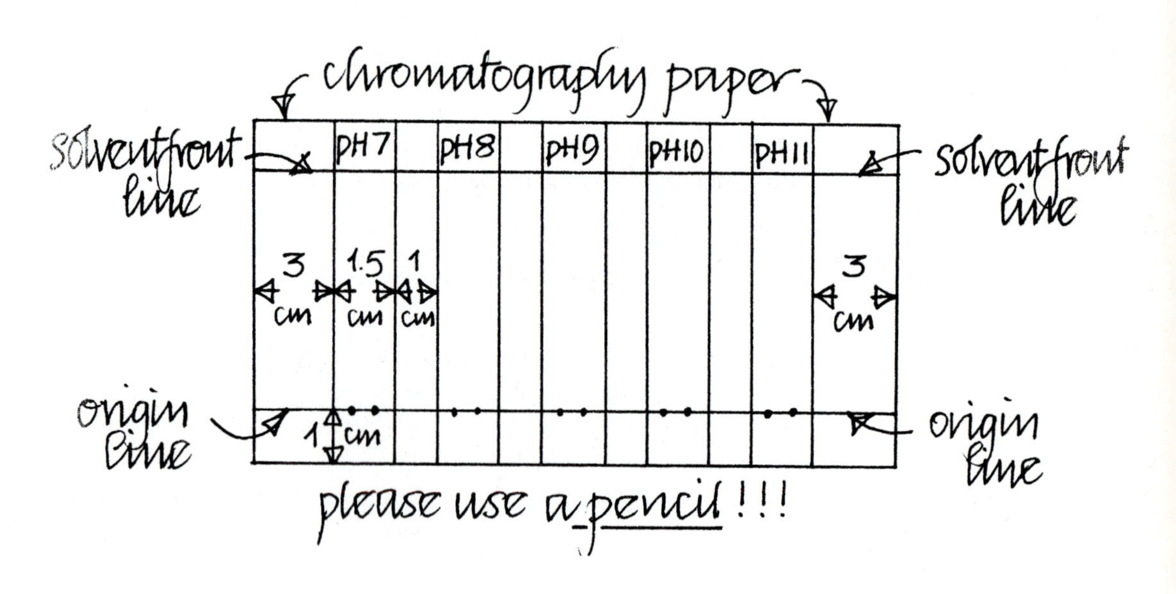

2. Clean one 24-well tray and take it to Reagent Central. Fill 1 well about 1/2 full with pH 7.0 buffer solution, another with pH 8.0 buffer solution, and so on. (You will have pHs 7.0, 8.0, 9.0, 10.0, and 11.0.)
3. Fold a clean piece of paper towel in half and place it on the table.
4. Lay the marked piece of chromatography paper on the towel.
5. With a clean thin-stem pipet, transfer some pH 11.0 buffer and wet the paper by moving the pipet down the pencil lines, as shown.

   NOTE: You want to completely wet the rectangle marked pH 11.0 with the buffer without letting the pH 11.0 solution soak into the area marked pH 10.0.

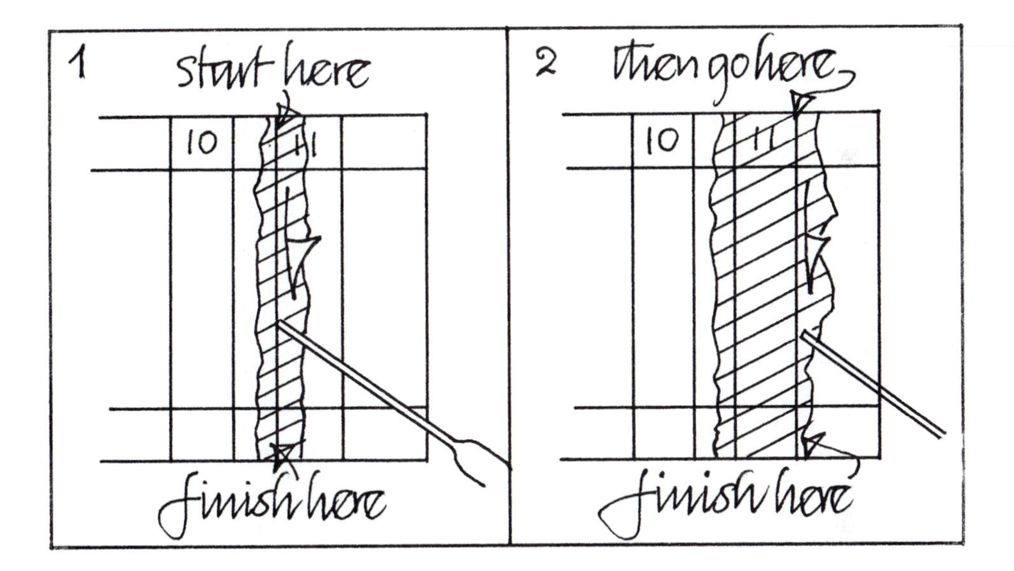

6. Wash the thin-stem pipet twice with water.
7. Transfer some pH 10.0 buffer from the well and wet the area marked pH 10.0 in the same manner as in Step 5.
8. Carry on with each buffer until you have finished with pH 7.0.
9. Wash the pipet and 1/2 fill it with water.
10. Wet the dry ends with water so that the whole piece of paper now appears uniformly wet.
11. Blot evenly and firmly with a folded paper towel.
12. Hang the paper on the drying rack (in still air!) and *note the time on a watch or lab clock.* Let the paper *semidry* for 3 minutes.
13. While the paper is semidrying, pour out the n-propanol/water mobile phase from the solvent tray into the special waste container. Dry the tray with a paper towel.

14. Put 2 squirts of hexane (C_6H_{14}) solvent into the tray. Cover the tray with the tank.

15. Obtain the 2 pens, one that contains a standard pure nicotine solution and the other that contains a cigarette extract.

    We have found that this is one of the best ways to store and apply very dilute nicotine solutions. The solutions in these pens also contain a dye that will enable you to see the precise location and size of a nicotine or a cigarette sample when it is applied to the chromatography paper.

16. At 3 minutes elapsed time, take the paper down and place it on a folded towel. Apply the standard pure nicotine solution to the left sample application mark of each pH area by holding the pen almost vertical and touching the paper. Keep the size of the applied sample small.

17. Now apply the cigarette extract to the right sample application mark of each pH area in the same manner.

18. Roll the paper into a cylinder and staple as before (Section A). Do not overlap the edges!

19. At *6 minutes elapsed time* from Step 11, put the cylinder, evenly and centrally, into the hexane in the solvent tray and replace the tank.

20. Watch the chromatography proceed. It will take a short time (only about 2 minutes!) *As soon* as the mobile phase reaches the solvent front line, remove the cylinder and place it on a dry paper towel. The hexane will evaporate quickly.

21. Open the cylinder and place the chromatogram on a paper towel.

22. Locate the nicotine spots by applying Dragendorff's reagent to each buffered area.

    NOTE: The nicotine spots will appear as orange colored spots.

    CAUTION: Do not get Dragendorff's reagent on your hands. If you do, wash with cold water and check with your instructor.

23. Hang the paper to dry. It will take about 30 minutes.

    Organize your time and begin to set up Sections C and D during the waiting period.

24. Lay the dried paper down and draw a circle around the leading edge of each orange nicotine spot.

    - Measure and record the R_f values for each nicotine spot. You may average the R_f values for the standard pure nicotine and the nicotine from cigarette extract spots for *each* pH.

    NOTE: If the moist buffered phase chromatograph did not give you a reasonable R_f versus pH profile (see Section C), you may repeat the experiment.

Section C. Calculation of the K_b of Nicotine from Chromatographic Data

Goal: *To use data obtained from a plot of R_f versus pH to calculate the base dissociation constant K_b for nicotine.*

Discussion: In order to carry out this calculation, it is important that you understand the derivation of the R_f-pH relationship for weak bases. The derivation is presented in the Background Chemistry section at the beginning of this laboratory module.

Experimental Steps:

1. Plot a graph of R_f versus pH of the stationary phase.
2. Draw a smooth curve through the points.

   The chromatograms at various pH values and the R_f versus pH plot should look similar to the following.

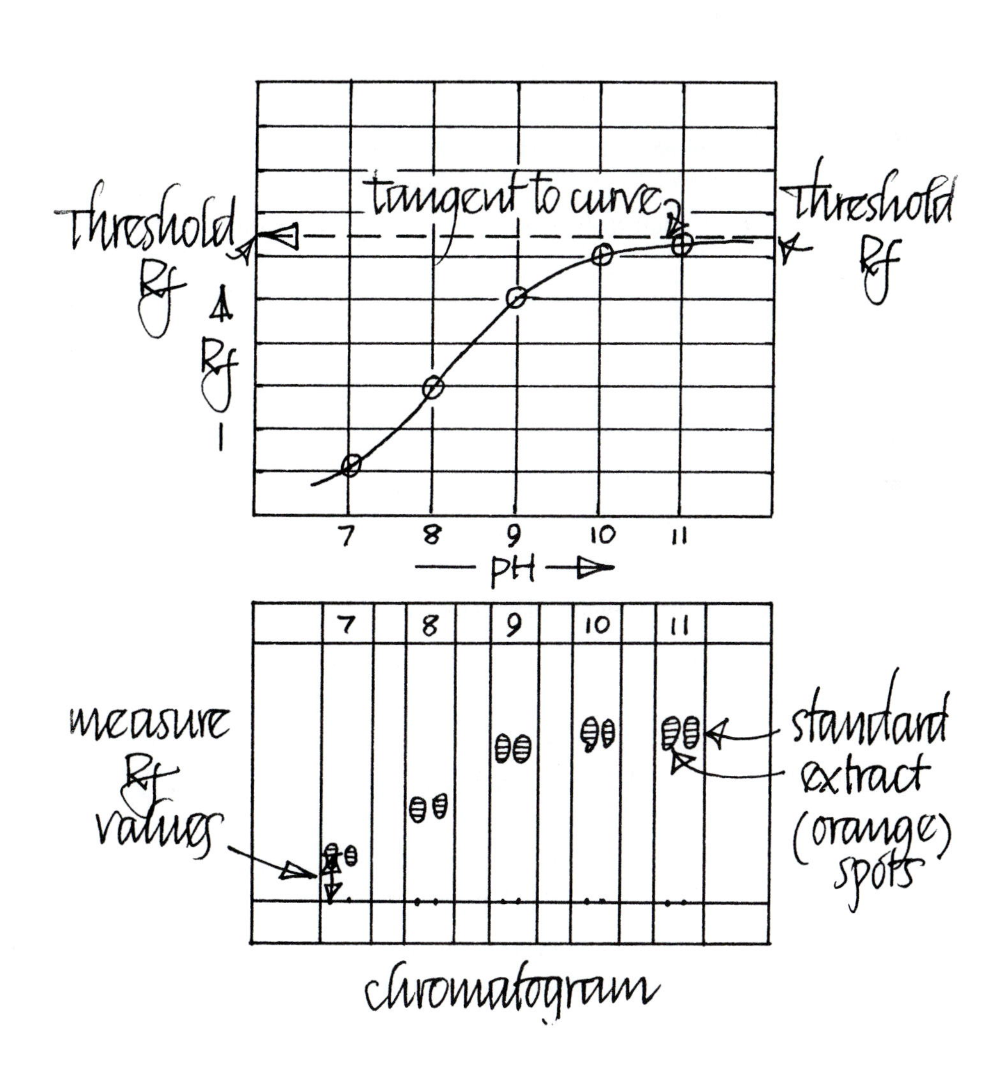

3. Find the threshold R_f value at which the R_f is pH independent. This is easily obtained by drawing a horizontal line as a tangent to the R_f versus pH curve at the flat part of the curve (at about pHs 10 and 11).

   At the threshold R_f, the nicotine is virtually all in the free base form (B) and, therefore, is acting as a nonelectrolyte. The chromatographic relationship for a nonelectrolyte is simply

$$R_f = \frac{kr}{kr + 1}$$

   which, on rearrangement, gives

$$kr = \frac{R_f}{1 - R_f}$$

   where R_f is the threshold R_f, k is the partition coefficient for nicotine between hexane and aqueous buffered stationary phase, and r is the phase ratio.

   - Calculate kr from the expression.

4. Look at the graph and select a pH value at which the R_f value is in the middle of the curved part of the curve. Draw a vertical line through the selected pH until it cuts the curve. Draw a horizontal line to the R_f axis.

   - Note the pH and R_f.
   - Calculate the $[H^+]$ for the selected pH.
   - Now use the expression derived earlier to calculate the K_a for BH^+, the conjugate acid of the free base B of nicotine:

$$K_a = \frac{[H^+]}{\frac{kr}{R_f} - 1 - kr}$$

   where R_f is the R_f value at the selected pH (and therefore $[H^+]$), and kr is the constant you calculated in Step 3.

   - Now for conjugate acids and bases, in aqueous solutions,

$$K_w = K_a\ K_b$$

   where K_w is the ion product of water, 1.0×10^{-14} (at 25 °C). Calculate K_b for nicotine.

   - At what pH of the stationary phase would you expect the nicotine to have an R_f value of zero?
   - Which form of nicotine, B or BH^+, would you expect to pass most easily through a lung cell wall into the blood stream?
   - How could the concentration of nicotine in tobacco be determined by using moist, buffered phase paper chromatography?

Section D. Preparation of a Liquid Chromatography Column

Goals:

(1) To construct a liquid chromatography system from straws. (2) To prepare a silica gel chromatography column that can be used to test chromatographic processes and separations.

NOTE: Please do not discard the various plastic materials used in making the chromatography apparatus. We will recycle them wherever possible.

Experimental Steps:

1. First build a stand for the chromatography column. This is easily done by using 2 straws and a 96-well tray (round-bottomed wells). Use a 1/4" office punch to punch a hole about 2" from the end of one straw and a hole in the middle of the other straw.
2. Push the first straw into well H–12 of the tray.
3. Push the second straw through the hole in the vertical straw. If necessary, cut the end of the straw at an angle to facilitate pushing it through the hole.
4. Obtain a column tip (the small plastic piece with a narrow hole). Place a *small* wad of polyester fiber into the tip.

   NOTE: Do not pack it too tightly or it will impede the flow of mobile phase.
5. Obtain a third straw. Push the column tip firmly into the end of the third straw.
6. Push the third straw through the hole in the horizontal straw.
7. Place a small cup under the column straw and squirt a stream of water into the straw to wash the walls and wet the polyester.
8. Use the plastic scoop (provided for you) to put 1 scoop of 100–200 mesh silica gel into a small cup.
9. Add about 15 mL of water to the cup and stir thoroughly with the slurry pipet. Suck up some slurry. Transfer the slurry to the column quickly!

   NOTE: The idea is to do this quickly to avoid settling of the large particles.

   NOTE: If you get an air block in the column, insert a slim straw and the block will disappear.
10. Transfer more slurry while the first batch is settling. Use the slim straw to work the packing in the column. Try to make a column silica gel slurry packing about 5–6 cm long.
11. Add water to the column until it is full (to the top).
12. Obtain a pump syringe and pull the plunger almost to the end, in order to *push* air not pull air!
13. Push the plastic tubing into the top of the column and hold it there firmly. Gently push the syringe plunger in and push water through the column.

CAUTION: ***Do not push the liquid level below the top of the packed column*** **or you will have to repack the column because severe channeling and cracking will occur.**

The liquid level will not go below the top of the packing if you don't push it. The pressure you used in Step 13 helps to pack the silica gel tightly, and this pressure reduces the void volume considerably. Reducing the void volume increases the efficiency of the column.

- Why?

14. Tap the straw at the top of the packing and rotate the straw so that the top of the silica gel stays flat.
15. Add 2–3 cm of water to the column and tap to flatten.
16. Place a punched circle of no. 3 paper (obtain from Reagent Central) into the top of the column. Push the circle down the column with a slim straw.

    NOTE: As it enters the water, the air bubble will disappear.
17. Push the paper until it is *flat* against the silica gel packing. Gently smooth with the slim straw while rotating the column.
18. Fill the column with water from your wash bottle and push water through with the syringe pump as in Step 13. *Do not push water below the paper circle.*
19. Allow the column to drain naturally, and the liquid level will stay at the level of the paper.

    You have just carried out a sophisticated slurry-packing procedure for the preparation of a liquid chromatography (LC) column of good efficiency. Note that the procedure appears to be a little difficult the first time, but once you are familiar with the technique, it takes about 2 minutes to pack a column. If you do make a mistake while making or running any of the columns, it's OK. Simply pump the mobile phase and any sample off the column, pull off the tip, wash the silica gel out, and repack the column.

Section E. Investigations of Column Parameters and Processes

Goal: *To examine the chromatographic characteristics of the LC column prepared in Section D.*

Discussion: You will use an unretained dye component to quantitatively measure band spreading as a function of band migration distance, and you will also measure the void volume of the column.

Experimental Steps:

1. Obtain 3 drops of 0.05 %wt bromocresol purple solution in well A–1 of your tray.
2. Cut off the top 2 cm of the column with scissors so that the thin-stem pipet will reach the paper circle.

3. Mark a point (with a fine permanent marker) close to the top of the straw. Call this point 1.
4. With a thin-stem pipet, suck up a small volume of bromocresol purple dye, following the techniques shown below. Very carefully, keeping the pipet vertical, lower the tip of the pipet until it just touches the paper circle in the middle.

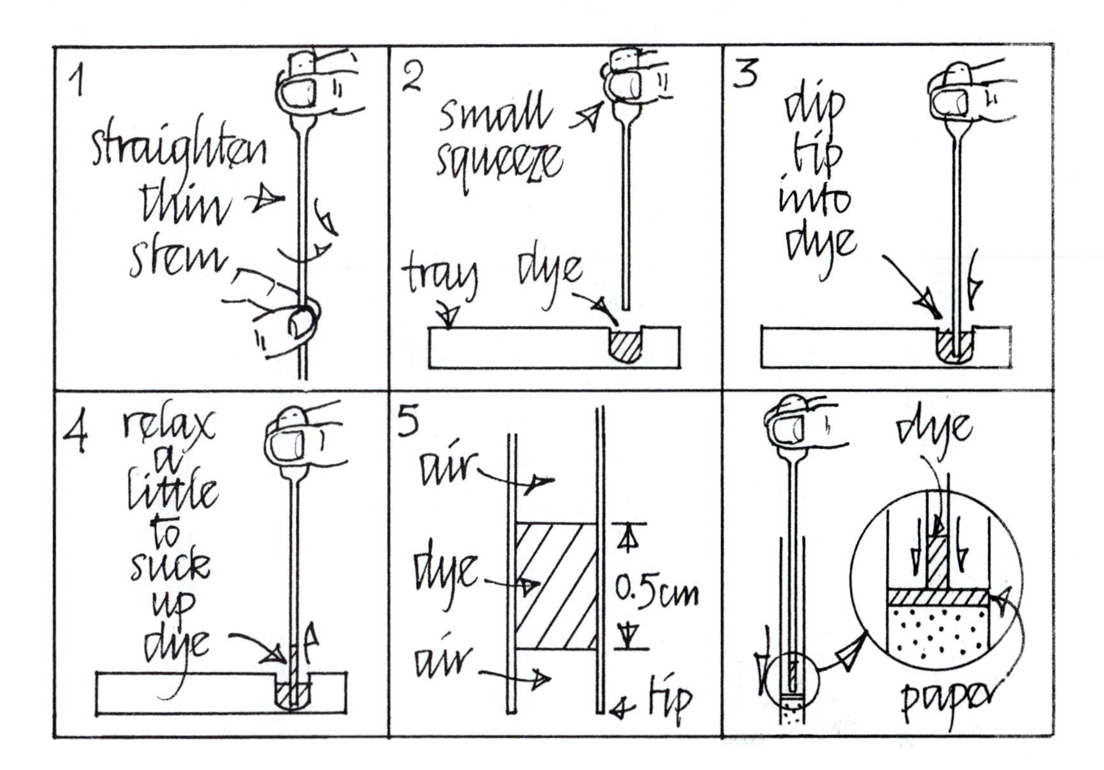

5. Very, very gently squeeze the pipet bulb until the dye is just soaked into the paper and *not* on the packing. Remove the pipet. Have a ruler ready.
6. Quickly fill the column with water to point 1 (use a wash bottle), and the chromatography will start.
7. Rotate the column 90° so that it is horizontal. The flow will stop (the water will stay in the column).
   - Measure the width of the band of color (W_b in mm) and measure the band migration distance — i.e., the distance from the paper circle to the middle of the band of color. Estimate both lengths to a mm.
8. Turn the column until it is vertical again, and the flow of mobile phase will resume.
9. Stop the flow by 90° rotation about every cm or so of band migration.
   - Measure and record the width of the band of color and the band migration distance.

10. When the band of color begins to enter the column tip, try to estimate the point in time when 1/2 of the band has come off the column and 1/2 of the band is still on. At this time make a mark on the straw at the water level. Call this point 2. Also estimate the width of the band of color as it leaves the column.

    The chromatographic progress of an unretained component through the column is shown in the following:

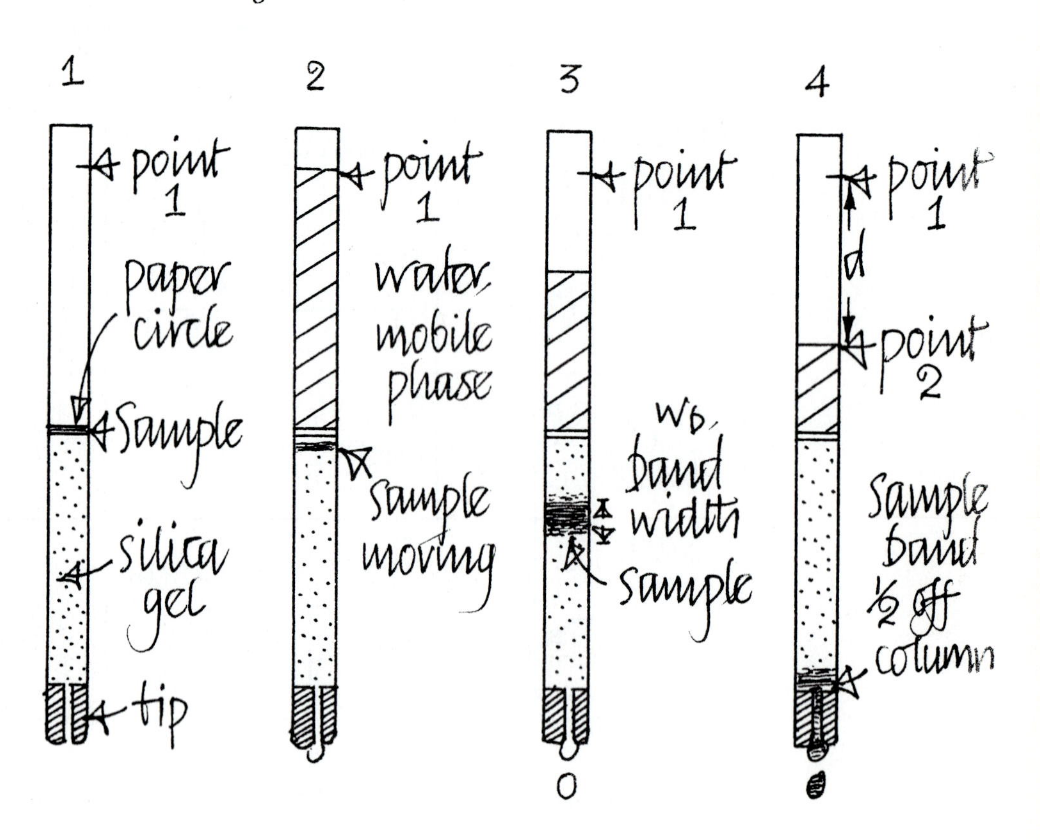

- Measure and record the distance d — i.e., the distance between point 1 and point 2 (in cm, estimate to a mm).
- Measure the length L of the actual column packing.

CALCULATIONS ON BAND SPREADING

- Plot a graph of band width W_b (vertical axis), versus band migration distance (horizontal axis).
- Compare your graph with the graph shown in Figure 18.7 which shows some typical data for the same type of experiment.

These plots show several significant trends. The *increase* in band spreading W_b gets smaller as the particle size of the silica gel decreases. Columns packed with very small particles are generally considered to be very efficient chromatographic systems.

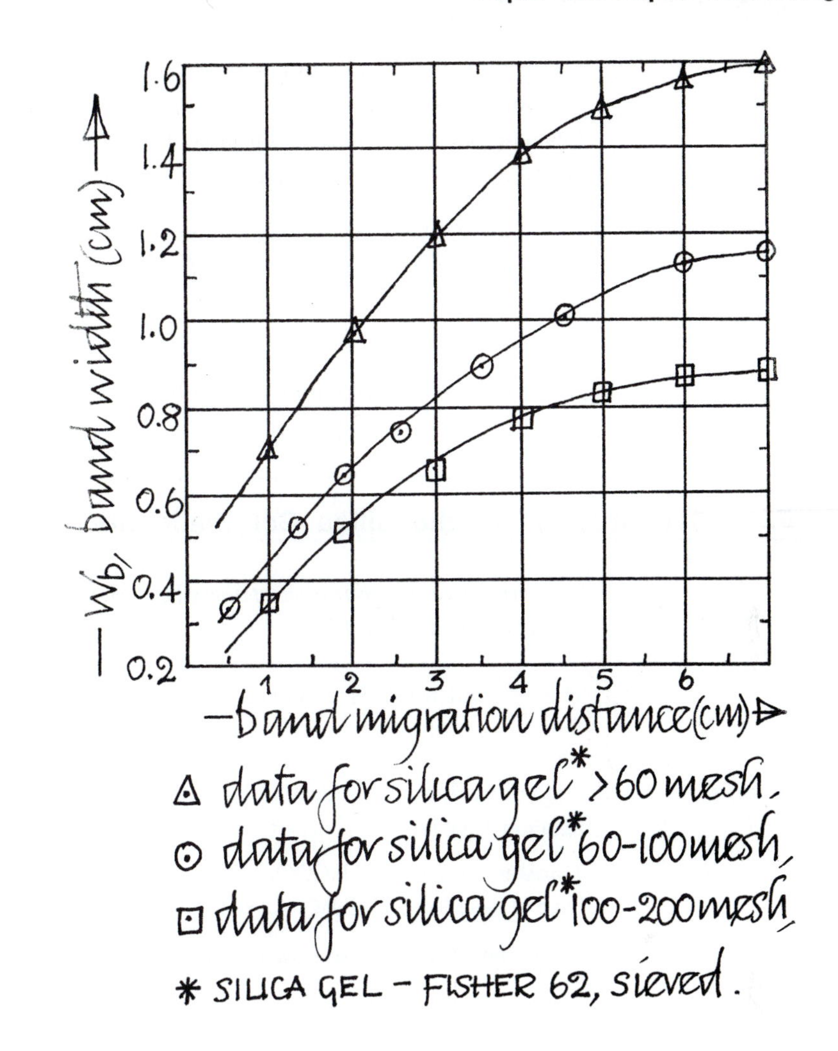

Figure 18.7 Band Width Versus Migration Distance Data for Silica Gels of Different Particle Size

CALCULATION OF VOID VOLUME (V_O), THE NUMBER OF THEORETICAL PLATES (N), AND THE HEIGHT EQUIVALENT TO A THEORETICAL PLATE (H).

V_O, N, and H are calculated from the experimental data obtained in Step 10 and the following discussion. The void volume can be calculated by assuming that the straw column is a cylinder of constant radius 0.3 cm and therefore

$$V_O = d \text{ cm} \times \pi \times (0.3)^2$$

- Calculate and report the void volume (in mL) for your column.

The number of theoretical plates N is given by

$$N = 16\left(\frac{V_O}{W_b}\right)^2$$

where V_o is the void volume and W_b is the width of the band of color as it leaves the column. V_o and W_b must be in the same units (i.e., mL), and, therefore, you need to calculate W_b in mL using the formula for the volume of a cylinder:

$$W_b \text{ (in mL)} = W_b \text{ (in cm)} \times \pi \times (0.3)^2$$

Calculate N. You can now calculate the height equivalent to a theoretical plate H, which is one of the most important efficiency factors in a chromatographic system:

$$H = \frac{L}{N}$$

where L is the length (in mm) of the silica gel column packing.

Section F. Derivatization of the Silica Gel Stationary Phase

Goal:

To carry out a derivatization of the silica gel in which the pH of the surface of the gel is changed from 6.8 to 4.5.

Discussion:

The derivatization is achieved by allowing a dilute acetic acid solution to flow through the column.

Experimental Steps:

1. Use your column from Section E. Use a 24-well tray to obtain 0.1 M CH_3COOH from Reagent Central.
2. Transfer the 0.1 M CH_3COOH to the column with a thin-stem pipet. Make sure you have at least 5 cm of CH_3COOH in the column.
3. Allow the CH_3COOH to flow under gravity until all the solution has gone through.
4. Pump 2 column lengths of water through the column to remove excess CH_3COOH. *Do not* pump the water level below the paper circle.

   The surface of the silica gel has now been protonated, and the surface pH is about 4.5.

Section G. Chromatography of Selected Synthetic Dyes

Goal:

To chromatographically separate a synthetic dye mixture into individual components using a derivatized silica gel column packing.

Experimental Steps:

1. Use good transfer technique and a thin-stem pipet to apply a small volume of the synthetic dye mixture to the paper circle on the column (as shown in the diagram in Section E, Step 4). *Do not overload the column.*
2. Clean the thin-stem pipet and use it to add a 30% v/v ethanol/water solution (the mobile phase) to the column (point 1). The chromatography will proceed.
3. Watch the chromatographic process.

- Draw a picture of the separation.

4. As the middle of each band of color comes off the column, make a mark on the straw at the mobile phase level.
5. Collect the effluent from each band of color in a separate well of a 1 x 12 well strip. This will allow you to collect *pure* fractions of each dye that may be used for spectroscopic examination.
6. Calculate the retention volume V_R for each of the dye components. The calculation of V_R may be carried out in the same manner as void volume in Section E.

Section H. LC of Beet Pigments

Goal:

To obtain a separation of the two major betacyanin pigments in natural beet (Beta Vulgaris) *juice.*

Experimental Steps:

1. Use the LC column that you used for the separation of the synthetic dye mixture. You can clean the column by pushing 3 column volumes of water through the column. Be careful not to push the liquid level below the paper.
2. Obtain a beet extract. This may be obtained easily by squeezing a small section of a fresh red beet or by using the juice from canned beets.
3. Carry out an LC separation on the juice and present samples of the 2 major pigments to your instructor.

Chapter 19

Gas Chromatography

"The second proposal, to look for halocarbons, was rejected as frivolous because it was 'obvious' that no apparatus existed sensitive enough to measure the few parts per trillion of chlorofluorocarbons I was proposing to seek."

James Lovelock, *The Ages of Gaia*

Introduction

Gas chromatography (GC) is one of the most powerful and widely used methods for the qualitative and quantitative analysis of volatile components in sample mixtures. Although a relatively new method — the first paper describing the use of a gaseous mobile phase was published by Martin and James in 1952 — GC is now used in all areas of science, medicine, and industry. Some typical analytical applications are trace hydrocarbons and other pollutants in air; petroleum refinery products; barbituates and other drugs in blood, breath, saliva, and urine samples; flavoring agents in foods; trace contaminants in beer, wine, and spirits; and pheromone sex attractants in insects.

Gas chromatography is a technique for the separation of volatile substances by percolating a sample mixture (in vapor form) in a gaseous mobile phase through a porous stationary phase contained in a long tube. The technique can be divided into two fundamental types: gas-solid chromatography (GSC) and gas-liquid chromatography (GLC). In GSC the separation is accomplished by passing the sample, in a carrier gas, over a solid stationary phase. The different components in the sample have different adsorption affinities for the stationary phase, and some are slowed down with respect to others. In GLC, components in the sample are separated by passing the sample and carrier gas over a stationary phase consisting of an inert solid support coated with a nonvolatile liquid. In GLC the distribution process between the mobile gas phase and liquid stationary phase is *partition*. GLC is by far the more versatile method because of the wide range of liquid stationary phases that are commercially available.

Background Chemistry

A general introduction to the theory of chromatography is given in the Background Chemistry section of Chapter 18, "Paper and Liquid Chromatography." If you are new to chromatography, it is probably advisable to read the section before you continue on into the Background Chemistry section of "Gas Chromatography."

A basic GC system consists of a carrier gas, a heated sample injection port, a separating column, and a detector. Commercially available instruments cost $5,000-$50,000 and often come with a dedicated computer for data collection, storage, and interpretation. The GC flow schematic is shown in Figure 19.1.

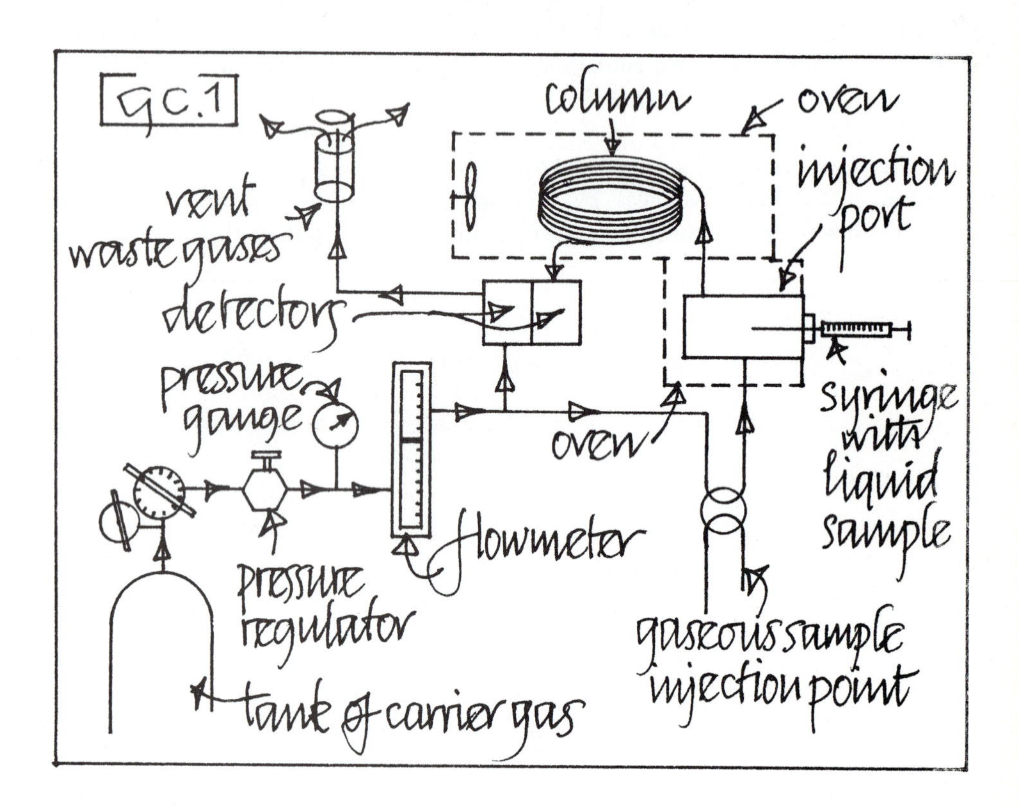

Figure 19.1 Schematic of a Typical GC

The carrier gas is usually a pure, inert gas (e.g., He, H_2, Ar, or N_2) stored in a pressurized tank. The flow rate of the mobile phase must be very carefully controlled in GC because the rates of migration of all components are dependent on it. Various pressure gauges, flow controllers, and meters accomplish exact carrier gas flow control.

The samples to be analyzed by GC may be gases, liquids, or solids. Solid and liquid samples must be volatilized; thus, they must be heated as they are introduced into the injection port. Generally, a very small sample volume is needed — on the order of 0.1 µL to 50 µL. The volatilized sample is swept onto the separating column by a flowing stream of carrier gas. The two main types of column in general use are shown in Figure 19.2.

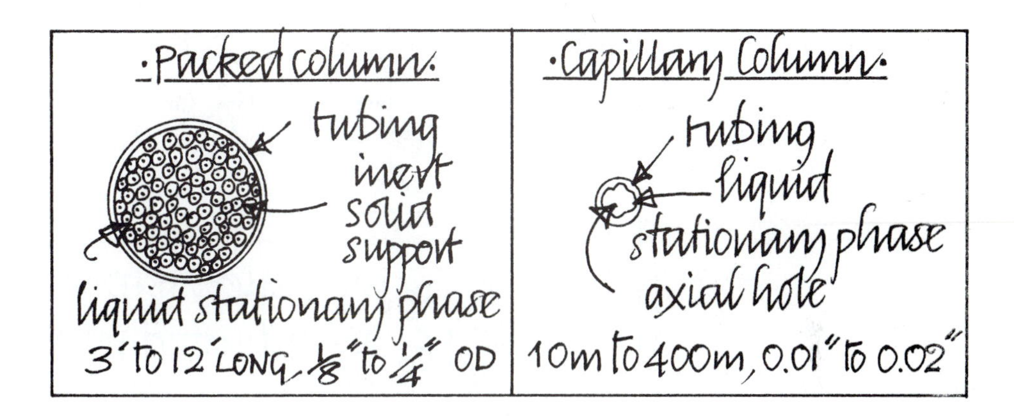

Figure 19.2 Two Types of GC Columns

Packed columns are relatively short because of the high pressure required to push the gases through the stationary phase. These columns are inexpensive and therefore widely used. Capillary columns are much narrower and can be much longer because of the hole all the way through the column. Capillary columns are tough to make and are expensive, although the increase in efficiency is worth the price, particularly for the analysis of very complex samples (e.g., gasoline).

Both types of columns are available with any one of several hundred different liquid stationary phases. Selection of the type of liquid stationary phase is based on the type of sample to be analyzed. The real power and flexibility of GC as a method of analysis rest on the fact that the stationary phase can almost be tailored at will to fit the separation problem. The choice is often made on the "like dissolves like" principle, or put in a more sophisticated way, the liquid is chosen on the basis of polarity index. The column is usually placed in an oven, the temperature of which can be raised or lowered (and monitored) in any predetermined manner. The separated components that leave the column are then quantitatively detected by a suitable detector or, in some instances, may be trapped and recovered. Again, one of the tremendous advantages of GC is the variety of sensitive, quantitative detectors that are available, e.g., thermal conductivity, mass spectrometry, and even live male insects (used as detectors of insect pheromones).

Many commercial GCs have three types of built-in detectors: *thermal conductivity* (TCD), *flame ionization* (FID), and *electron capture* (ECD). The detector output (signal) is usually fed to a strip chart recorder or to a dedicated computer (are there any other kind?). A typical GC chromatogram is shown in Figure 19.3.

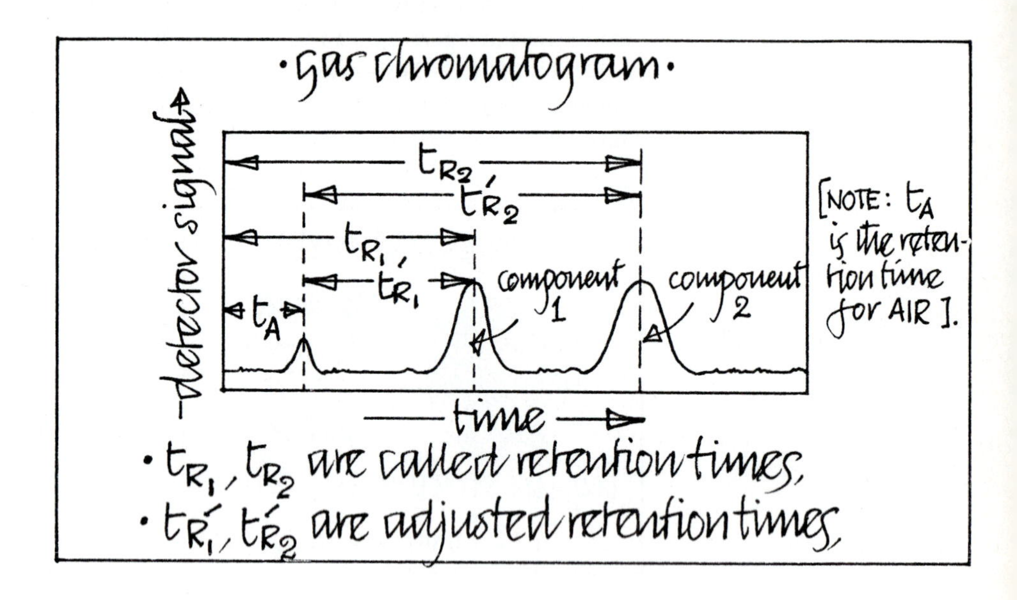

Figure 19.3 A GC Chromatogram

The various components do not have R_f values, in the same sense as in paper chromatography, because the components actually come *out* of the gas chromatograph, and the mobile phase is continuously flowing (compare PC). In GC the retention parameter is called the *retention time* (t_R) and is the time that elapses between the injection of the sample and when the center of the component band is detected by the detector. Almost always, the injection of the sample into a gas chromatograph results in air being injected. Air components (O_2 and N_2) are generally unretained — i.e., have no affinity for the liquid stationary phase — and quickly appear in the detector. The time between sample injection and the detected air peak is called the *retention time of air* (even though it is not retained by the stationary phase). The detectors also produce a concentration profile that, with a suitable calibration line, can be used to quantitatively measure the amount or concentration of any sample component.

A SMALL-SCALE GAS CHROMATOGRAPH

The Department of Chemistry could not afford to buy a gas chromatograph for each of you (in spite of the current tuition trends!). However, you can build your own working GC for about $0.25. The GC design was developed by the author over a three-year period (between 1974–1977). The column stationary phase is constructed of a glass or straw tube filled with dry Tide detergent ("you can trust Tide!") and uses natural gas as carrier gas. The GC detector is rather unusual in that the principle has been known for 90 years and is only now beginning to be used in some modern instruments. You will be constructing a Beilstein detector — named after a famous German chemist, who incidentally didn't discover the principle; a Swedish chemist, Berthollet, did!

The detector is a copper coil that is placed in a small flame generated at the column exit by burning natural gas. One of the real reasons for using natural gas (methane) as a carrier gas is that it comes from a highly controlled valve and has a convenient useful line pressure of about 6–7 ounces per square inch. The normal gas tap serves as a fine-tuning regulator for the carrier gas flow. The methane is then used as a fuel source for the small-scale premix burner and Beilstein detector.

Normally in commercial GC instruments, the injection port is actually a little oven that quickly converts the liquid sample (often dissolved in a volatile solvent) into vapor. In your GC the samples are halocarbon vapors and, thus, heating is not necessary. The port is made from ordinary Bunsen burner tubing (latex tubing). Surprisingly, the latex is a self-sealing material that will take repeated injections without leaking. The separating column is a short (20 cm x 0.8 cm) tube of soft glass packed with activated Tide detergent. Tide detergent is a complex mixture of about twelve ingredients formulated to get clothes clean. However, from the GC viewpoint, it consists of an inorganic solid ($Na_5P_3O_{10}$, sodium tripolyphosphate) coated with polar, high molecular weight, organic surfactant. The organic surfactant probably serves as the liquid stationary phase in GC. The Tide must be activated before use in order to remove the perfume and some water. Activation is carried out by placing a large tray of the detergent in an oven at 150 °C for about 4 hours. The particle size of the powder is not ideal, but it is in the correct range to allow a reasonable carrier gas flow through the column (provided the column is not too long). The actual process of putting the Tide into the tube (packing the column) is critical to achieving successful separations. If the column is packed too tightly, there will be no carrier gas flow, and if the column is packed too loosely, channeling will occur, giving rise to poor or no separations. Of course, the larger the Tide column, the longer are the retention times for components and the greater is the band spreading of each component. A column length of 20–50 cm seems to be reasonable for the gas pressures encountered in most schools and universities.

It is worth noting that there are several reports in the scientific literature of the use of solid detergents as stationary phases in gas chromatography. In the GC that you will be building, the Tide-packed column will be used at ambient temperatures. The halocarbon samples that you will be separating have a relatively high vapor pressure at room temperature, and elevated column temperatures are not necessary. The Tide column will separate the sample mixture into individual halocarbon components that then must be sensed by some type of detector. You will be using a Beilstein detector, which is a sensitive device for the detection of compounds containing halogen atoms, but which is not very sensitive, or does not work at all, for other substances.

The *Beilstein detector* is a sensitive, selective GC detector that emits visible light when separated halocarbon components go through it. A quantitative analysis may be obtained by using some type of photodetector (e.g., a CdS cell or photodiode) to transduce emitted light into an electronic effect. Of course, a calibration line is required for each component because the detector and transducer response depend on the chemical structure of the detected substance. A Beilstein detector consists of a copper wire coil that is placed in the relatively cool part of a small flame. As the flame plays over the copper (and the surface copper(II) oxide), the surface reacts with free electrons in the flame and is kept clean and reactive. This phenomenon is very beautiful to watch because the black surface ripples with a golden sheen as the flame reduces Cu^{2+} back to Cu°. If a halogen-containing vapor is burned in the flame, highly reactive halogen atoms are formed which then quickly react with the fresh copper surface to form volatile copper halides. The halides then rapidly react with OH radicals in the flame to give various copper species (e..g., $CuOH^+$) that are thermally excited by the heat of the flame. The excited copper species emit green-blue light as they return to the ground state. Emission of green-blue light is a definite indication that a halogen-containing component has arrived in the detector. By the way, the Beilstein effect has been used for many years in devices for detecting leaks in air conditioners and refrigerators!

Once you have built the GC, the best way to learn about the technique is to get involved in investigating a practical problem. The next section presents an interesting environmental sciences application that involves the analysis of industrial and commercial products and addresses the general problem of halocarbons in the environment.

HALOCARBONS AND THE ENVIRONMENT

A large number of chemical compounds containing carbon-halogen bonds have been made by the chemical industry and have proved to be extremely useful in industry, commerce, and agriculture. The uses are varied and include such applications as pesticides, hydraulic fluids, electrical transformer fluids, solvents, aerosol propellants, anesthetics, refrigerants, air conditioner fluids, foam expanders, plastics, etc., etc. Some specific examples follow.

Insecticide:

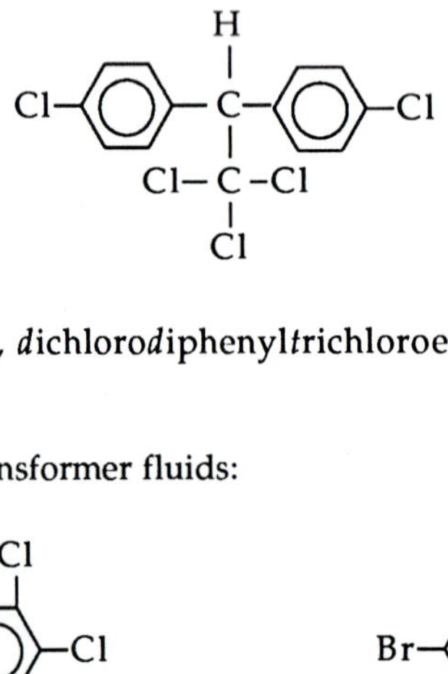

H
Cl– C –Cl
Cl–C–Cl
Cl

DDT, *d*ichloro*d*iphenyl*t*richloroethane

Hydraulic fluids and transformer fluids:

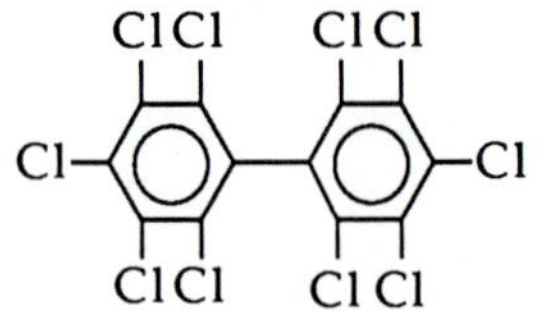

ClCl ClCl
Cl– –Cl
ClCl ClCl

PCB, *p*oly*c*hlorinated*b*iphenyl

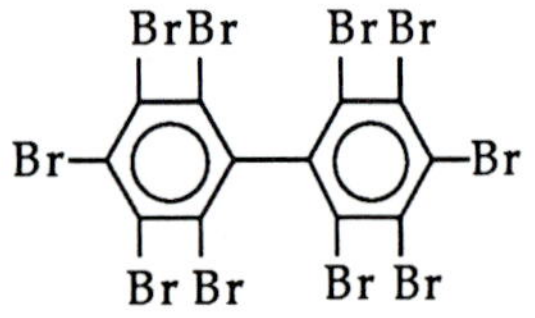

BrBr BrBr
Br– –Br
BrBr BrBr

PBB, *p*oly*b*rominated*b*iphenyl

Solvents:

$CHCl_3$ (H–C bonded to three Cl) — Chloroform

CCl_4 (Cl–C bonded to three Cl) — Carbon tetrachloride

Solvents, propellants, refrigerants, foam expanders, air conditioner fluids, etc.:

Freon 11	Freon 12	Freon 13	Freon 14	Freon 113
$CFCl_3$ (F–C with Cl, Cl, Cl)	CF_2Cl_2 (F–C with F, Cl, Cl)	CF_3Cl (F–C with F, F, Cl)	CF_4 (F–C with F, F, F)	$Cl_2FC{-}CF_2Cl$ (Cl–C(Cl)(Cl)–C(F)(F)–F)

The use of many of these halocarbons has been restricted, and some of them have been banned outright because of environmental health problems. The environmental problems have arisen mostly from the chemical and biological properties of these halocarbons. Carbon-halogen covalent bonds are very stable, and these compounds do not break down easily under most environmental conditions. This chemical longevity, together with the fact that many of these compounds are fat soluble, means that halocarbons stay around a long time and concentrate up food chains. (For one interesting aspect, see Chapter 8, "An Introduction to Acids and Bases.")

In this module, it is the smaller molar mass halocarbons (e.g., the Freons) that are of interest. The Freons, sometimes called chlorofluorocarbons (CFCs), are very stable, volatile, and cheap compounds. These characteristics make them valuable as refrigerants, propellants, and foam expanders. Their stability means that they persist in the air — the troposphere — for a long time (half-life is about 75 years; for Freon, 12), eventually drifting into the stratosphere. Unfortunately, the stratosphere is a sink for Freons because at this high altitude, the high-energy ultraviolet (UV) light from the sun causes even stable compounds to break down. In a complex sequence of photochemical reactions, the Freons are decomposed by light into chlorine atoms:

$$CCl_3F + h\upsilon \rightarrow Cl + CCl_2F$$

$$CCl_2F_2 + h\upsilon \rightarrow Cl + CClF_2$$

The chlorine atoms catalyze the destruction of ozone (O_3) in a series of free radical reactions, one of which is

$$Cl + O_3 \rightarrow ClO + O_2$$

Recently, it was discovered that each year, in September and October, the stratospheric ozone layer over Antarctica shrinks drastically and in many places disappears completely. Again, it appears that chlorofluorocarbons are the culprit. Unusual chemical reactions on the surface of polar stratospheric ice clouds generate chlorine from chlorofluorocarbons, and the ozone is destroyed. The maintenance of normal ozone concentrations in the stratosphere is critical to life on earth. The stratosphere ozone cycle removes most of the high-energy UV light that would otherwise reach the earth's surface. The consequences of a reduced ozone concentration in the stratosphere may vary from increased skin cancers to severe chromosome damage and dramatic climate changes.

A recent global conference of CFC producers, users, and scientists agreed that the problem of ozone destruction by chlorofluorocarbons is a real and severe environmental threat. In an unprecedented decision the conference announced CFC production restriction and eventual phase-out of CFCs altogether. This decision has stimulated several large companies to begin research on developing new CFC substitutes for refrigerants and foam-expanding agents. In spite of all the recent furor, it is important to note that most of the CFCs that have *ever* been produced are still present in the atmosphere. The ozone problem is not going to go away for many decades, if ever.

Gas chromatography has played a key role not only in the initial identification of CFCs in the atmosphere, but also in the current research on ozone "holes" in the polar regions. In fact, it could be said that the whole CFC story started about 20 years ago in an English country garden! In 1970 an independent scientist named James Lovelock, who now works in a barn-turned-laboratory in Cornwall, England, invented a new type of GC detector called the *electron capture detector*. He attached the detector to a GC, and his first sample for analysis was English garden air. The GC chromatogram revealed two previously unidentified peaks that were shown to be Freons 11 and 12. Lovelock did a rough calculation based on the Freon concentrations measured by the GC experiment and came to the conclusion that all of the Freons ever manufactured were still present in the atmosphere — the rest is history.

It is strange but true that the discovery of the CFC–global ozone problem relied on the invention and use of a GC detector. In this laboratory module, you have the opportunity to build your own sophisticated GC equipment, investigate its limitations, and carry out analyses on various samples containing halocarbons.

Pre-Laboratory Quiz

1. What is gas chromatography?

2. In GLC what is the name of the distribution process that occurs between the mobile and stationary phase?

3. In GC what is the retention parameter called?

4. What carrier gas will you use in your GC?

5. Give a brief description of the stationary phase you will be using in your packed-column GC.

6. How does a Beilstein detector work?

7. Give the chemical formula for a Freon.

8. Give 2 chemical reactions that show how CFCs destroy ozone.

9. Who invented the electron capture detector?

10. Give one major use of a Freon.

Laboratory Experiments

Flowchart of the Experiments

Section A. The Construction of a Gas Chromatograph

Section B. Measurement of the Retention Time of Air and the Gas Flow Rate

Section C. Measurement of the Retention Times of Halocarbons

Section D. GC Separation of Halocarbon Mixtures

Section E. GC Analysis of Industrial Products

Section F. Optimizing a GC System: The Van Deemter Plot

Section G. Quantitative Analysis by GC with Photodetection

Requires one three-hour class period to complete

CAUTION: In this series of experiments, you will be using natural gas as a carrier gas in GC. Be aware that natural gas is flammable and in certain circumstances can be dangerous. Please remember that burning gas is hot, and so are metal objects (such as detectors and windbreaks) that come into contact with flames. Please sign out and in for the syringe sample injector you use during the experiment.

Section A. The Construction of a Gas Chromatograph

Goal:

To build a working small-scale gas chromatograph that is capable of separating several halocarbons.

Before You Begin:

You are about to build the odd-looking device, actually a small-scale gas chromatograph, shown in the diagram below. As you work through Section A, refer to the diagram to compare with your own construction.

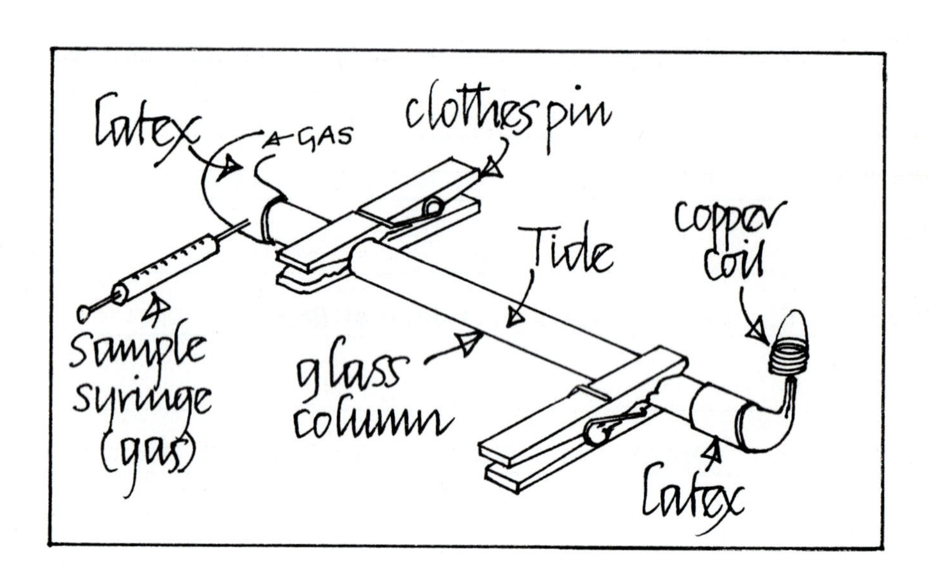

Small-scale gas chromatograph

Experimental Steps:

1. Obtain the following from Reagent Central: 1 x 20 cm piece of 8 mm glass tubing; 1 x 2 ft and 1 x 2 cm pieces of latex tubing; 1 glass Pasteur pipet; small bundle of polyester fiber; 1 piece of copper wire; 1 glass cutter or scorer; 1 box matches; 2 clothespins; 1 small cup of activated Tide detergent; 1 plastic scoop; and a windbreak made from a beverage can.

2. *Do not wash* the 20 cm piece of glass tubing or it will take a long time to dry. Place a *small* plug of polyester fiber inside one end of the glass tube.

3. Insert a small cork at the end of the tube with the plug.
4. Scoop up some of the activated Tide detergent and place the end of the scoop into the vertically-held glass tube.
5. Deliver the detergent at an even rate, tapping the tube gently as you fill it.

   NOTE: If the scoop gets blocked, remove and invert it over the cup and tap. Do not attempt to poke it out or it will really clog.
6. Refill the scoop and keep pouring and tapping until the tube is completely filled with Tide. Remove the scoop.
7. Keep the tube vertical and very gently bounce the tube on the table (at the cork end). The Tide will settle a little.
8. Add more Tide until it is about 0.5 cm from the end.
9. Place a plug of polyester fiber into the end of the tube to keep the Tide in.

   You have now prepared a Tide-packed gas chromatography column. Since this is possibly the first time that you have attempted this high technology endeavor, there might be some probability that the column packing is not perfect. Don't worry — diagnosis of "sick" columns is easy, and repacking takes only a few minutes. It is important that you handle the column carefully. Place it gently on the table and try not to bump it too much. Now you can construct the burner and Beilstein photoionization detector.

Beilstein Burner Construction

10. Obtain a glass Pasteur pipet. Place the part where it narrows down in a small burner flame (or even a match flame). Keep rotating it until it starts to bend slowly. Stop rotating and let it bend, under gravity, until it forms a right angle. You can help it a little if you like!
11. Place the pipet on the table to cool.
12. Hold the large end firmly on the table and use a scorer (or file) to scratch and cut off the thin end so that you are left with a tip 2–3 cm long, as shown below.

    CAUTION: Glass can be dangerous so consult your instructor if you are uncertain about this step.

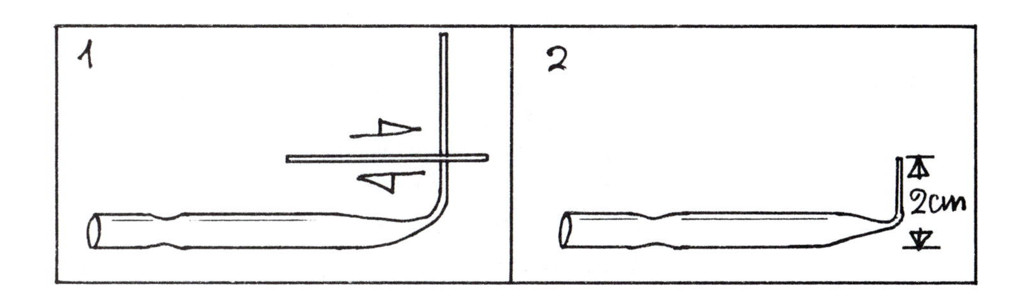

13. Likewise, cut off the larger diameter end about 2-3 cm from the bend with a scorer or file. This technique is not easy! If you have problems check with your instructor. Save the cutoff part.

14. Fire polish the sharp, large-diameter end. You have now made the burner.

GC Detector Construction

15. As demonstrated below, hold the straight, cutoff part of the pipet saved in Step 13 and *tightly* wind a copper wire coil 10 turns. As you wind, keep your thumb tightly on the end and put tension on the wire as you wind it. Leave a tail of copper wire about 3 cm long.

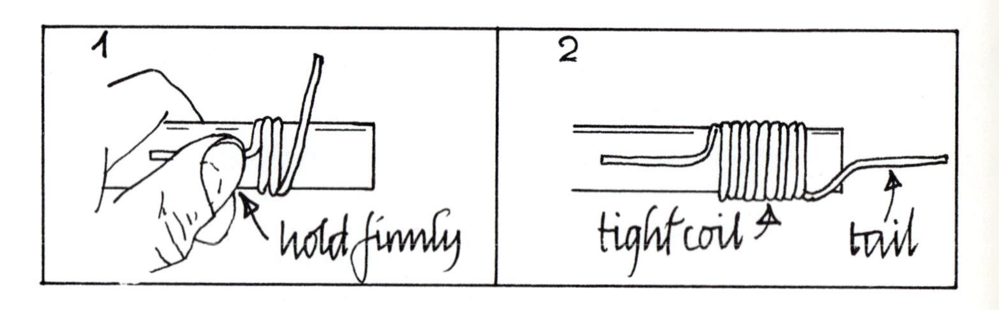

16. Slip the tight coil off the glass tube and adjust the tail so that it bends and then is positioned down the axis of the coil.

17. Cut the tail off about 1 cm from the coil.

18. Now slide the tail into the narrow part of the glass burner.

    Presto! You have created burner and detector! Now you are ready to assemble the gas chromatograph.

Gas Chromatograph Construction

19. Push one end of the latex tubing onto the natural gas tap and carefully push the other onto the column. Try to avoid pulling the polyester wad out.

20. Push the small piece of latex on the other end of the column.

21. Attach the burner to the small piece of latex. Rotate it if necessary to ensure that the column will lie naturally on the table with the burner vertical.

22. Clip 2 clothespins onto the column as stabilizers.

23. Place a windbreak made from a beverage can around the burner so that the flame exhaust can exit through the tab hole in the top. The can also acts as a flame stabilizer.

    Congratulations. You have just made a small-scale, packed-column gas chromatograph with a latex injection port and a Beilstein burner photoionization detector. The carrier gas is natural gas.

Section B. Measurement of the Retention Time of Air and the Gas Flow Rate

Goal: *To "age" the Beilstein detector and measure the retention time for unretained air.*

Discussion: The t_R for air (retention time for air or air peak time) may be used to calculate the linear gas velocity of the carrier gas.

Experimental Steps:

1. Turn the gas tap full on. Wait about 5 seconds, strike a match, and hold the flame to the top of the coil.
2. Adjust the gas tap so that the flame is about 0.5–1.0 cm above the top of the coil. Let the heat from the flame "age" the coil.

   The visible blackening of the copper surface is due to $2Cu + O_2 \rightarrow 2CuO_{(s)}$.
3. Wait about 30 seconds, then remove the coil with a pair of tweezers and hold it for a moment to cool.
4. Place it carefully on the table.

   NOTE: Aged detectors are fragile.
5. Relight the burner flame if necessary.

   NOTE: From now on do not touch the gas tap. The flow is set.
6. Obtain a plastic, graduated 1.0 mL syringe from your instructor. Your instructor will have you sign for it.
7. Pull the plunger out to the 0.5 mL mark.
8. Push the needle into the latex injection port within a cm or so of the end of the column. Smoothly and quickly, push the plunger in.

   Watch the flame carefully and note that after a very short time, the flame will dip smaller and then go back to its regular size. The dip corresponds to a change in air-fuel ratio as the injected air arrives in the burner.
9. Repeat Step 8.

   - Measure the elapsed time from the injection to the flame dip.

   The elapsed time is the retention time for air — i.e., the time it takes for the injected air to travel unretained through the column and into the flame.

   - Carry out at least 3 measurements of the retention time for air and record your data.
   - Calculate the gas flow rate (linear gas velocity) by

   $$\bar{u} = \frac{L}{t_A}$$

   where $\bar{u}$ is the average linear gas velocity (in cm s^{-1}), L is the length of the GC column (in cm), and t_A is the retention time for air (in seconds). In any GC experiment the carrier gas flow rate must be measured because *all* retention times for components will be dependent on it. The faster the carrier gas is flowing, the faster the sample components will move through the column.

NOTE: The carrier gas flow rate is sometimes measured in a different way (with a soap bubble flow meter) and is reported as a volumetric flow (i.e., mL per minute). If you have time, you might want to build a simple soap-bubble flow meter. Check with your instructor.

Section C. Measurement of the Retention Times of Halocarbons

Goal:

To measure the GC retention times for a series of organic compounds containing carbon-halogen covalent bonds.

Experimental Steps:

1. Using tweezers slip the copper coil back into the burner. Relight the burner if necessary.

   NOTE: It is very important that the flame be only on top of the coil. When the burner and detector are optimized, the flame will be steady, on top of the coil, and not very luminous. If the flame is burning from the glass tip or inside the coil, the coil will get too hot and the Beilstein effect will not work well. Try and adjust the coil with your tweezers, but do not change the gas flow rate!

2. At Reagent Central you will find 25 mL conical flasks stoppered with rubber septa. Take back to your place one of each of the 5 flasks labelled Freon 11 (CCl_3F), Freon 12 (CCl_2F_3), dichloromethane (CH_2Cl_2), chloroform ($CHCl_3$), and carbon tetrachloride (CCl_4).

3. Pump the syringe plunger several times to clean the syringe with air.

4. Pull the syringe out to the 0.1 mL mark.

5. Start with one of the Freons. Insert the needle into the septum.

   NOTE: Do not tilt the flask — you are going to withdraw *vapor*, not liquid!

6. Push the plunger in and pull it out to the 0.1 mL mark. Remove the syringe and needle.
7. Stick the needle into the GC injection port. Be ready to begin timing.
8. Inject the vapor, using good injection technique, and immediately begin timing.
9. Watch the detector flame.

   - Record the elapsed time from injection to:

     (a) The first appearance of a green-blue color in the flame

     (b) The maximum intensity of green-blue flame

   The first time you try this, everything generally happens too fast to get decent results (it took me 7 times!). Obtain another Freon sample and try again.

   You have just measured the times shown in the diagram below.

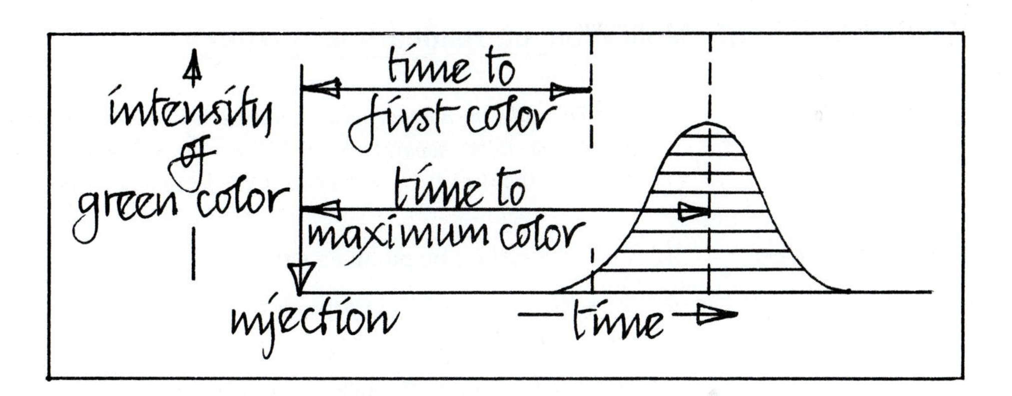

   The elapsed time from injection to the maximum green-blue flame color is the retention time t_R for that halocarbon. Measuring the time to the first appearance of green-blue flame color will allow you to calculate W_b, the band width of the halocarbon peak.

10. Repeat the above measurements (Steps 3 through 9) on the halocarbons that are available to you.

    Use more vapor for some of the less volatile halocarbons — e.g., 0.2 mL CH_2Cl_2, 0.3 mL $CHCl_3$, and 0.5 mL CCl_4. You may use the same syringe provided that, after injecting each sample, you *remove the plunger entirely, replace, and pump air several times to pump any residual vapor from the needle.*

    - Measure and record the retention time for each halocarbon.
    - Record the retention time for air and flow rate.

    If you think that for some reason the flow rate is changing (e.g., the building gas pressure might change), then redetermine the retention time for air.

    - Calculate the W_b values for each halocarbon peak.

HINT: Assume that the peak is Gaussian, i.e., symmetrical. Note that the units of t_R and W_b (in this GC experiment) are seconds.

- Calculate an approximate value for the number of theoretical plates N and the height equivalent to a theoretical plate H for one of the halocarbons (e.g., dichloromethane).

Use the expressions suggested earlier for these calculations:

$$N = 16\left(\frac{t_R}{W_b}\right)^2 \quad \text{and} \quad H = \frac{L}{N}$$

where t_R is the retention time (seconds), W_b is the width of the GC peak (in seconds), and L is the length of the column (mm).

Section D. GC Separation of Halocarbon Mixtures

Goals:

(1) To make a homogeneous mixture of several halocarbons and to obtain a gas chromatographic separation of all of the components. (2) To measure halocarbon retention times and W_b values in order to determine peak resolution.

Experimental Steps:

1. Clean the syringe by pumping with air.
2. Stick the needle into the septum of the Freon 12 flask. Pull the plunger out to the 0.1 mL mark.
3. *Leaving the plunger set,* stick the needle into the Freon 11 septum. Remove 0.1 mL of Freon 11 by pulling the plunger out to 0.2 mL. In the same way (in the same syringe), take 0.1 mL CH_2Cl_2 and 0.3 mL CCl_4.

   NOTE: *Do not contaminate* the individual halocarbons in the flasks by pushing the plunger in at any time.

   Be ready to record the elapsed times to the first and maximum green-blue flame color.
4. Inject the mixture into the gas chromatograph.
   - Record the elapsed times and describe what happens.
   - Why does the width of the peak W_b increase as the t_R increases?

   One important parameter in any chromatographic systems is the *resolution* obtained between two successive components, which is defined by

$$\text{Resolution} = \frac{\text{band migration difference between 2 components}}{\text{component spreading}}$$

- Calculate the resolution obtained between Freon 12 and Freon 11. Note the diagram below.

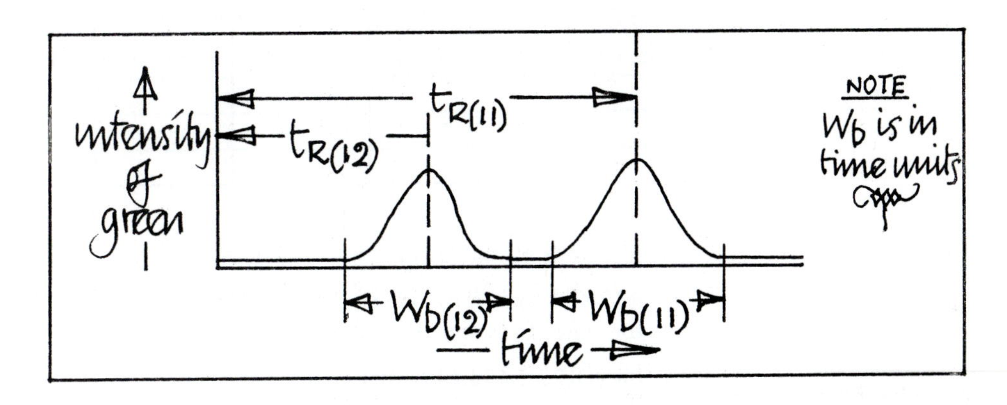

$$\text{Resolution between Freons 11 and 12} = \frac{t_R(11) - t_R(12)}{\dfrac{W_b(12) + W_b(11)}{2}}$$

- What does a resolution of 1.0 tell you about the peaks in a chromatographic separation?

Section E. GC Analysis of Industrial Products

Goal: *To use the gas chromatograph and your acquired knowledge of halocarbon separations to carry out a qualitative analysis of an industrial product containing halocarbons.*

Discussion: Your instructor will provide you with an industrial product containing one or more of the halocarbons that you have studied in the GC laboratory module. Analyze the product by GC and report your result. Your instructor will then give you a brief product description indicating the uses and specifications of the product.

- How would you analyze a solid sample?

Section F. Optimizing a GC System: The Van Deemter Plot

Goal: *To measure retention times and W_b values for dichloromethane at several different linear flow rates (of carrier gas).*

Discussion: A plot of height equivalent to a theoretical plate (H) versus linear flow rate ($\bar{u}$) is called a *Van Deemter plot.* This plot may be used to select flow rates at which optimum separations can be achieved.

Before You Begin:

Make a table in your lab notebook similar to the one shown below. In Step 6 you will be asked to record your data and calculational results in this table.

Retention Time for Air	$\bar{u}$ (cm s^{-1})	Time for 1st t_R (s)	Appearance	W_b (s)	N	H (mm)

Experimental Steps:

1. Make a new 3-turn copper wire coil detector for your gas chromatograph.
2. Adjust the gas flow rate so that the retention time for air is about 2–3 seconds. Leave the flow rate set at this value.
3. Inject 0.2 mL CH_2Cl_2 vapor.
   - Time the first and maximum appearance of green-blue color and record.
4. Decrease the flow rate slightly by turning the gas tap and again determine the retention time for air at this new flow rate. Leave the flow rate set at this new rate.
5. Inject 0.2 mL CH_2Cl_2 vapor.
   - Measure and record times as before.
6. Carry out one more set of time measurements at a slightly lower rate. Try to obtain about a 1 second difference in the retention time for air at the different flow rates.
   - Record your data and calculational results in the table you made in your lab notebook.
   - Measure and record the length of Tide packing in the column (L) in mm.
   - Make a graph of H (vertical axis) versus $\bar{u}$.

   This plot is called a Van Deemter, Von Klinkenberg, and Zuiderweg plot, or Van Deemter plot. The plot will often show a minimum in H at some flow rate. A minimum in H means a maximum in chromatographic efficiency. The flow rate at which this occurs is called the *optimum flow rate.*

Section G. Quantitative Analysis by GC with Photodetection

Goals:

(1) To add a quantitative emitted-light detector (a cadmium sulfide cell and digital multimeter) to the GC. (2) To use the instrumental combination to obtain a calibration for the quantitative analysis of dichloromethane.

Experimental Steps:

1. Obtain a cadmium sulfide (CdS) cell light detector from your instructor.
2. Build a straw holder and stand for the cell.

NOTE: Replace the 3-turn coil with the original 10-turn coil.

3. Position the cell horizontally and at a convenient height about 4 cm away from, and pointed directly at, the detector flame. Use the cutout can as a windbreak and flame stabilizer.

4. Connect the CdS cell to a multimeter and set the meter to measure resistance. If you can, adjust the sensitivity of the meter to give reasonably stable readings.

5. Inject 0.6 mL CH_2Cl_2 vapor into the GC and measure the maximum emission of light — i.e., minimum resistance on the meter — as the halocarbon goes into the flame and causes the emission of green-blue light.

   - Record the measurement in your lab notebook.

6. Now inject 0.5 mL CH_2Cl_2 vapor into the GC. Determine the minimum resistance on the meter for this volume of CH_2Cl_2 vapor.

   - Record the minimum resistance on the meter for this volume of CH_2Cl_2.

7. Perform Step 6 again with each of the following amounts of CH_2Cl_2: 0.4 mL, 0.3 mL, 0.2 mL, 0.1 mL, and 0.05 mL.

   - Plot mL CH_2Cl_2 vapor versus minimum resistance on the CdS cell. Draw a smooth curve through the experimental points.

   If you know the temperature of the CH_2Cl_2 liquid/vapor sample, the ambient atmospheric pressure, and the following vapor pressure expression, then you can calculate the exact number of moles of CH_2Cl_2 injected into the GC. The vapor pressure expression is

$$\text{Log}_{10} P = \left(-0.2185 \times \frac{A}{T}\right) + B$$

   For CH_2Cl_2, A = 7572.3 and B = 8.1833; for 25 °C, T = 298 K (valid from –70 °C to 41 °C). Then the vapor pressure of CH_2Cl_2 is 428 Torr. You can now use the ideal gas law PV = nRT to calculate the number of moles.

   Once you have obtained a calibration line at a particular flow rate with a particular detector and CdS cell, etc., etc., then you can carry out a quantitative analysis for that substance on the GC system.

8. Obtain an unknown from your instructor and have a go.

Chapter 20

Surface Chemistry: Bubbles and Films

Introduction

The calming effect of oil on rough water has been known for thousands of years. Benjamin Franklin reported some of the earliest scientific experiments on the effect of oily films on water. His famous experiment on the pond on Clapham Common was described thus: "... the oil, though not more than a teaspoonful, produced an instant calm over a space of several yards square, which spread amazingly, and extended itself gradually 'til it reached the lee side, making all that quarter of the pond, perhaps half an acre, as smooth as a looking glass."[1]

The strange and unusual behavior of liquid surfaces has stimulated chemists and physicists alike to try to find explanations for the spreading of oil on water, the calming effect of oil on rough water, etc. The study of the chemistry of surfaces has proved to be so fruitful and of such practical importance that it has become a separate discipline called surface chemistry. The real challenges in the surface chemistry of liquids involve, for example, research in the dynamics of the interactions of the oceans with the atmosphere, research on biomembranes in plants and animals, and research on the environmental consequences of large-scale oil spills in the ocean.

The containment of oil spilled in the ocean, or indeed in any aquatic environment, is extraordinarily difficult. The removal of oil spread over ocean and land surfaces is even more difficult. Nowhere were these problems more evident than in the recent massive oil spill from the supertanker Exxon-Valdez in beautiful Prince William Sound off the coast of Alaska. The March, 1989 Exxon-Valdez disaster tragically demonstrated the enormous risks inherent in the large-scale production and transportation of fossil fuels. The calming effect of oil on rough water turns into environmental nightmare when a teaspoon of oil becomes ten million gallons of Alaskan crude!

[1]Cited in *The Compete Works of Benjamin Franklin*, vol. V, J. Bigelow (ed.), G.P. Putnam and Sons, New York, 1887, p. 253.

Background Chemistry

A *surface* may be defined as the place where two different phases meet. The interfacial region between gases and liquids, immiscible liquids, liquids and solids, and immiscible solids may all be called surfaces. Surface chemistry is a rich and fascinating area of science because most chemical transformations take place at interfaces. Extreme changes in many chemical and physical properties occur at surfaces. Atoms, molecules, and ions at and around a surface exist in a very different environment than those deep within the bulk material. The inequality of forces and the close proximity of different species lead to an enormous variety of unusual and useful chemistry. The emphasis in this series of laboratory experiments is on the surface chemistry of liquids, particularly the interface between aqueous solutions and air. A knowledge of the water-air interface is central to understanding the beauty of bubbles, much of the biology of plants and animals, and the way our water planet works.

First let us consider the interface between bulk pure water and air. A water molecule that is in the interior, remote from the surface, is completely surrounded by other water molecules. The water molecules are attracted to each other by intermolecular hydrogen bonds that are continuously forming and breaking as the molecules collide. On a time-averaged basis, each water molecule experiences a symmetrical force field that has the same magnitude in all directions. The direction in which any water molecule moves is determined solely by momentum exchanges with colliding molecules and is therefore random.

The molecular picture near the water-air interface is very different. A water molecule that is approaching the interface in its random thermal motion experiences an increasingly unsymmetrical force field. As it moves closer to the surface, the number of water molecules on its interior side remains the same, but the number of exterior water molecules decreases rapidly. The water molecules in the dynamic surface layer are continually being attracted inward and, of course, are continually being replaced by molecules diffusing from the interior. The surface layer, which is perhaps two or three molecules thick, is therefore in a continual state of tensile strain, and the restoring force is called the surface tension. For a flat surface the *surface tension* is defined as the force acting parallel to the surface and at right angles to a line of unit length anywhere in the surface. The SI unit of surface tension is *newton per meter* ($N\ m^{-1}$). The magnitude of the surface tension of a pure liquid in equilibrium with its vapor and air is dependent on the magnitude of the intermolecular attractions in the liquid. Water has a relatively high surface tension because of its relatively strong hydrogen bonding, whereas ethanol has a much smaller surface tension because of its weaker hydrogen bonding.

One of the most important practical consequences of surface tension is that a liquid spontaneously tries to shrink (minimize) its surface. A drop of any liquid in equilibrium with its vapor and in the absence of external forces spontaneously assumes the form of a sphere. This shape corresponds to the minimum surface area for a given volume of liquid. Work must be done on a drop to increase its surface area; therefore, the surface molecules are in a state of higher free energy than those in the bulk interior of the liquid.

It is important to emphasize that the phenomenon of surface tension is a consequence of two molecular properties of liquids: *intermolecular attraction* and *rapid molecular motion*. The thin surface layer is undergoing constant depletion and replenishment as molecules move out of and into it. The average residence time of a water molecule in a water surface layer is about a microsecond (10^{-6} s at 25°C). Unfortunately, there are many descriptions of surface tension in the literature that are misleading because they do not emphasize the molecular dynamics. One common analogy for surface tension is that the surface of a liquid is like a "skin." The skin is said to be like that of the stretched rubber skin of an inflated balloon. This picture is a bad analogy because the molecules in the rubber membrane remain fixed in the surface in a continuously strained condition, whereas in a liquid surface there is always tremendous fluidity.

The introduction of other substances into the water-air system complicates the chemistry considerably. These multicomponent systems, however, are most interesting because they represent most of the actual, practical examples of surface chemistry. It is generally impossible to predict theoretically what will happen at the surface in complex systems. Often the best approach to complex systems is to carry out experiments and then try to explain and interpret the results. One special type of binary liquid system that merits special attention is the presence of insoluble oily materials at the water-air interface. This type of system is important in food processing (e.g., salad dressings), crude oil production, oil spills, biological membrane formation, fabric cleaning, and the survival of flora and fauna. The behavior of oily materials at aqueous surfaces depends very much on the chemical structure of the material. Nonpolar substances consisting of carbon and hydrogen (e.g., motor oil) are immiscible and not soluble in aqueous solutions. These substances are said to be *hydrophobic*, or water-hating, substances. Polar substances (e.g., ethanol) that can interact strongly with water are said to be *hydrophilic*, or water-loving, substances. The balance of the hydrophobic and hydrophilic characteristics in a particular molecule determines to a large extent its surface chemistry. A good example is motor oil, which in the unused state is a homogeneous mixture of hydrophobic hydrocarbons. Used motor oil, however, has been drastically oxidized by high-temperature engine reactions with air to become partially hydrophilic in character. The difference in the surface activity of the oil upon water is quite dramatic.

The experimental breakthrough in the study of oily materials on water came in 1891 when Fräulein Agnes Pockels (in her kitchen!) developed an original and elegant method of manipulating oils on water. Her technique involved a trough filled to more than the brim with water and the use of "barriers" that were swept across the surface to control the extent of an oil film. Pockel's letter to Lord Rayleigh[2] stimulated a century of extraordinarily fruitful research in surface chemistry. The trough method she developed was refined, quantified, and extended by Rayleigh, Langmuir, Blodgett,

[2]Pockels, A., *Nature* 43, *1891*, p. 437.

and many others, and has achieved a recent renaissance with the advent of computer-controlled barrier systems. Rayleigh proposed that the films were only one molecule thick; Langmuir made a giant step in showing that long-chain fatty acid films were oriented with the hydrophilic acid group in the water and with the hydrocarbon tail stuck vertically out into the nonpolar air. The cumulation of all the intense research on the surface properties of molecules with hydrocarbon tails and polar heads has been the basis for understanding the chemistry and biochemistry of detergents and biological membranes.

Soap molecules have played a central role in the development of aqueous surface chemistry. These molecules are salts of long-chain acids. A typical example is sodium palmitate (made from palm oil):

```
   H H H H H H H H H H H H H H H    O
   | | | | | | | | | | | | | | |   //
H- C-C-C-C-C-C-C-C-C-C-C-C-C-C-C-C
   | | | | | | | | | | | | | | |   \
   H H H H H H H H H H H H H H H    O⁻Na⁺
```

The long hydrocarbon tail of the molecule is nonpolar (hydrophobic), and the anionic head is quite polar (hydrophilic). In general, soap molecules are called *surfactants*, — i.e., they are surface-active agents. They can be anionic (as in sodium palmitate) or nonionic or cationic, depending on the charge on the tail. The dissolution of a surfactant in water produces a series of bizarre molecular arrangements. The hydrophobic tail is "squeezed out" by water molecules that are strongly hydrogen bonded, yet the polar tail is strongly bonded to water. The result is cooperative and oriented gathering (aggregation) of molecules both in the bulk aqueous solution and at the surface, as shown in Figure 20.1.

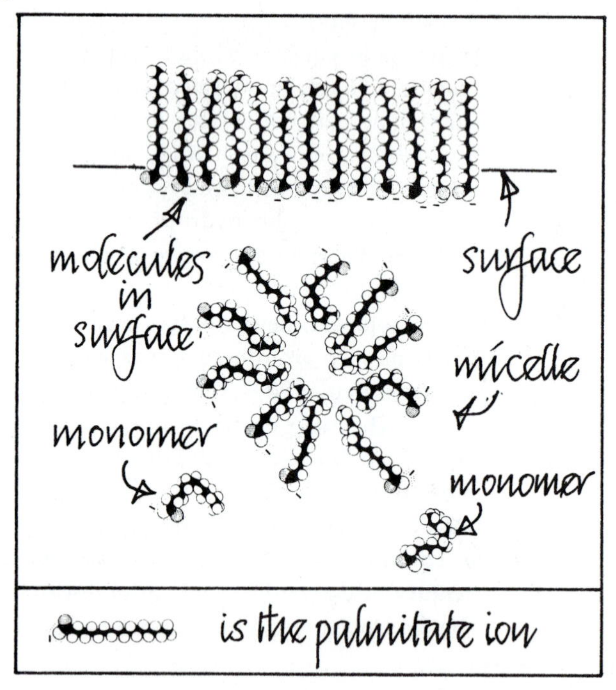

Figure 20.1 Aggregation of Palmitate Ions

The surface tension of aqueous solutions of surfactants is much smaller than that of pure water. The molecular explanation is that the hydrophobic tails resist moving into the aqueous interior, whereas the hydrophilic heads strongly attract water molecules and slow down the return of the water molecules into the interior. Surfactant molecules reduce the aqueous surface tension while at the same time maintaining the flexibility and tremendous fluidity of the interface. A lowered surface tension produces bubbles, films, and froths in agitated solutions containing soap molecules.

When air is blown or shaken into a soap solution or when a wire frame is dipped and pulled out of a soap solution, beautiful soap films and bubbles are formed. Films and bubbles are bilayers of surfactant molecules, sandwiching water and counter ions, as shown in Figure 20.2.

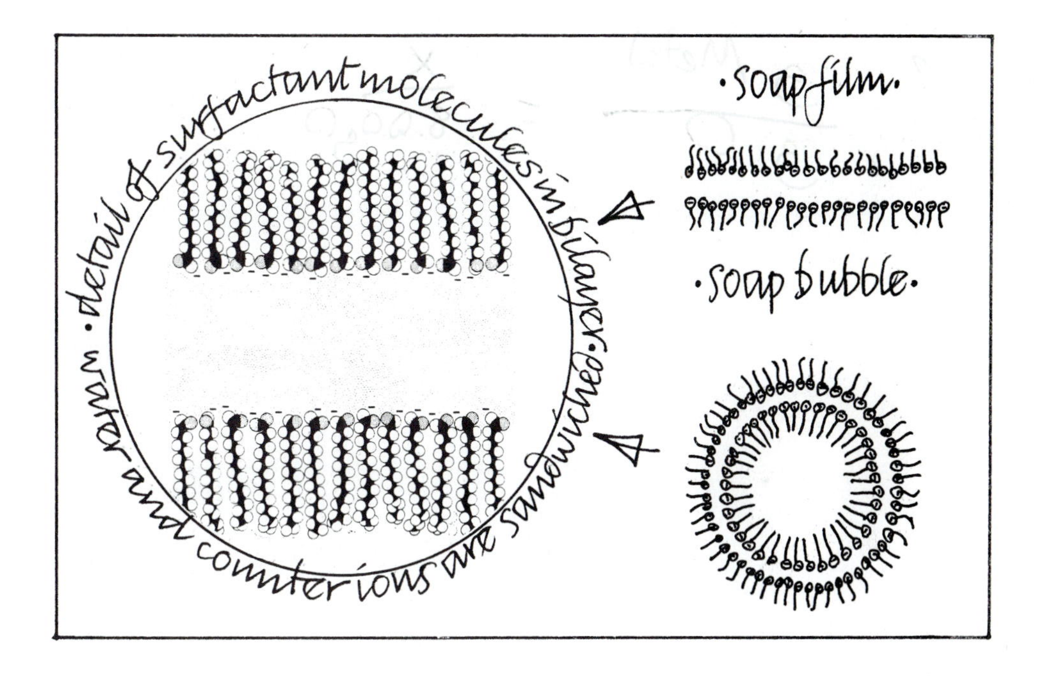

Figure 20.2 Bilayers in Soap Films and Bubbles

The gravitational effect on the bilayer gradually drains molecules downwards toward the bottom of the film, and the film begins to thin. Beautiful interference colors play on the surface and eventually disappear in blackness. The bilayer has become thinner than light itself. Suddenly the film breaks as the negative heads come close and repel each other. Drops fly away at 50 miles an hour . . . and the bubble is gone forever.

Most of what we know about soap bubbles was discovered by the remarkable Belgian scientist J.A.F. Plateau, whose great work, *Statique Expérimentale et Théorique des Liquides*, was published 30 years after he became blind. We owe a debt to Plateau's wife, whose diligent observations allowed Plateau to discern the fundamental geometries of bubbles and films. We now know that the geometry, and most other properties, of soap bubbles and films is a direct consequence of the coupling of intermolecular attraction and dynamic molecular motion at the surface of liquids. The path of an individual soap molecule is totally unpredictable. The collective action of billions of molecules completely determines the macroscopic shape of bubbles and films. The perfect economy of space and the wonderful architecture of soap films is just one of the many amazing phenomena exhibited by chemical surfaces. In this laboratory module you have an opportunity to explore the world of surface chemistry and perhaps even make a new architectural contribution to the collection of "beautiful rooms."

? $\frac{\text{g Metal}}{\text{g O}} = \frac{x}{8.00\text{ g O}}$

Laboratory Experiments

Flowchart of the Experiments

Section A.	The Measurement of Surface Tension by Capillary Rise Techniques Part 1. The Surface Tension of Water Part 2. The Surface Tension of Ethanol Part 3. The Surface Tension of Soap and Detergent Solutions

Section B.	Langmuir-Blodgett-Pockels Techniques and Oil Spill Chemistry

Section C.	Some Properties of Soap Films

Section D.	Minimum-Distance Networks and Soap Film Computers

Section E.	Three-Dimensional Systems: Bubbles and Films on Frames

Section F.	Free and Captive Bubbles and So Froth

Section G.	Fun with Big Bubbles

Requires one three-hour class period to complete

Section A. The Measurement of Surface Tension by Capillary Rise Techniques

Goals:

(1) To observe and measure the rise of liquids up a capillary tube. (2) To be able to calculate the surface tension of a liquid from an appropriate mathematical relationship.

Discussion:

Measurements will be made on several pure liquids (e.g., water and ethanol) and on solutions containing surface-active solutes.

Section A. Part 1. The Surface Tension of Water

Before You Begin:

Surface phenomena are extraordinarily dependent on the presence of small amounts of surface-active materials. Skin oils and many other contaminants can dramatically affect the results of surface tension measurements (particularly by this technique). Absolute cleanliness is necessary in order to achieve success in this series of experiments. Wash your hands before you start and make sure that all soap is removed — soap is a surface-active agent. All your apparatus should be cleaned and rinsed with distilled water. The capillary tube is particularly susceptible to contamination with oils and grease. Plastic wash bottles containing wash liquids (e.g., ethanol and distilled water) are at Reagent Central.

Experimental Steps:

1. Construct a straw clamp and stand using 2 straws, scissors, and a 1/4" office punch. Use a 96-well tray (RB) as a base.
2. Thoroughly clean a 24-well tray and rinse well with distilled water. Slap the tray on a clean paper towel to remove drops of water.

   NOTE: Avoid touching the sides of the wells with your fingers — skin oil will destroy the experiment.
3. Clean the narrow capillary tube by washing with a stream of ethanol followed by several washes with distilled water. Make sure that all the ethanol is washed out of the inside of the capillary tube. Handle the tube with a clean piece of dry paper towel.
4. Touch the end of the capillary tube to the clean towel and allow any liquid in the inside bore to be sucked out and absorbed by the paper. Do this at both ends. If there is still a liquid slug in the middle, shake the tube gently and then touch the paper towel to the end of the capillary.
5. Place the tube into the straw clamp. Do not allow the tube to touch any surface other than the straw.
6. Fill 1 well of the 24-well tray about 3/4 full with distilled water.

7. Lift the stand and clamp over the well so that the capillary tube gently enters the water. Make sure that the end of the tube is at least 0.5–1 cm under the surface of the water.

   The water will rise up the capillary tube.

8. The next step is *important* and *not easy to do.* Hold a cutoff microburet over the hole at the top end of the tube so that good contact is made. Very, very gently squeeze the bulb in order to suck or blow the column of liquid up or down a little.

   This technique will ensure that the liquid is really at equilibrium and at the maximum height it will rise to.

9. As illustrated below, carefully measure the height of the column of liquid (h) in the tube from the level of the liquid in the well.

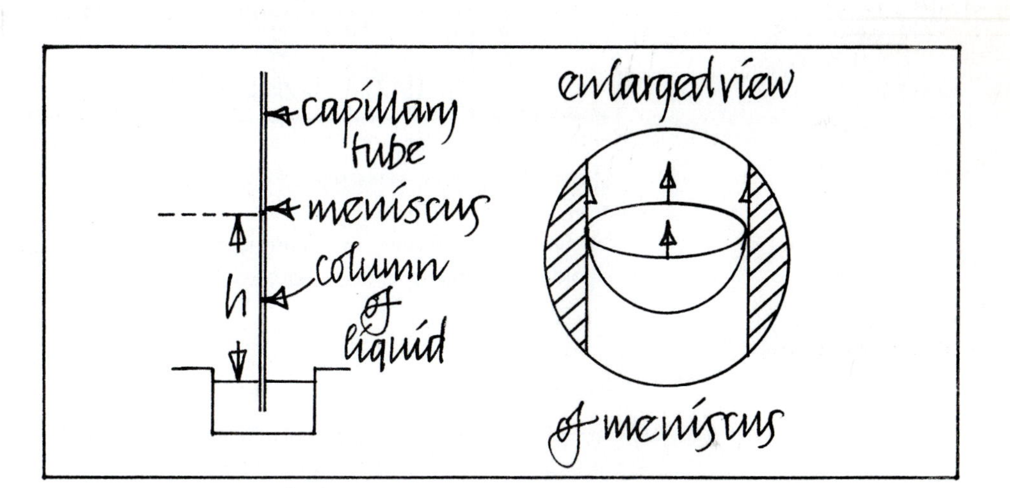

- Record the measurement.

10. With a clean piece of paper towel, remove the tube, touch the end to remove the residual column of liquid, and replace it in the liquid in the well.

11. Repeat Steps 7 through 9.

    - Do you think that the capillary tube must be vertical? Why?
    - Would the water rise if the capillary were made out of a plastic, such as polyethylene?

Calculation of the Surface Tension of a Liquid

A liquid will rise up a narrow tube only if the liquid actually "wets" the tube wall — i.e., the contact angle is 180°. The surface tension of the liquid acting at the top of the column of the liquid in the tube holds the column up against the force of gravity pulling the liquid column down.

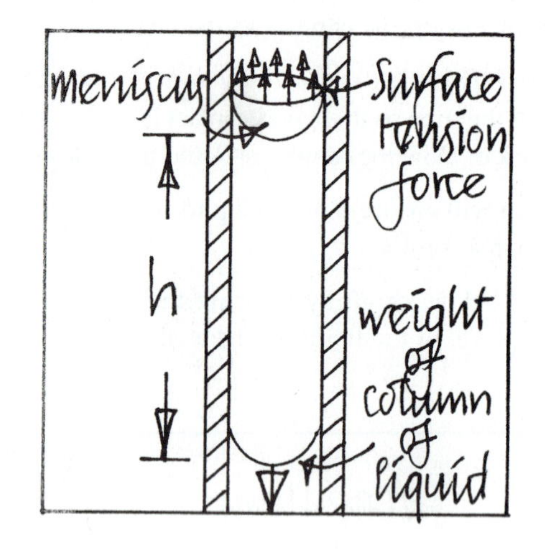

Let

r = radius of tube (cm)

γ = surface tension (dyne cm^{-1})

D = density of liquid (g cm^{-3})

g = acceleration due to gravity, 980 cm s^{-2}

h = capillary rise (cm)

Then

Surface tension force holding column up = circumference of liquid column × surface tension

$$= 2\pi r\gamma$$

Weight of liquid = volume of column × density × acceleration due to gravity

$$= \pi r^2 hDg$$

At equilibrium, the two forces are equal; thus,

$$2\pi r\gamma = \pi r^2 hDg$$

Rearranging and cancelling,

$$\gamma = \frac{rhDg}{2}$$

and the unit of surface tension γ is dyne cm^{-1}. To convert this value into an SI unit, multiply by 10^{-3}, and the unit is newton per meter (N m^{-1}). The literature value for the surface tension of pure water (against air) at 25 °C is 72 dyne cm^{-1}, or 0.072 N m^{-1}.

Ask your instructor for the radius of the tube and use 1.00 g cm^{-3} for the density of water.

- Calculate the surface tension of water from the average of your capillary rise measurements.

- Would you expect the surface tension to increase or decrease if the temperature were raised? Why?

NOTE: If you do not know the radius of the capillary tube, then you can carry out the capillary rise experiment and *assume* the value for the surface tension of water (given earlier). The calculation will then enable you to find the average tube radius. You have calibrated the tube and can now use it to measure the surface tension of other liquids.

Section A. Part 2. The Surface Tension of Ethanol

Experimental Steps:

1. Clean the capillary that you used in Part 1, only this time wash it thoroughly (inside and out) with ethanol from a wash bottle.
2. Wipe with a dry paper towel. Remove any residual ethanol by contacting the end of the tube with the dry towel. Do the same at the other end if necessary.
3. Dry a well in the 24-well tray with a microtowel and fill it about 3/4 full with ethanol (C_2H_5OH).
4. Carry out 2 measurements of the capillary rise of ethanol using the same technique used for water (Part 1).
   - Record the measurements.
   - Calculate the surface tension of ethanol given that the density of 95% by volume ethanol is 0.82 g cm^{-3}.
   - In your record give a brief discussion of why the surface tension of ethanol is different from water. Give particular emphasis to the intermolecular forces in each system.
5. If you have time, it might be interesting to make a solution with a mole fraction of 0.5 water and 0.5 ethanol and measure its surface tension.

Section A. Part 3. The Surface Tension of Soap and Detergent Solutions

Experimental Steps:

1. Select 1 of the soap or detergent solutions that are available at Reagent Central.
   - Make a note of the brand name, type, and concentration of the soap or detergent.
2. Fill a clean well about 3/4 full with the selected solution.
3. Wash the capillary tube with ethanol and distilled water.
4. Carry out 2 measurements of the capillary rise for the same soap or detergent solution using the standard technique.
   - Calculate the surface tension of the soap or detergent solution. Assume that the density of the solution is 1.0 g cm^{-3}.

- In your laboratory record, give a brief explanation of your surface tension result.
- Would you expect the surface tension of an inorganic salt solution (say, 1 M NaCl) to be less than or greater than that for pure water? Explain.
- Do you think that changing the gas above a surface (say, by replacing air with dinitrogen) would change the surface tension?
- Do you think that the surface tension is related to the charge (i.e., anionic, neutral, or cationic) on a surface-active agent?

Section B. Langmuir-Blodgett-Pockels Techniques and Oil Spill Chemistry

Goals:

(1) To explore the surface activity of a variety of natural, synthetic, and modified hydrocarbon oils using a Langmuir-Blodgett-Pockels trough. (2) To investigate several strategies for oil spill containment and cleanup.

Before You Begin:

The observation of surface phenomena, in this instance on a water surface, requires considerable care and thought. The room lighting will often determine the appropriate viewing angles at which to see the various phenomena. In the first experiment explore all the possibilities. Once you first see the activity at the water surface, subsequent observations are very easy.

Experimental Steps:

1. Clean a plastic petri dish thoroughly with soap and water. Make sure that all the soap is rinsed away.
2. Fill the petri dish bottom about 3/5 full with room temperature water.
3. Place the dish onto a white surface (no lines in the background) in a position on the table where the surface can be viewed easily.
4. Try looking at the water surface from different angles until you can see the occasional specks of dust floating at the surface.

   NOTE: Once you have established a good viewing position, place the dish in roughly the same place for all experiments and make a mental record of where you are in relation to the light and the dish. The dish and water system is a simple Langmuir-Blodgett-Pockels trough (LBPT).
5. Obtain a container of lighter fluid.

   CAUTION: Make sure there are *no flames* in the room!
6. Open the container and use the pipet to drop 1 drop of the fluid into the center of the water surface in the dish.
7. Observe and make notes. Several interesting things are going to happen in a relatively short period of time. Watch carefully (in a good viewing position) until the hydrocarbon fluid disappears.
   - Record your observations!

NOTE: If everything went by too fast, all you have to do is place another drop onto the surface and repeat the experiment.

At this stage it is perhaps worthwhile to form a discussion group with your peers (and the instructor). Try to formulate explanations and interpretations of the complex sequence of phenomena you have just observed.

To help you do this, here are some thoughts. Lighter fluid is a homogeneous mixture of many hydrocarbons ranging from small molecules (e.g., C_3H_8) to large ones (e.g., C_9H_{20}). The fluid mixture is obviously volatile — it disappeared! The mixture consists of a range of compounds of different molar mass, each of which has a different vapor pressure (at the temperature of the water in the dish). The smaller molecules will evaporate faster than the larger molecules; therefore, the fluid composition changes as the fluid evaporates. The fluid is less dense than the water and is immiscible with water. The series of colored spectra that eventually appear in the fluid layer form when the layer thickness approaches the wavelength of visible light.

8. If you have time you might want to explore the effect of temperature on the surface behavior of hydrocarbons.
   - What would happen on ice-cold water?
   - How about the effect of substances dissolved in the water?
   - Would a seawater surface exhibit the same characteristics?
9. You can investigate the effect of waves on the fluid layer in the following way. Hold a cotton swab vertically close to the edge of the dish and dip it in and out of the water in a periodic motion.
10. Clean the dish and again fill it about 3/4 full with water.
11. Obtain a sample of unused (new) motor oil. Transfer some oil to the surface by touching a plastic stem (or toothpick) to the oil and then to the surface.
   - Describe and explain the behavior of the oil. ~~spreads over surface~~ stays on surface
   - Why does the oil form a lens? hydrophobic
   - Does the oil disappear? yes
   - What happens if you make waves?
12. Design some simple experiments to explore the best way of removing the oil from the surface.
   - Would your idea work on a large scale?
   - What would happen if the water temperature were almost 0 °C?
   - Why won't a cotton swab pick up the oil very easily? Try it!
13. Remove the oil (by your best method) and clean the dish thoroughly with soap and water.
14. Place the dish on the plastic surface against a white background.

15. Fill the dish with water until the water "piles up" above the plastic edges of the dish (as illustrated below).

16. Slip a straw over a glass rod. Hold the rod at both ends (use both hands). Place the rod and straw horizontally onto the edge of the dish and in the piled-up water. Pull the rod over the dish keeping the straw touching the edges of the dish. As you do this the surface will be swept clean, as shown below.

If you leave the rod "parked" on the surface at the end of the sweep, any surface contamination that was swept in front will stay enclosed. The sweep technique (invented by Pockels) is an extremely effective method of ensuring a clean surface.

17. You can now remove the contamination by quickly dabbing the end surface with a clean, dry paper towel.
18. Obtain the used-oil sample.
19. Using a thin plastic stem, transfer a *small* drop to the center of the water surface in the dish.
20. Observe carefully what happens.
    - Draw a picture in your record.

- Compare the behavior of the used versus new oil.
- Describe the color of the oil (view at a good angle!).

21. Now you need to do two things at once! Keep watching the oil at an angle at which you can see the color and push the rod and straw over the surface for about 3–4 cm. Then pull it back and push again (keeping it on the edges of the dish).
    - Describe what you see in your notebook.
22. You now have an oil slick that has spread out over a considerable area. Design a way in which the slick can be removed from the surface. Experiment with microbooms (use cut pieces of straw or cotton or fishing line) to try to contain the slick.

    If you sweep it, you are still faced with the problem of removing the swept oil. If you were in an ocean environment, the boom would have to be rather long!
    - Describe your design in your lab notebook.
23. Make waves and observe the behavior of the oil slick.
    - Record your observations.
24. Obtain a detergent solution and dip a clean plastic stem — i.e., *not* contaminated with oil — into it. Transfer the small volume of detergent solution to the middle of the slick.
    - What happens?
25. If you have time, you might want to investigate the effect of ocean composition, temperature, waves, wind, and ships on an oil slick. To make a new slick, all you have to do is empty the dish, clean it thoroughly with soap and water, wash your hands thoroughly, and start again.

Section C. Some Properties of Soap Films

Goal: *To use a soap solution to investigate some of the chemical and physical properties of bubbles and films.*

Experimental Steps:

1. Obtain a piece of aluminum wire from Reagent Central.
2. Bend the wire to make a smooth ring (about 6 cm in diameter) with a handle at right angles to the plane of the ring. The ring should easily fit into the bottom of a petri dish.
3. Obtain soap solution from Reagent Central and fill the dish about 3/4 full — this is a special solution specifically formulated for making good bubbles and films. It is labelled as such.
4. Holding the ring by its handle, dip the wire ring into the soap solution and gently remove it.
5. Tilt the ring at an angle so that you can see the film that has formed on it. Use your hand lens to make observations on the nature of the film.

- How thick do you think the film is? Is there any way of measuring the thickness of a soap film?
- What are those weird wriggly areas?
- How does the film get around the wire?
- Are there any colors? Describe them.
- Do the colors change with time?
- How long does it take for the film to collapse?
- How does it collapse?

6. Dip again and make another film. Poke a straw through the film.
   - Does it collapse?
   - Is there a film inside the straw now?
7. Move the straw in a circular motion in the film.
   - Describe what happens.
   - How do the soap molecules move out of the way so quickly?
   - Can a film be made on any type of material, e.g., a rubber band or a plastic ring?
   - In your record draw several pictures to show the ring being lifted from under the surface of the soap solution and out into the air and explain why the film is a thick bilayer. Use the head and tail stick representation for individual soap molecules.
   - Is the film on the wire ring the minimum surface area for a circle? Explain.
8. Tie a piece of thread very loosely across the wire ring so that it is attached on opposite sides of the ring.
9. Dip the ring with thread into the soap solution in the dish. Make sure that the thread is thoroughly wetted.
   - Describe the result.
10. Poke the thread with a straw.
    - Describe its action.
11. Break the film on one side of the thread with your finger or a dry glass rod.
    - Interpret the action of the forces that must be acting on the thread.
12. Dip the ring and thread again and arrange the thread so that it forms a looped circle in the middle. Now, break the film inside the thread loop (but not outside).
    - Describe and explain what happens.
13. Poke at the thread with a straw without breaking the outside film.
    - What happens when you push the loop in toward the center of the ring?

You have seen in these simple experiments that soap films always try to form minimum-area surfaces. The reason is that soap and water molecules in the top and bottom surfaces of the film are more strongly attracted to soap and water molecules inside the film than to the air molecules (O_2 and N_2) outside the film. At the molecular level the film is a chaotic mass of continuously moving molecules that are being pulled into the inside of the film.

In the next few experiments you can investigate minimum-distance networks and soap film computers.

Section D. Minimum-Distance Networks and Soap Film Computers

Goals:

(1) To be able to solve minimum-distance problems. (2) To deduce some of the mathematical relationships that govern soap network systems.

Before You Begin:

Before beginning to work with actual soap films, think about the following puzzle. There are 3 points that are spaced equally apart, forming an equilateral triangle. The puzzle is to connect all 3 points such that the sum of the lengths of all lines is the smallest possible — i.e., a minimum distance between all 3 points. The author has tried two ways, one of which is much shorter than the other. You can easily see if you have solved the puzzle by using a ruler to measure the total line length, as shown below.

These types of problems have become known as Steiner problems, after the nineteenth-century mathematician Jacob Steiner, who made a study of them. The soap film approach to solving Steiner problems was made famous in the 1940s by the mathematician Richard Courant. OK, let's solve the puzzle with a soap film computer!

Experimental Steps:

1. Empty the soap from the dish into the waste soap container. Rinse the dish. Turn it over and place it on the plastic surface.

   You are going to use it as a support for a glass sheet.

2. Obtain 2 glass or plastic sheets from Reagent Central. Place one of the sheets onto the dish (it will hang over the end). Form an equilateral triangle with 3 thumbtacks

placed about 1–2 cm apart in the middle of the glass sheet. Place the second sheet onto the tacks.

3. Grasp both sheets and press so that you can lift and carry them without losing or moving the tacks — no mean feat!
4. Dip the entire sandwich into the bucket of soap and then withdraw it. If you do this cleanly and avoid the froth, you will be presented with an answer to Steiner's 3-point problem.
5. If you are careful (and lucky) you can lay the sandwich back down on the dish and use a ruler to measure the total line length and prove the answer is correct.
6. You might want to experiment with 3 point problems in which the triangle is not equilateral. Some typical minimum distances are illustrated below.

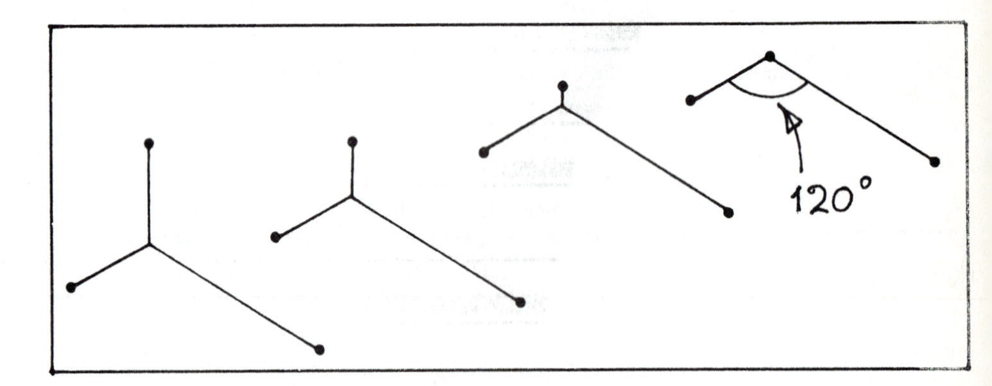

There is always a triple film junction with 120° angles, and in the limiting case, the triple junction degenerates and joins the tack that forms the 120° vertex.

7. Add a fourth thumbtack to see if you can create a 4-point (square) system that can be solved with a soap film computer. Some possibilities are,

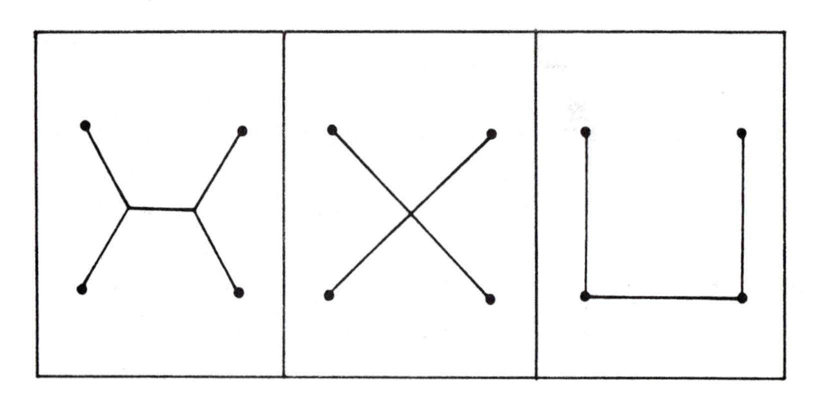

You can see that there are 2 networks that have an equal probability of being answers. The soap system will inevitably adopt one of these 2 minimum networks. It is impossible to predict in advance which one will appear! You can actually push the network into the other minimum form by gently blowing on the intersection with a straw. The network will move through a point where 4 films actually come together and then adopt the more stable, minimum network. It is interesting to note that soap networks connecting an irregular quadrilateral can adopt minimum forms that when blown on adopt a shorter network. Both forms are minimum networks, but one is more minimum than the other!

- How could these minimum networks be shown in graphical form?

The mathematical formula governing the topology of Steiner's problems is that the number of points N_1 minus the number of 3-way junctions N_3, equals 2:

$$N_1 - N_3 = 2$$

The Steiner problems involve two-dimensional film networks, but it is possible to solve three-dimensional problems using the fact that soap films always tend to form a minimum area, as you will see in Section E.

Section E. Three-Dimensional Systems: Bubbles and Films on Frames

Goal: *To extend the soap film computer concept to solve minimum-area problems in three-dimensional systems.*

Discussion: Observations and measurements will be made on several wire polyhedra.

Experimental Steps:

1. At Reagent Central there are numerous soldered wire frames forming both regular and irregular polyhedra. Select some of the simpler arrangements to begin with (e.g., the tetrahedron or cube).
2. Fill the large cup (or beaker) with soap solution and take it back to your place in the laboratory.
3. Use a towel to sweep any froth of small bubbles from the top of the soap solution.
4. Dip the frame (the entire polyhedron) into the soap solution and withdraw it slowly. The soap film computer (the frame) will adopt the minimum area bounded by the particular polyhedron that you are using.
5. Select several more complex polyhedra from Reagent Central and carry out Steps 3 and 4 with each. Selectively break a film and watch how the minimum surface responds. Redip to obtain the analog answer and use a straw to blow on various places in the films.

- Draw pictures of the minimum areas adopted by 3 different polyhedra. For each area, note how many films meet and where they meet — i.e., at points or lines — and note the angles formed in the process.
- Describe whether you can obtain alternative minima.
- Note the colors in the films. Describe color changes as the films drain.

NOTE: The minimum area inside a tetrahedral frame is

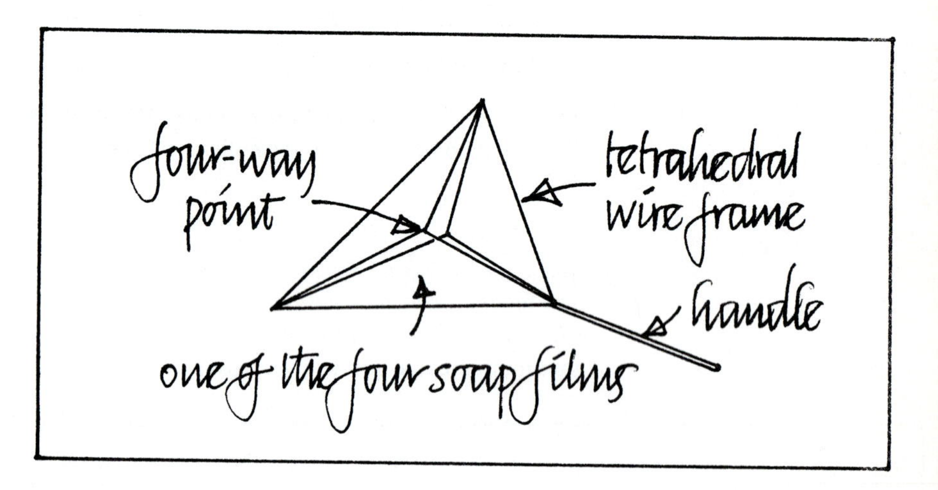

It is especially interesting to note that at the 4-way point, the angles are exactly 109° 28' 16". You will see later that this angle precludes the making of a regular froth of regular cells.

6. Generate a little froth on top of the soap solution and slowly dip a frame. If you are lucky, as you dip and withdraw the frame, you might catch a bubble in the middle of the minimum area.
7. Use a straw to enlarge the bubble (by blowing into it).
   - Describe the intersecting points and angles, etc.
8. If you have time try constructing an irregular structure from a piece of wire. How about a helix form?
   - Can you predict the minimum area for the structure that you made?

Section F. Free and Captive Bubbles and So Froth

Goal: *To investigate some of the characteristics of free and captive bubbles and examine the pressure and area relationships in different sized bubbles.*

Before You Begin: Remember to draw pictures of what you see in the following experiments.

Experimental Steps:

1. Let's start with a very simple experiment and some simple questions. Blow a few free-floating bubbles.
   - Describe how you did that!
   - Why are free bubbles spherical in shape?
   - What determines the size of a free bubble?
   - Is there any limit to the size of a free bubble?
   - Why do bubbles fall?
2. Pour a little soap solution into a petri dish until the bottom of the dish is just covered. Poke the end of the straw into the soap and blow down the straw to make a captive bubble.
   - Describe its shape.
3. Poke the straw into the bubble and blow gently.
   - What happens to the bubble?

   Now here is a very interesting question,
   - Is the air pressure inside a large bubble greater or smaller than the air pressure in a small bubble?

   The answer is smaller — which perhaps seems slightly paradoxical in view of the fact that the bubble got larger when you blew into it! Everyone knows, for example, that if you pump air into an automobile or bicycle tire, the tire expands and the pressure increases. In contrast, though, when you blow air into a bubble, the bubble expands and the pressure *decreases*. The explanation for this apparent paradox is that tires are far less elastic than soap bubbles, and the rubber allows stresses to build up. In contrast, a soap film never exerts a force of more than about 0.066 N m^{-1} (which is 2 times the surface tension of a soap solution because a soap *film* has 2 surfaces). The force is about the same for thin and thick bubbles and for small and large bubbles. A bubble simply expands until the force exerted by the air inside exactly balances the force within the film.
4. Test the pressure theory by blowing another hemispherical cap bubble on the dish.
5. Now place the straw about 0.5 cm away from the side of the bubble, blow with constant force, and observe the resulting curvature in the bubble.

   You are applying pressure to the bubble and it adopts a much larger curvature — i.e., smaller diameter curve.
6. Poke the straw into the soap solution in the dish and blow a captive bubble that occupies about 1/2 the dish. Remove the straw.
7. Blow another captive bubble in the other half of the dish.

- Describe what happens when the bubbles touch. Explain your observations by invoking an important property of soap films.

8. Poke the straw into one of the bubbles.

   - Why doesn't the bubble leak air out and go down?

   HINT: This is a trick question.

9. Now blow into the bubble to increase its diameter. Observe the curvature of the film that joins the two bubbles.

   - Why does the curvature go in that direction?
   - What angles are formed at the intersection of the bubbles?

10. Dip the straw in soap solution and lift it out. Keeping the straw vertical, blow gently down the straw to form a bubble on the end of the straw. Now stop blowing (and remove your mouth).

    - Explain why the bubble goes down.

11. Repeat Step 10, except this time when you stop blowing, quickly and smoothly place your index finger over the end. The bubble will stay. Now gently lower the bubble until it touches the soap in the petri dish. You can repeat the experiment as many times as you like — the change is subtle, but definitely real.

    - Explain what happens and particularly note any change in size.

    When the spherical bubble changes to a captive hemisphere, there is no change in the amount of air. The volume of the spherical bubble must therefore equal the volume of the hemisphere, i.e.,

$$\frac{4}{3}\pi r_S^3 = \frac{1}{2}\left(\frac{4}{3}\pi r_H^3\right)$$

where r_S and r_H are the radii of the sphere and the hemisphere, respectively. Upon rearrangement and cancellation,

$$r_S = \left(\frac{1}{2}\right)^{1/3} r_H$$

$$= 0.79\, r_H$$

which means that the radius of the sphere is 79% of the radius of the hemisphere.

A very simple and effective method of keeping bubbles around for a long time in order to make observations is to blow the bubbles in the dish and then place the top on the petri dish.

11. Design an experiment to study the shape and angles of a multiple-bubble froth.

    - Can you see any pentangular dodecahedra?

Section G. Fun with Big Bubbles

Goal: *To have Fun!*

Before You Begin

Your instructor will demonstrate amazing technical skill by attempting to produce world-record bubbles. The author has produced bubbles that are about 12 ft in diameter. All you need is the big bubble machine, which may be purchased at your local toy store for a few dollars. (Or you can make one yourself — check with your instructor.)

Experimental Steps:

1. Play with the lab's big bubble machine!
2. Can you put a bubble inside a bubble? Try it!
3. Can a person get inside a bubble?

Chapter 21

Natural Products Chemistry: Anthocyanins as Food Dyes

Roses are red, Violets are blue...How come?

Anonymous

Introduction

Natural products from plants are the source of an array of substances that are essential to the well-being and survival of humans. Proteins, carbohydrates, fats, minerals, and vitamins are familiar to most people. However, plants also synthesize an enormous number of compounds, called *secondary plant products,* that are usually found only in relatively small amounts. The plant functions of these secondary products are not understood very well, although some are known to be part of a plant's defense mechanism or are attractors of pollinators. The study of natural products chemistry, particulary of secondary products, has yielded a rich and varied supply of medicinals (antibiotics, antitumor agents, analgesics, and stimulants); agricultural products (insecticides); natural dye stuffs (indigo); products for the chemical industry (gums and resins); and a wide spectrum of flavorings and essential oils. Natural colors have long held an important position in the dyeing industry and have brought joy to scientists and nonscientists alike for their beauty in food, wine, and flowers.

One of the first important scientific studies of natural colors was reported by Robert Boyle in 1664, in his book *Experiments and Considerations Touching Colors.* Boyle used the color changes of plant extracts to differentiate acidic, basic, and neutral substances (for more details, see Chapter 8, "An Introduction to Acids and Bases"). His favorite indicator was syrup of violets, which he consistently used to identify neutral substances. In 1784 James Watt reported the first use of a red cabbage extract as a universal indicator, and this extract was again highly recommended by Michael Faraday in his 1827 laboratory manual, *Chemical Manipulations.* By the beginning of the twentieth century, the sap-soluble red and blue natural plant colors had been recognized as being similar in chemical structure and were given the collective name of *anthocyanins.* The other major classes of natural colors, which were also recognized at about this time, are the green chlorophyls (all green plants), the yellow, orange, and red carotenoids (such as carrots, lemons, oranges, tomatoes, and red peppers), and the orange and brown tannins (tea, coffee, and roots).

The anthocyanins constitute one of the most important and widespread group of coloring matter in plants. The intensely colored, water-soluble anthocyanins are responsible for nearly all the pink, scarlet, red, purple, violet, and blue colors in the petals, leaves, and fruits of higher plants. They also occur in roots and tubers, but have never been found in animals. In the living plant, anthocyanin color is usually very stable, although transient colors can occur in young spring leaves and in senescing flowers. Autumnal colors are mainly due to anthocyanins in combination with other yellow and brown pigments. Flower color, shape, scent, and nectar are all major factors in attracting bees, butterflies, birds, ants, and other pollinators to plants to ensure fertilization. Bees are particularly attracted to blue flowers, and humming birds to orange and red flowers. Anthocyanin color is an important aid to seed dispersal (by birds) in strawberry, cherry, currant, and many other berries. Both fruit and flower color give immense aesthetic pleasure to people the world over. Deliberate selection of color varieties among garden plants and horticultural crops has been practiced for a very long time.

Modern anthocyanin chemistry began with the complete structural elucidations of anthocyanins by Richard Willstätter in Germany (1912–1935) and with the laboratory synthesis of many anthocyanins by Sir Robert Robinson in England (1924–1938). In the last 50 years, there has been increased interest in many aspects of anthocyanin chemistry and biochemistry. Research has revolved around many fundamental and controversial questions:

- How can a single class of compounds be responsible for the incredible color variety expressed in leaves, fruits, and flowers?
- What makes blue flowers blue?
- Is it possible to grow a naturally blue rose?
- How do plants synthesize anthocyanins?

- What are the functions of anthocyanins in plants?
- What chemical reactions are responsible for the color changes observed with acids and bases?
- What is the chemical explanation for color changes in aging red wines?

Getting answers to these and other questions in the field of natural colors (and in natural products chemistry in general) is not an easy task. The key steps in understanding both the *in vitro* and *in vivo* behavior of anthocyanins involve extraction from the natural source, separation of the complex mixtures, purification, identification, and, finally, a laboratory synthesis.

Anthocyanins have been ingested by humans and other species for thousands of years in the form of leaves, roots, fruits, juices, extracts, and fermentations, apparently without any serious problems. Many fruit juices — e.g., grape, blueberry, pomegranate, and cranberry — contain relatively high concentrations of anthocyanins, and there appear to have been no reports of adverse health effects due specifically to anthocyanins. The numerous instances of the coloration of urine resulting from the ingestion of red beets, which was at first attributed to anthocyanins, is now known to be caused by the completely unrelated betacyanin pigments in beets. About 10% of the U.S. population lacks the necessary enzyme to break down these rather unusual nitrogen-containing pigments.

In spite of the long history of ingestion, very little is known about the nutritional function (if any) or even the biochemical fate of anthocyanins in the human diet. There is some evidence that anthocyanins may help to preserve foods and also play a role in strengthening collagenous tissue. Certainly, the attractiveness of foods containing natural colors is an important nutritional consideration. Unquestionably, color is one of the first characteristics of foods perceived by the senses and is indispensible as a means for rapid identification and acceptance of foods. Already there have been a number of successful attempts to incorporate anthocyanins in commerically processed food products. Concentrated fruit juices are perhaps the best example, along with the use of hibiscus flowers as coloring material for herb teas. Recently, red grape skin extracts have been used as "artificial" coloring for fruit yogurt, cheesecake, and other dairy products. The commercial use of natural colors in produced foods is minor compared to the synthetic color additives that make up more than 90% of food dyes on the market. Synthetic dyes are now widely used as major coloring additives in sausage, beverages, gelatine desserts, cereals, candy, ice cream, sherbert, snack foods, pharmaceuticals, and pet food.

You are probably aware of the considerable controversy surrounding the use of preservatives and flavor and color additives in foods. The U.S. Food and Drug Administration has in the past placed bans on several widely used synthetic dyes. In 1976 the color additive FD and C red no. 2 was banned on the basis of several rather controversial studies that reported that the dye was a potential carcinogen in laboratory animals. The regulatory basis for the ban is the Delaney Amendment, which states that an additive cannot be listed for any use in foods, drugs, or cosmetics if it is found to induce a cancer when ingested by humans or animals. The ban on red no. 2 dye left only seven food dyes listed as safe: FD and C yellow no. 5, yellow no. 6, red no. 3, blue no. 1, red no. 40, blue no. 2, and green no. 3. Several of these (red no. 3, red no. 40, and blue no. 2) are suspect and are currently being studied in animal feeding studies. The possibility of carcinogenic, teratogenic, or mutagenic properties associated with these synthetic dyes has stimulated research into the use of natural colors, particularly the anthocyanins, as food, drug, and cosmetic dyes.

The overall objective in this natural products chemistry module is to examine the question, Can naturally occurring anthocyanins be used as nontoxic food, drug, and cosmetic dyes? Several large corporations in France, the United States, and Japan, are spending very large amounts of money trying to answer this question. Do you suppose that, based upon what you've learned thus far about small-scale techniques and your own creativity, we can study the same question as effectively but for much, much less?

There are several important criteria in developing answers to questions about the use of natural colors in commercial products. Reliable sources of inexpensive plant materials — e.g., grape skins and red cabbage — must be available, together with the appropriate technology for the extraction, purification, and formulation of the anthocyanin product. More importantly, laboratory studies of the chemistry of anthocyanins must be carried out to provide a fundamental knowledge of the behavior of dyes in simple aqueous solution and in the final product environment. Finally, the decisions concerning the replacement of existing synthetic dyes with anthocyanins must rest on a total assessment of product performance and consumer acceptance. In this laboratory module you will be investigating many of these aspects in order to try to evaluate whether anthocyanin extracts could be a major source of nontoxic food, drug, and cosmetic dyes.

Background Chemistry

Anthocyanins are organic compounds that are usually found in the aqueous sap of the vacuole of epidermal plant cells. These compounds have a complex structure consisting of an aromatic three-ring molecular region, one or more attached sugar molecules, and sometimes acyl groups attached to the sugar molecules. A good example is the reported structure for one of the eight anthocyanins in red cabbage (*Brassica oleracea var capitata rubra*), which is illustrated in Figure 21.1.

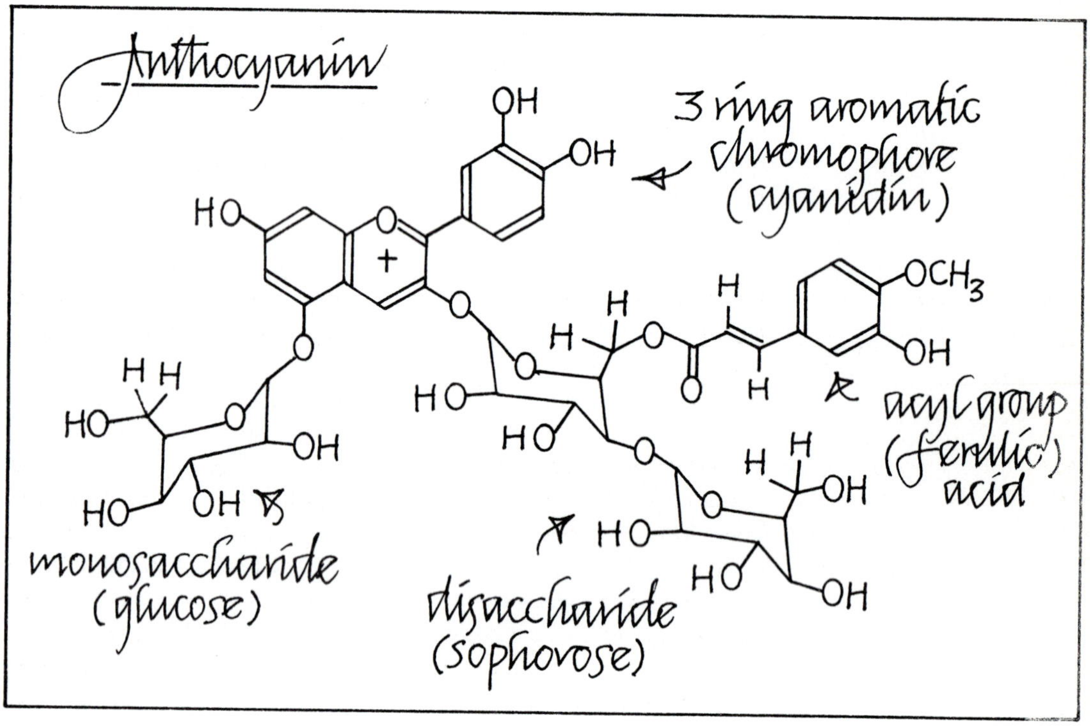

Figure 21.1 One of the Eight Anthocyanins in Red Cabbage

In spite of the tremendous variation of anthocyanin structure found in nature, most anthocyanins have been found to contain one of the following three-ring aromatic molecular regions (called *anthocyanidins*), with a variety of attached sugar molecules:

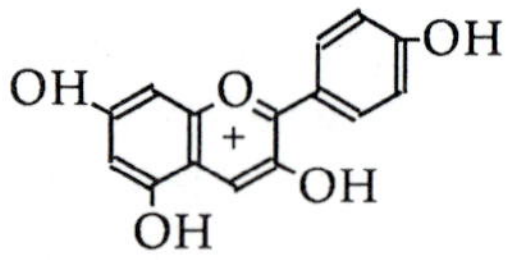

Pelargonidin 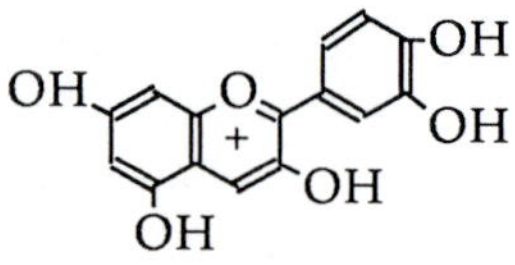 Cyanidin 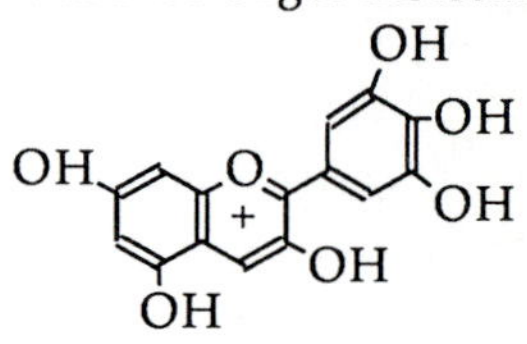 Delphinidin

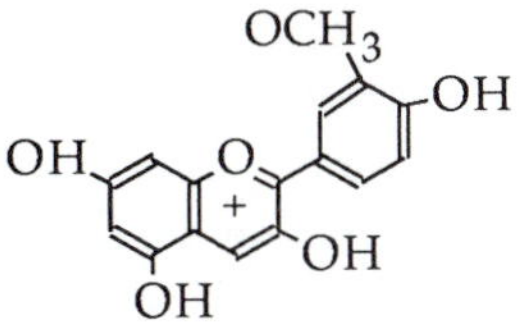

Peonidin 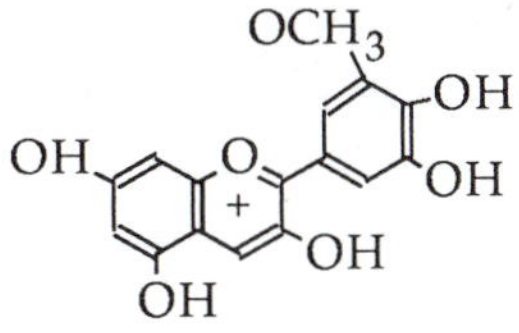 Petunidin 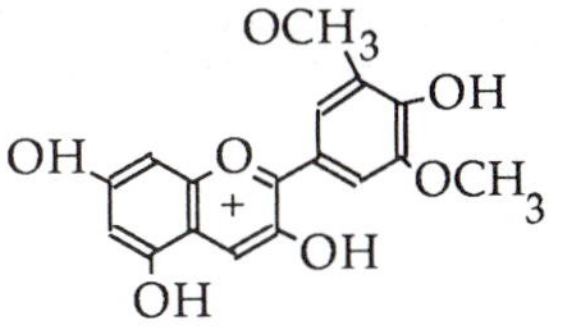 Malvidin

The incredible range in the color of flowers depends mainly on which particular anthocyanidin is present in the anthocyanin(s) in the flower. Orange geranium and red raddish contain a pelargonidin-type anthocyanin, whereas red and purple roses and cabbage contain cyanidin types, and blue larkspur contain delphinidin-type anthocyanins. The anthocyanidin part of the anthocyanin molecule is the part that absorbs visible light and is therefore responsible for the color of a flower seen in visible light. The parts of the molecule that absorb light are called *chromophores,* and the other parts of the molecule — e.g., –OH groups, sugars, and acyl groups — that serve to fine-tune the color are called *auxochromes.*

The *in vivo* expression of flower is under genetic control, and many factors can apparently modify the anthocyanin color. The nature, concentration, and spatial location of anthocyanins in a plant can dramatically affect the color. Flowers with small amounts of delphinidin-type anthocyanins look blue, whereas large concentrations can give a black hue (e.g., eggplant skin). The pH of the petal sap in flowers can vary in development and can also be different in different genotypes. The young flower buds of "Heavenly Blue" morning glory are a deep reddish-purple. As the flowers open, the color changes to a beautiful sky blue. This color change is associated with a definite increase in petal sap pH from about 5.6 to 7.2. The presence of other substances that are present in the cell — e.g., metal ions Al^{3+} and Fe^{2+} and co-pigments flavones, tannins, and pectins — can also modify the anthocyanin color.

The search for a chemical explanation for the dramatic *in vitro* color changes of anthocyanin extracts reported by Robert Boyle in the seventeenth century has continued for over 300 years and is still the object of much research. Recent work has shown that aqueous solutions of anthocyanins undergo a variety of structural transformations when the pH of the solution is changed and that the kinetics of these transformations is a major factor in determining the strange color sequences. Extraction of anthocyanin from plant material with a dilute aqueous solution of a strong acid inevitably produces a red-colored extract. The red color is due to a protonated form of the anthocyanin, called a *flavylium salt* (for actual structures and schematics, see Figure 21.1). If the pH of a red acidic solution is changed to about pH 8, a dramatic color change from red to blue or bluish purple is observed. The change is due to an acid-base reaction in which the acidic red form loses a proton to give a neutral form called an *anhydro base.* You might recall that acid-base reactions are extremely fast, and this reaction is no exception. However, if the pH of a red acidic solution is changed to about 4 or 5, a slower change to a colorless solution occurs. This strange reaction is thought to occur by water attacking the red form to give a colorless *carbinol* in which the chromophore is no longer capable of absorbing visible light. The situation becomes even more complicated if the pH of a red solution is changed to pH values greater than about 10. More protons are pulled away from all the forms, and there is a very slow breaking open of the ring (containing oxygen) to produce a yellow anthocyanin form called a *chalcone*. The rate of color formation and disappearance is dependent on the specific value of the pH *change* and on how the pH change is accomplished. The whole sequence of transformations is shown structurally and in an abbreviated schematic form in Figure 21.2.

All the color changes discussed thus far were obtained by adjusting the pH of aqueous solutions in a laboratory context (*in vitro* or *in plastico*). Considerable controversy surrounds the exact nature of the form of anthocyanins in a plant cellular environment (*in vivo*). An obvious explanation of the variation in flower color would be to propose that simple pH changes were responsible. However, it is now known that the intracellular pH of most petal sap ranges from about 4 to about 6.5, pH values at which the major *in vitro* form is colorless! It is highly probable that more complex intramolecular and intermolecular interactions are at work and that the expressed color of each flower is perhaps specifically defined by its own unique environment.

One interesting aspect of all of these color transformations is the proposed use of anthocyanin extracts as nontoxic food, drug, and cosmetic dyes. The use of natural extracts currently appears to be a far better economic route than the synthesis of anthocyanins in the laboratory. However, it is worth mentioning that several Japanese and at least two U. S. companies are investigating synthetic approaches to anthocyanins. Red grape skins contain several acylated anthocyanins that perhaps could be used as food dyes and are currently thrown away. Several large companies are very interested in "recycling" grape and other anthocyanins as coloring agents in carbonated beverages and processed foods. Recently, a patent was issued for the use of "Heavenly Blue" morning glory anthocyanins in cosmetics. The key factor in these commercial applications is an understanding of the factors that control the color form of anthocyanins in a particular product.

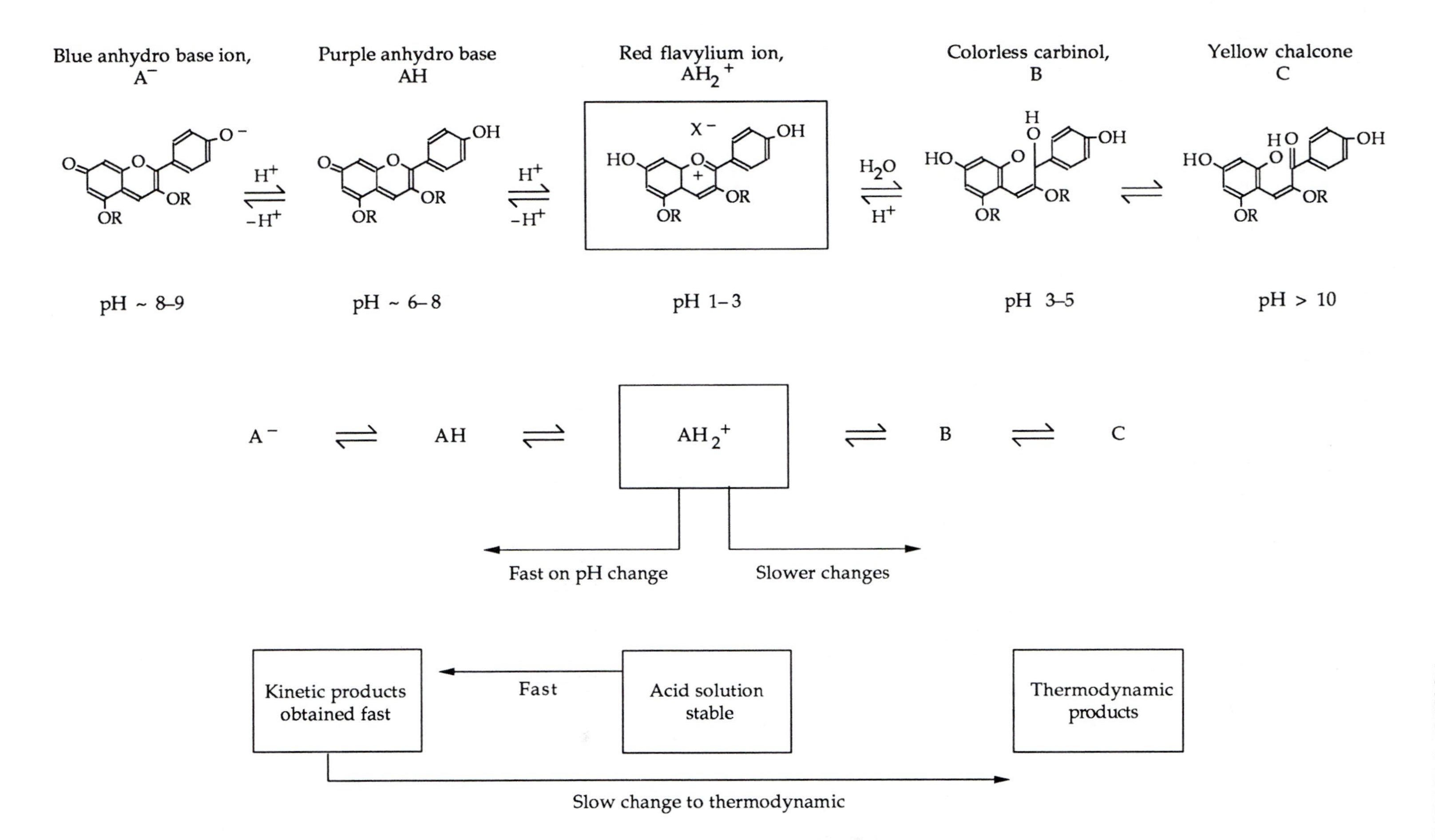

Figure 21.2 Structural Changes of Anthocyanins with Changes of pH

Pre-Laboratory Quiz

1. Which British scientist classified acidic and basic substances by reactions with colored plant extracts?

2. Which natural plant extracts can be used as universal acid-base indicators?

3. Give the general name for the coloring material in tomatoes.

4. Give a definition of an anthocyanin.

5. Give 1 function of anthocyanins in plants.

6. Why has there been controversy about the use of synthetic color additives in food?

7. What is an anthocyanidin?

8. Which anthocyanidin type is present in the anthocyanin of geranium?

9. What is the name for the form of an anthocyanin that is stable in very acidic solution?

10. What evidence suggests that simple *in vitro* anthocyanin chemistry is considerably different from *in vivo*?

Laboratory Experiments

Flowchart of the Experiments

Section A. The Extraction of Plant Dyes

Section B. Anthocyanin Purification by Chromatography

Section C. The Solution Chemistry of Anthocyanins

Section D. Spectroscopy of Anthocyanins

Section E. Anthocyanin Synthesis

Section F. The Evaluation of Anthocyanins and Betacyanins as Food, Drug, and Cosmetic Dyes

Requires one three-hour class period to complete

Section A. The Extraction of Plant Dyes

Goals:

(1) To investigate the extraction of plant dyes from flowers, fruits, and leaves. (2) To assess the suitability of the extraction processes for the commercial production of natural food, drug, and cosmetic dyes.

NOTE: The materials that give color to plants are called by many different names — e.g., dye and dyestuffs, pigments, colorant, coloring material, etc. — in the scientific and technical literature. For consistency, the word *dye* is used throughout the experimental sections of this laboratory module.

Experimental Steps:

1. At Reagent Central you will find several types of plant material — e.g., fresh red cabbage, flower petals, and fruits (when in season). Use scissors or a paper punch to cut off small pieces (about 1 cm^2) of each of the plant samples and place them in wells of a 24-well tray.
   - Make a note of which plant samples you have taken.
2. Take them back to your place.
3. Inspect the plant material carefully with your hand lens.
   - Describe the nature of the plant dye, e.g., are the dyes in the epiderm or all through the tissue?
   - How easy is it to obtain juice from the plant material?
4. First work with red cabbage. Cut off 3 small pieces and place each piece in a well of a 1 x 12 well strip. Place the pieces in adjacent wells in order to make comparison easier.
5. Add 3 drops water to one piece, 3 drops of white vinegar to another, and 3 drops ethanol to the third.
   - Record the well location of the added liquids.
6. If you have flower material, repeat Steps 2 through 5 for small samples.
7. Stir each sample.
   - Record any color change in the sample or the liquids.
8. If you have access to a microwave oven or a conventional hot plate, you can examine the effect of heat on the sample. If you use a microwave oven, the 1 x 12 well strip can be placed directly in the oven. Consult with your instructor for power and time settings.

   CAUTION: ***Do not*** **use any device with flames to heat the ethanol solution. It will undoubtedly catch fire!**
9. If you have a conventional hot plate, use small containers (not the 1 × 12 well strip) to examine the effect of heat on the samples. Bring the samples to a gentle boil for not more than 1 or 2 minutes.

10. Remove from the microwave oven or hot plate and examine the products. Allow to cool for a few minutes.

11. Use a pipet to transfer 1 drop of each extract to a filter paper.

12. Make a straw drying stand and let the solvents partially evaporate.

    - Describe what happens.
    - Make a tabular summary of the characteristics of each extraction.

    The three extractions that you have just carried out are completely different in their effect on the plant material and the dye contained in it.

    - Which of the extraction processes chemically changed the anthocyanins in the plant materials?
    - What are the products of these changes? (Consult the Background Chemistry section.)
    - What are some of the compounds other than dyes that are probably extracted from the plant material in these processes?

    The ethanol extraction is rather unusual in that the anthocyanins form colorless compounds in the presence of excess ethanol. However, evaporation of the ethanol restores the original color almost completely.

    Let us choose the alcohol extraction as perhaps the best of these three extraction processes. Describe how you would extract 800 kg of red cabbage with ethanol and produce a concentrated anthocyanin extract.

    - Could you use another alcohol (e.g., methanol) in the extraction? Why not?

Section B. Anthocyanin Purification by Chromatography

Goals: *(1) To remove unwanted impurities (e.g., sugars) from plant anthocyanin extracts by means of column chromatography. (2) To recover and test the purified anthocyanin fraction.*

Experimental Steps:

1. Construct a column chromatography system by making a straw stand and clamp in a 96-well tray. Use a plastic column tip (with a small wad of polyester fiber in the tip) on the end of a straw as the chromatography column. (See Chapter 18, "Paper and Liquid Chromatography.")

2. Rinse the straw with water. Allow to drain to a small cup or beaker.

3. Transfer a neutral aqueous slurry of fine-mesh resin to the column and try to make the column about 5 cm long.

4. Add a little water to the column and then push a wetted 1/4" circle of filter paper (already punched out for you) onto the top of the resin. Use a slim straw to guide the circle down and to gently tamp it onto the top of the resin.

5. Add a little more water and allow the water to drain under gravity to make sure that there is good flow through the column and that the water level stops at the paper circle.

   NOTE: Capillary action will stop the column from draining completely. It is important that the liquid level not go below the circle or channeling will occur.

6. Cut about 4 cm off the column straw.
7. Obtain 3 drops of concentrated red cabbage extract from Reagent Central. Use a well from the 96-well tray.
8. Use a thin-stem pipet to transfer 1 drop of the extract to the top of the column. Allow it to flow onto the packing.
9. Wash out the thin-stem pipet and fill it about 1/2 full with 0.1 M HCl/ethanol solution. Add the solution to the column straw.
10. Allow the anthocyanins to elute under gravity and collect the colored fraction in a well of a 24-well tray.

    Save the anthocyanin fraction for Section C.

11. A good way to show that the sugars in the red cabbage extract do not interact with the cation exchange resin is to add 1 drop of the extract to a vial. Add a very small volume of resin slurry and shake to carry out a simple batch extraction.

    It will take a few minutes for the anthocyanins to adsorb onto the resin.

12. When the color has disappeared from the aqueous phase, remove a few drops of the liquid and test for reducing sugars (e.g., glucose) by means of the Benedict or Fehling tests. Your instructor will give you directions as to where to obtain the solutions for these tests.

Section C. The Solution Chemistry of Anthocyanins

Goal: *To investigate the solution chemistry of selected anthocyanins, with a particular emphasis on acid-base, redox, and complexation reactions.*

Discussion: A comparison of the aqueous solution chemistry of anthocyanins, betacyanins, and synthetic FD and C dyes will also be carried out. The chemical information gathered in these investigations will be used to assess the prospect of using anthocyanins and perhaps betacyanins as food dyes. You might also wish to try other anthocyanins.

Experimental Steps:

1. Obtain a microburet of anthocyanins (red cabbage or red rose extracts), one of betacyanins (red beet extract), and one of a red FD and C dye (e.g., red no. 3 or red no. 40).

   NOTE: A convenient method for carrying out many reactions and making comparisons is to use 1 x 12-well strips or a 96-well flat-bottomed tray.

2. First examine the acid-base chemistry of anthocyanins. Go to Reagent Central and deliver 2 drops of each buffer solution (there are pH 1 through 11 solutions) to different wells in a strip (or tray). Take the strip back to your table.

3. Deliver 1 drop of the red cabbage extract (or the red rose extract) to the pH 1 buffer. Stir.

   - Record what happens.
   - Compare the color with the color of unreacted extract.

4. Continue in this manner until you have examined the effect of each pH buffer on the extract.

   Note particularly any immediate color changes and color changes that occur over a longer time period. If you are in doubt about the chemistry you might want to go back to the simple scheme for the pH-jump chemistry of anthocyanins given in the Background Chemistry section.

5. Use the same technique to examine and compare the acid-base chemistry of beet extract and synthetic FD and C dye.

   The author has discovered that a very convenient way of making a collection of anthocyanin extracts (in fact a sort of library of anthocyanins) is to express a plant or fruit juice directly onto a filter paper or soak a filter paper in a concentrated aqueous ethanolic plant extract. The paper is then hung up to dry or warmed very gently for a short time in a microwave oven. Once the paper is dry, the anthocyanin is stabilized in its natural form and can be kept for a long time. An office punch can be used to punch out dozens of convenient "reaction ready" anthocyanin circles. Store the circles in a stoppered vial or bottle and you can do anthocyanin chemistry anywhere, anytime.

6. Check with your instructor to see if out-of-season exotics are available — e.g., pomegranate juice, passion flower extract, Christmas cactus flower extract, etc. One of these will be available in the form of paper circle anthocyanins. Use tweezers to select a few circles and place them on the plastic surface. Repeat Steps 2–4 on the circles.

   One drop of reagent dropped onto a circle is usually sufficient to carry out the reaction. Excess reagent is removed with a cotton swab.

7. Examine the effect of oxidizing and reducing agents on all of the dyes by reacting 1 drop of each dye with 1 drop 3% H_2O_2 solution (oxidizing agent) and 1 drop 0.5 M Na_2SO_3 (reducing agent).

   Acid-base and redox reactions on *in vivo* anthocyanins can be carried out by placing small pieces (~ 0.5 cm^2) of plant material into a petri dish and then diffusing in a gaseous reagent. To examine the effect of H^+ on red cabbage anthocyanin *in vivo*, poke a pin several times into a small piece of red cabbage (to break throught the epicuticular wax) and place it (and any other plant pieces) into a petri dish.

8. Drop 1 drop 3 M HCl into the dish, close to the plant materials. Replace the cover. Wait.

- Record what happens. Use your hand lens to view the plants.

9. The effect of bases can be examined by removing the HCl with a cotton swab and placing a drop of 2 M NH_3 into the dish.

   - Summarize the fundamentals of the solution chemistry of anthocyanins, betacyanins, and synthetic dye(s) in table format. You will need this information to interpret the results you get in Section E.

Section D. Spectroscopy of Anthocyanins

Goals:

(1) To obtain a visible absorption spectrum of one of the anthocyanin species present in the acid-base equilibria scheme. (2) To compare the wavelengths of maximum absorption of anthocyanins.

Before You Begin:

The experimental procedure for the operation of a visible region spectrometer will depend on the type of spectrometer that is available to you. If you are in doubt about how to use the instrument, check with your instructor and read the instruction manual. The instructions given in this section are for a spectronic 20, Bausch and Lomb instrument.

Experimental Steps:

1. Select an anthocyanin extract from those that are available. Decide on which form of the anthocyanin you want to take the spectrum of. From a practical point of view, you can choose the flavylium ion form (in acid solution) or the anhydro base form (in slightly basic solution).
2. Use a thin-stem pipet to transfer 10 drops of anthocyanin extract to a clean spectrometer tube (or cuvet).
3. If you wish to make the flavylium form (red form), add about 3–5 mL of 0.1 M HCl to the extract and stir. If you wish to make the anhydro base form, add about 3–5 mL of pH 8.0 tris buffer (not the buffer used in Section C) to the extract and stir.
4. Fill a second, clean spectrometer tube about 1/2 full with water to act as a blank solution.
5. Turn the spectrometer on and wait about 2 minutes for the instrument to warm up.
6. Set the wavelength control at 390 nm by rotating the wavelength control. Make sure that the sample compartment lid is closed.
7. Rotate the zero control until the needle on the meter is on the zero mark of the black scale.

   NOTE: Parallax errors should be avoided by moving your head so that the needle and its reflection in the scale mirror are superimposed.
8. Open the lid and push the tube containing the blank sample down as for as it will go.
9. Rotate the tube so that the mark on the tube and the mark on the sample compartment coincide. Close the lid.

10. Standardize the light control by turning it until the meter needle is on the zero mark on the red scale (absorbance).
11. Remove the blank.
12. Push the tube containing the anthocyanin sample into the compartment. Close the lid.
13. Read *absorbance* directly from the absorbance scale. If the color is too intense, dilute the solution with acid or buffer (see Step 3) until you have a solution with an initial absorbance of about 0.2 (at 390 nm).
14. Scan the visible spectrum from 390 nm to 620 nm at 20 nm intervals.
15. Zero the instrument (repeat Steps 6 through 9) between each absorbance reading at each wavelength.
    - Record the absorbance reading at each wavelength.
    - Plot a graph of absorbance versus wavelength. Draw a *smooth* curve through the points.

    The graph, which is called the *visible absorption spectrum,* identifies the particular anthocyanin and its chemical form in solution.
    - Determine the wavelength of maximum absorption (λ_{max}) from the absorption spectrum.

    Your instructor will provide you with the λ_{max} values for other anthocyanins in various forms.

Section E. Anthocyanin Synthesis

Goals: *(1) To carry out a simple synthesis of an anthocyanin from rutin, a naturally occurring flavonol. (2) To chemically characterize the reaction product as an anthocyanin.*

Experimental Steps:

1. Place a small amount of rutin (about 50 mg) into a well of a 24-well tray.
2. Add 10 drops of 1 M HCl and 4 drops of ethanol.
3. Stir to make a suspension of the rutin.
4. Add 2 or 3 small pieces of magnesium ribbon.
5. Wait for the magnesium to start dissolving in the HCl.
    - What is the chemical equation for the reaction of HCl with $Mg_{(s)}$?
    - What gas is evolved?
6. Let the reaction go for a few minutes. Stir occasionally.
    - Record what happens as the reaction proceeds.
7. Filter the solution using a small-scale filtration system and retain the filtrate.

8. Design a simple set of experiments to show that an anthocyanin was produced in the synthetic reduction of rutin.

Rutin is found in many plants, especially the buckwheat plant (*Fagopyrum esculentum* Moench., *Polygonaceae*), which contains about 4% (dry weight basis). The structure and reaction is

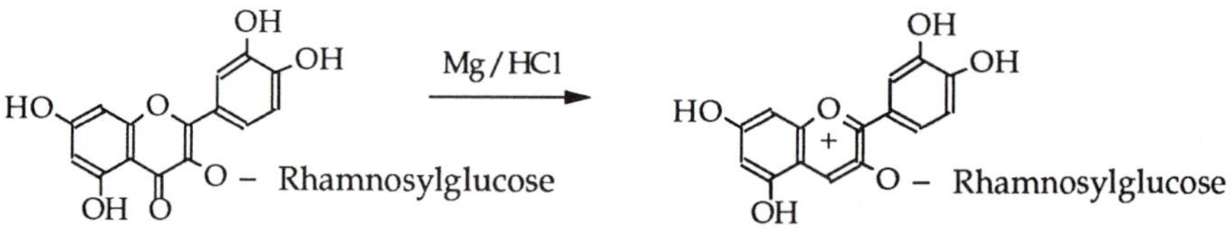

OH OH HO O O - Rhamnosylglucose OH O —Mg/HCl→ OH OH HO O + O - Rhamnosylglucose OH

The mechanism for the reaction is not known. The synthesis, however, has attracted a lot of attention from entrepreneurs interested in large-scale production of anthocyanins. Several patents have recently been issued that claim that the yield of anthocyanin is increased substantially in the presence of ultraviolet light. The obvious advantage of this method of production is that anthocyanins can be made cheaply, in one step, from such ubiquitous flavonols as rutin.

Section F. The Evaluation of Anthocyanins and Betacyanins as Food, Drug, and Cosmetic Dyes

Goals:

(1) To incorporate anthocyanin and betacyanin extracts in a variety of commercial products. (2) To evaluate the dye products for chemical compatibility, stability, and color.

Before You Begin:

I am going to ask you to design the experiments and the evaluations. However, here are a few suggestions to help you get started.

(a) Incorporate the anthocyanin or betacyanin extracts in small volumes of simple, available products — e.g., a colorless carbonated beverage, plain yogurt, unflavored plain gelatin dessert, cornstarch, *hot* herb tea, and generic face cream.

(b) Use small quantities of materials.

(c) The wells of a 24-well tray make excellent reaction vessels. You can heat or cool a well by filling the surrounding volume with hot water or ice-cold water.

(d) A suitable control dye would be one of the FD and C reds (e.g., red no. 3 or red no. 40).

(e) Comparisons are best made by arranging the various samples so that they are close to each other. Trays and strips are ideal containers for these comparisons.

It would be a good idea to pool your efforts with your peers. Check with your instructor and arrange an organized research program.

CAUTION: At this point of the research into the use of natural dyes in products, it would seem premature to try out the product. *Do not eat, take, or apply* the dyed samples.

Chapter 22

Zinc Links: Coordination Chemistry and Nutritional Deficiency

Introduction

The elements essential to human survival cannot be synthesized; they must come from plants and other animals. They must be in the diet! The study of the human need for trace elements is now at a very interesting and exciting point. Of the 106 (or so) known elements, 11 make up the bulk of all matter, and only 17 trace elements have thus far been shown to have biological function in animals. The present state of knowledge allows an evaluation of only a few of these in human nutrition; they are fluorine, chromium, manganese, iron, cobalt, copper, zinc, selenium, molybdenum, and iodine. In this laboratory module, we are going to investigate some of the unusual characteristics of one of the few water-soluble minerals, zinc.

Zinc is an essential element for both plants and humans. It is known to be present in at least 20 enzymes, such as carboxypeptidase A and B, carbonic anhydrase, and alcohol dehydrogenase, which are involved in most major metabolic pathways. Zinc is also a critical component of DNA polymerase, and without it protein synthesis and cell division do not proceed normally. The biologically available body pool is small, and the turnover rate is quite fast, primarily because of the water solubility of zinc compounds. Zinc deficiency symptoms appear almost immediately, and recent data suggest that zinc-containing enzymes involved in nucleic acid synthesis and degradation are very sensitive to zinc intake.

Some of the first cases in which a severe deficiency of zinc in humans was established were reported by Prasad in 1966 and involved a study of eleven male, Iranian dwarfs. The symptoms were dwarfism at the age of 20, infantile sex organs, and a lack of mental acuity. In these cases Prasad found that the zinc deficiency arose not from a lack of zinc in the diet, but from a lack of absorption from the gastrointestinal tract. The normal diet of the Iranian villages from which the dwarfs came consisted mostly of unleavened bread, which contains phytate, a substance that prevents the absorption of zinc. The habit of eating clay (geophagia) was found to be very common among villagers in the Shiraz region. Clay is also a substance that strongly binds zinc. A zinc supplement of 100 mg of zinc sulfate per day was fed to the patients, and the result was astounding. The dwarfs gained height and within about one year started developing normally. Marginal zinc deficiency was described in a survey of apparently healthy children in Denver, Colorado, by Hambridge in 1972. The original diagnosis arose partly as a result of a check on school children who had very poor appetite and did not eat their school lunches. The poor appetite was the result of a loss of taste acuity. Other symptoms were low levels of zinc in hair ($< 0.70\ \mu g\ g^{-1}$) and suboptimal growth. An increase in the daily intake in zinc in these children brought about an almost immediate, marked improvement.

Many other clinical disorders have now been linked with possible zinc deficiencies — e.g., slow post-operative healing of wounds and burns. Most burn centers and many hospitals now use dietary zinc supplements to promote the growth of healthy tissue. Burn patients, in particular, have so much stress that significant zinc losses occur via the urinary pathway and the fluid exuding from the burned areas. Other symptoms of zinc deficiency include skin stretch marks, joint pains, sex and endocrine problems, and male growth lag. It has also been established that women have a different type of zinc metabolism, and women who take birth control pills generally have lowered serum zinc concentrations. Dermatologists have recently concluded that white spots in fingernails (leukonechia) may indicate zinc deficiency.

A critical question is, What factors lead to zinc deficiencies? Zinc compounds occur in the soil as soluble salts and, thus, can easily be washed away. Many glaciated areas in the northern United States have soils that are deficient in zinc. The sandy soil in Florida has a very low zinc content, and zinc salts are routinely added to fertilizers. Plants that grow on zinc-deficient soils (and animals that eat such plants) will be low in zinc content. Spontaneous zinc deficiency has been diagnosed in farm and feedlot animals in large sections of the United States and has necessitated routine zinc enrichment of animal feeds with zinc supplements. The extra dietary zinc has increased feed efficiency up to 25% in hogs and poultry and is one of the reasons for the relatively low cost of pork and chickens in the United States. Zinc salts have also been found to be very effective in treating some specific disorders in farm animals. In New Zealand a disease caused by a fungus toxin found in certain pasture grasses produces liver damage, severe abnormal sunburn, and pregnancy difficulties in sheep and cattle. Field trials with sheep and milking cows have shown that large doses of oral zinc sulfate significantly reduced toxin-induced liver damage.

Many types of food processing can result in severe zinc losses in foods. Milling processes in wheat flour can remove up to 80% of the initial zinc. Zinc losses through the canning process amount to about 60% in canned legumes, 83% in canned tomatoes, 70% in canned carrots, and almost 90% in green beans. Frozen peas have less zinc than fresh peas because the surface layer of trace metals is removed with EDTA to produce a brighter green color when the peas are cooked. The zinc content of cooked foods is presented in Table 22.1 so that you might assess your own intake of this essential mineral.

Let us look at a sample calculation of the zinc content of a typical dinner. One-half pound of raw steak, when cooked, will weigh approximately 120 g. The zinc content of beef is 60 mg per 1000 g of meat. The cooked steak will thus contain

$$\frac{120\text{ g}}{1000\text{ g}} \times \frac{60\text{ mg}}{1} = 7.2\text{ mg of zinc}$$

A portion of green beans would weigh approximately 100 g. The zinc content is 2 mg per 1000 g of beans. The portion will contain

$$\frac{100\text{ g}}{1000\text{ g}} \times \frac{2\text{ mg}}{1} = 0.2\text{ mg of zinc}$$

A normal portion of potatoes is 100 g. The zinc content would be

$$\frac{100\text{ g}}{1000\text{ g}} \times \frac{3\text{ mg}}{1} = 0.3\text{ mg of zinc}$$

The dinner would provide a maximum of 7.7 mg of zinc. Cooking processes, such as boiling potatoes, can remove zinc, and the actual zinc content might be considerably less than 7.7 mg. The average daily zinc content of a mixed diet consumed by an American adult is 10 to 15 mg, for children, 5 mg, and for adolescents, 13 mg. The current RDA (recommended dietary allowance) for zinc for an adult is 15 mg per day.

The turnover of body zinc has been calculated from isotope studies to be in the range of 6 mg to 10 mg per day. This very high turnover rate is good evidence that zinc is bound rather loosely in the protein complexes in which it is utilized in the body. Recent studies on two of the zinc-containing enzymes mentioned earlier, carbonic anhydrase and carboxypeptidase, seem to confirm the presence of relatively weak zinc protein bonds. Ingested zinc is absorbed at several locations in the gastrointestinal tract, although the major site of absorption appears to be in the second portion of the duodenum. The mechanism of zinc absorption from the gut is not known. However, the process probably involves the formation of a small, low molar mass, zinc chelate (perhaps even a zinc-peptide complex). Zinc that is not absorbed is carried through the intestines and excreted in the feces.

Many factors have been found to affect the absorption of zinc in humans and other animals. The biological availability of zinc from food is known to be a major factor in absorption. Biological availability is primarily controlled by the strength of the chemical bonding between Zn^{2+} and natural complexing agents in food and by the formation of insoluble, nondiffusible zinc complexes. Most of the well-documented instances of zinc deficiency have occurred in animals whose main dietary protein intake was obtained from plant sources (and seeds in particular). Zinc from plants is not as biologically available as zinc derived from animal protein. The difference is caused by the presence of large amounts of phytates in seeds. Phytates are sodium, calcium, magnesium, and other metal ion salts of phytic acid (see the Background Chemistry section for more details) that form insoluble and nonabsorbable zinc phytate complexes in the gastrointestinal tract. There now remains little doubt that phytate in animal

Table 22.1 **Zinc Content of Various Types of Food**

Dairy Products	ppm	grams	
whole egg	15	48	
egg yolk	15	17	
egg white	0.2	31	
raw milk	3.5	244	(1 cup)
skimmed milk	3.0	246	(1 cup)
butter	1.5	10	
Average	9.0		

Meat	ppm	grams	
beef	60	120	(cooked)
lamb	50	100	(cooked)
chicken thigh	30	80	(cooked)
chicken breast	10	100	(cooked)
beef liver	6	75	(2 slices)
Average	9.0		

Vegetables	ppm	grams
peas	40	100
carrots	20	100
potatoes	3	100
cauliflower	2	100
cabbage	8	100
corn	3	100
tomato	2.5	100
green beans	2	100
legumes (average)	11	
roots (average)	3.4	
leaves (average)	1.7	

Seafood	ppm	grams	
atlantic oyster	1000	100	(raw)
hard clams	200	100	
Average	18		

Fruits	ppm	grams	
peaches	1	100	
pears	1	100	
bananas	3	150	(1 medium)
cantaloupe	1	100	(1/4 melon)
lemon	2	50	(1/2 lemon)
orange juice	1	150	
grapefruit juice	1	100	
Average	1.7		

Cereals	ppm	grams	
wheat bran	140		
wheat germ	130		
unpolished rice	8	150	(1 cup)
polished rice	3	150	(1 cup)
whole wheat bread	10	23	
whole rye bread	13	23	
white bread	1.5	23	
Average	18		

Nuts	ppm
whole nuts	34
peanut butter	20
Oils and Fats	
Average	8.4
Beverages	
alcoholic	0.9
non-alcoholic	0.2

The unit ppm is milligrams of zinc per 1000 g of food. The weight in grams of edible portions commonly used is given so that you may calculate zinc content in daily meals. The *Average* of some food groups is given although only a few items are listed.

diets markedly decreases the biological availability of zinc. Foods with high levels of phytate include soy products, wheat and rice bran, rye and corn products, barley, peanuts, sesame oil meal, mustards, infant cereals, legumes, and many other foods derived from seeds.

All the evidence presented thus far about the intake, turnover, and loss of zinc indicates that a significant fraction of the U. S. population is perhaps marginally sufficient in zinc. A small decrease in dietary zinc or a slight decrease in biological availability could quite easily (and quickly) lead to zinc deficiency. Many nutritionists have recommended that dietary zinc levels be increased, particularly in diets rich in phytate, in order to ensure an adequate availability. Supplemental zinc is inexpensive and is available in most multiple-vitamin-plus-minerals formulations on the market. Studies have shown that almost all of the various zinc salts — e.g., zinc oxide (ZnO), zinc sulfate ($ZnSO_4$), zinc gluconate ($Zn(C_6H_{11}O_7)_2$), and zinc EDTA — are equally effective. The *Physicians Desk Reference* lists several zinc dietary supplements, all of which contain zinc sulfate. Many claims have been made by food faddists that chelated zinc (usually zinc gluconate or glycinate) is "more easily assimilated" or is "more natural" than simple inorganic salts. However, there is little evidence to back up these claims, and it is worth pointing out that many of these so-called natural zinc supplements are extremely expensive. Ingested zinc is relatively nontoxic if the diet contains adequate copper and iron. Most animals appear to tolerate levels of the order of one gram per kilogram of body weight. Most of the reported adverse side effects (diarrhea) of large dosages are probably associated with the irritative effect of the high anion concentration. Finally, it is important to emphasize that the extent to which marginal zinc deficiency occurs in humans and animals is unknown because there is no specific test for zinc deficiency. Most experts recognize that the zinc nutritional requirements of animals and the recommended dietary allowances are simply best estimates made on the basis of very limited data.

Background Chemistry

The ability of metal ions to bind with many naturally occurring organic compounds has a profound effect on the availability and mobility of the metal ions. Many types of biochemical substances in the soil, including acids, peptides, proteins, polysaccharides, and humic compounds, form stable combinations with zinc. It is well established that zinc is strongly immobilized by stabilized organic matter, whereas smaller biochemical intermediates generally solubilize zinc. These zinc binding agents have a tremendous influence on the availability of zinc to plants and other biological systems. The fundamental study of the nature and strength of the chemical bonding between zinc and electron-rich compounds is part of a fertile and interesting area of chemistry called *complexation*, or *coordination, chemistry*.

Many metal ions, particularly polyvalent metal ions, react with electron pair donors to form complexes. The electron pair donors, called *ligands*, must have one or more lone pairs of electrons that can be donated to a metal ion to form one or more coordinate covalent bonds. A *complex* is therefore any chemical species that contains one or more coordinate covalent bonds. Similarly, a *coordinate covalent bond* may be defined as a covalent bond in which the shared pair of electrons comes from one of the bonded atoms. The study of complexes and complexation reactions is often referred to as coordination chemistry and, alternatively, as *Lewis acid-base chemistry*. A *Lewis acid* is an electron-deficient species that can accept and share electrons and form coordinate covalent bonds. Most of the Lewis acids found in aqueous solution are metal cations. *Lewis bases* (ligands) are usually anions or species with one or more lone pairs of electrons — e.g., dimethylglyoxime Cl^-, H_2O, NH_3, OH^-.

Consider the dissolution of a water-soluble zinc salt, such as zinc sulfate, to form an aqueous solution. The zinc sulfate dissociates and produces hydrated ions:

$$ZnSO_{4\,(s)} \longrightarrow Zn^{2+} + SO_4^{\;2-}$$

$$Zn^{2+} + 4\,H_2O \rightleftharpoons Zn(H_2O)_4^{\;2+}$$

Note that this complexation equilibrium reaction, unlike other types of equilibria, is written as a formation reaction rather than as a dissociation. Aqueous solutions of metal ions are Lewis acids that are literally swamped by a Lewis base — water. The hydrated zinc ion ($Zn(H_2O)_4^{\;2+}$, tetraaquazinc (II) ion), is a tetrahedrally shaped ion with four coordinate covalent bonds between zinc and the water molecule ligands. Each water molecule forms one coordinate covalent bond by sharing a lone pair on the oxygen atom with the zinc ion. A ligand that can form only one bond to a metal ion is called a *unidentate* ligand. The maximum number of ligands that can attach to a central metal ion is called the *maximum coordination number* and is often twice the ionic charge. Zinc almost always has a maximum coordination number of 4, and its chemistry is dominated by the formation of tetrahedral complexes.

The addition of a stronger Lewis base, e.g., ammonia (NH_3), to an aqueous solution of zinc ion will result in the sequential displacement of water molecules from the hydrated ion by ammonia molecules:

$$Zn(H_2O)_4{}^{2+} + NH_3 \rightleftharpoons [Zn(H_2O)_3NH_3]^{2+} + H_2O$$

$$[Zn(H_2O)_3NH_3]^{2+} + NH_3 \rightleftharpoons [Zn(H_2O)_2(NH_3)_2]^{2+} + H_2O$$

$$[Zn(H_2O)_2(NH_3)_2]^{2+} + NH_3 \rightleftharpoons [ZnH_2O(NH_3)_3]^{2+} + H_2O$$

$$[ZnH_2O(NH_3)_3]^{2+} + NH_3 \rightleftharpoons [Zn(NH_3)_4]^{2+} + H_2O$$

Equilibrium constants, called *stepwise stability constants*, can be written for each of the equilibria in the above sequence of complexation reactions:

$$K_1 = 3.9 \times 10^2 = \frac{[[Zn(H_2O)_3NH_3]^{2+}]}{[[Zn(H_2O)_4][NH_3]}$$

$$K_2 = 2.1 \times 10^2 = \frac{[[Zn(H_2O)_2(NH_3)_2]^{2+}]}{[[Zn(H_2O)_3NH_3]^{2+}][NH_3]}$$

$$K_3 = 1.0 \times 10^2 = \frac{[[ZnH_2O(NH_3)_3]^{2+}]}{[[Zn(H_2O)_2(NH_3)_2]^{2+}][NH_3]}$$

$$K_4 = 5.0 \times 10^1 = \frac{[[Zn(NH_3)_4]^{2+}]}{[[ZnH_2O(NH_3)_3]^{2+}][NH_3]}$$

The large value of K_1 (390) quantitatively indicates the formation of a stronger coordinate covalent bond between NH_3 and Zn^{2+} than between H_2O and Zn^{2+}. Notice that as the displacement of water by ammonia proceeds, the K values get smaller. This may be explained by the fact that there are statistically fewer water molecules to replace as the reactions proceed. The overall reaction for the formation of $[Zn(NH_3)_4]^{2+}$ can be written as

$$[Zn(H_2O)_4]^{2+} + 4\,NH_3 \rightleftharpoons [Zn(NH_3)_4]^{2+} + 4\,H_2O$$

which is the sum of the individual stepwise complexation reactions. *The overall stability constant* (β_4) for the overall reaction is:

$$\beta_4 = K_1K_2K_3K_4 = 4.1 \times 10^8 = \frac{[[Zn(NH_3)_4]^{2+}]}{[[Zn(H_2O)_4]^{2+}[NH_3]^4}$$

The two ligands for Zn^{2+} considered thus far, H_2O and NH_3, are both unidentate and share only one lone pair per ligand with the zinc ion. Many ligands are capable of donating two or more lone pairs of electrons per ligand species to a metal ion and are said to be *multidentate* ligands. Complexation reactions between multidentate ligands and metal ions form complexes called *chelates*. Chelates are considerably more stable than the corresponding complex formed with unidentate ligands. The bidentate analog of ammonia is 1,2-diaminoethane, which contains two lone pairs of electrons per molecule and takes part in a chelation reaction

$$[Zn(H_2O)_4]^{2+} + 2\,H_2N{-}CH_2{-}CH_2{-}NH_2 \rightleftharpoons [Zn(H_2N{-}CH_2{-}CH_2{-}NH_2)_2]^{2+} + 4\,H_2O$$

$$\beta_2 = 2.1 \times 10^{10} = \frac{[Zn(en)_2^{2+}]}{[Zn(H_2O)_4^{2+}][en]^2}$$

where en is an abbreviation for 1,2-diaminoethane. The formation of the beautifully ordered ring chelate structure releases four disordered water molecules, and the decrease in entropy drives the reaction. The overall stability constant for the formation of the chelate is about two orders of magnitude greater than that for the analogous unidentate complex, indicating much greater stability. In general, it is found that multidentate ligands form more stable complexes than do similar unidentate ligands, a phenomenon which is called the *chelate effect*.

Zinc dietary supplements are often in the form of zinc gluconate, in which the ligand is a glucose derivative called gluconic acid. The zinc gluconate chelation reaction forms an extremely water soluble, weak complex:

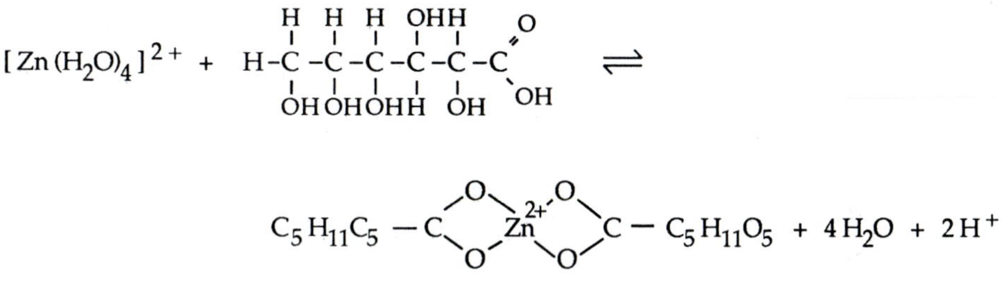

The overall stability constant has a small value (about 10^2) which reflects the rather weak bonds between the zinc ion and the gluconate ligands. Presumably, the common use of zinc gluconate as a dosage form for supplemental zinc is because of high water solubility and the gastrointestinal compatibility of the gluconate species. It is interesting to note that even very stable zinc chelates have been used as clinical dietary zinc supplements. Ethylenediaminetetraacetic acid (commonly abbreviated EDTA) reacts as a tetradentate ligand with zinc ion to form a very stable chelate with a stability constant of 3.2×10^{16}. In spite of the large value of the stability constant for ZnEDTA, the chelate apparently provides a good supply of zinc when it is administered in the diet.

One of the most important naturally occurring chelating agents for zinc is phytate. *Phytate,* which is prevalent throughout the plant kingdom, is a term that is loosely applied to phytic acid and phytic acid salts of calcium, magnesium, and other metal ions. Phytic acid is myo-inositolhexaphosphoric acid ($C_6H_{18}O_{24}P_6$) and has the structure

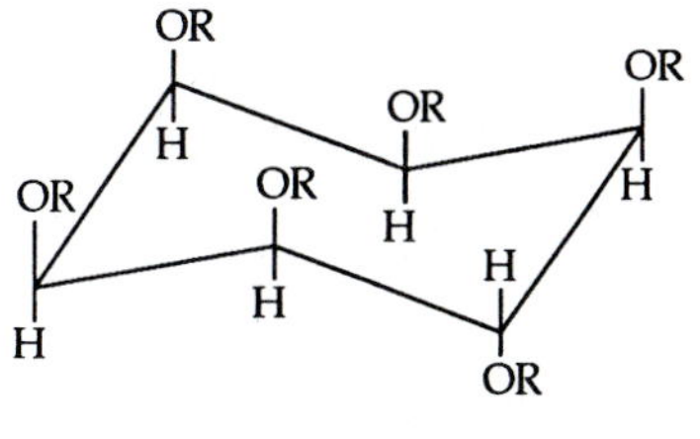

where R = PO_3H_2

Phytic acid is a strong Lewis base because of the six phosphate groups which are rich in lone electron pairs. Phytic acid complexes numerous metal ions, in various combinations and to different extents. Cation binding is pH dependent, and at a pH of 7.4, phytate binds metal ions to form chelates with stability constants in the following order: $Cu^{2+} > Zn^{2+} > Ni^{2+} > Co^{2+} > Mn^{2+} > Fe^{3+} > Ca^{2+}$. The stability and the solubility of phytate chelates in the gastrointestinal environment will determine the biological availability of metal ions. It is now well established that at the pH in the human intestine (pH 6), calcium and zinc both bind to phytate to form an insoluble chelate. The picture is further complicated by the synergistic effects that have been found when the stoichiometry of the phytate chelates are varied. Phytate coordination chemistry may perhaps be the dominant factor in the biological availability of zinc in the human diet. In the last decade the global food system has experienced an influx of many new plant food sources into food webs. Many of these plant sources represent significant concentrations of dietary phytate. The consequences for the nutritional status of populations with a marginal zinc sufficiency might be profound.

The study of the role of zinc in nutrition is made possible by the availability of analytical methods for the accurate determination of zinc concentration in agricultural, biological, and pharmaceutical samples. One of the most widely used methods for zinc analysis is a technique called atomic absorption spectrophotometry. Unfortunately, the instrumentation involved is rather expensive, difficult to operate, and unavailable in this chemistry course. However, coordination chemistry comes to the rescue with a colorimetric zinc analysis based on the formation of a blue zinc chelate with a commercially available dye called Zincon. The dye, in aqueous solution at pH 9, selectively forms a stable, highly colored 1:1 chelate with zinc ion:

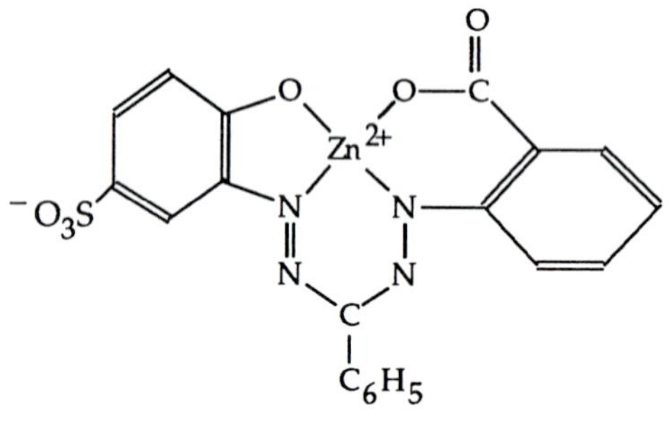

This reaction can be made the basis for a quantitative analysis for zinc by preparing a set of known concentration standards as a calibration series. The analysis of unknown zinc-containing samples is then carried out by a visual comparison with the calibration set or by an instrumental colorimetric procedure.

ADDITIONAL READING

1. Prasad, A. S., "History of Zinc in Human Nutrition," in A.S. Prasad, I.E. Dreosti, and B.S. Hetzel (eds.), *Current Topics in Nutrition and Disease*, vol. 7, Alan R. Liss, Inc., New York, 1982.

Ananda Prasad is the father of the zinc deficiency field and has some fascinating stories about the early diagnoses of zinc deficiencies. The rest of the volume also has many other readable and interesting articles.

2. *Recommended Dietary Allowances*, 9th ed., National Academy of Sciences, Washington, D. C., 1980.

Excellent. A "must read" for anyone interested in nutrition and health.

Pre-Laboratory Quiz

1. Give the name of an enzyme that contains zinc.

2. What are some symptoms of zinc deficiency?

3. What is the RDA for zinc (for an adult in the United States)?

4. What type of food is high in zinc?

5. What is the major site of zinc absorption in the human body?

6. Give the chemical formula for one Lewis base.

7. What is the maximum coordination number for the zinc ion?

8. Give a definition of a chelate.

9. What is the name of one of the most important naturally occurring chelating agents for zinc?

10. What type of bonding occurs between a metal ion and a ligand in a chelate?

Laboratory Experiments

Flowchart of the Experiments

Section A.	The Preparation of Calibration Standards for the Quantitative Colorimetric Determination of Zinc

Section B.	Zinc Determination in Unknown Samples

Section C.	The Chelation of Zinc by Phytate and Other Naturally Occurring Substances

Section D.	A Quantitative Instrumental Colorimeter for the Determination of Zinc

Section E.	Calculation of Your Daily Zinc Intake

Requires one three-hour class period to complete

Section A. The Preparation of Calibration Standards for the Quantitative Colorimetric Determination of Zinc

Goals: *(1) To serially dilute a standard zinc solution. (2) To carry out a reaction with Zincon dye to prepare a set of colorimetric calibration standards.*

Experimental Steps:

1. Clean two 1 x 2 well strips and remove any remaining water with a cotton swab.
2. Use good transfer technique to fill a large-drop microburet with standard zinc solution (10.0 mg L^{-1} Zn^{2+}).
3. Carry out a serial dilution of the standard zinc solution by delivering 1 drop to the first well, 2 drops to the second well, etc., of one of the 1 x 12 well strips. Use quantitative delivery technique and stop after you have delivered 10 drops to the tenth well.
4. Wash the microburet twice with water and fill it with water.
5. Deliver 9 drops of water to the first well (which already contains 1 drop of standard zinc solution), 8 drops of water to the second well, and so on.
   - Calculate the concentration of Zn^{2+} (in mg L^{-1}) of the solutions in each of the 10 wells.
6. Wash the microburet with water and squeeze the bulb several times (vigorously) to expel any water.
7. Suck up the liquid in the first well and gently expel it back into the well in order to ensure good mixing.
8. Suck it up again and use quantitative transfer technique to deliver 1 drop of the dilute zinc solution to the first well of the other clean 1 x 12 well strip.
9. Continue as in Steps 7 and 8 until you have transferred 1 drop from each well in the first strip to each well in the second strip.
10. Wash the microburet with water and use good transfer technique to fill it with Zincon solution (100 mg L^{-1} Zincon dye).
11. Add 1 drop of Zincon to each well of the second strip.
12. Add 2 drops of buffer pH 9 to each well and stir.
13. Hold the strip up and look through the *top* against a white background.
    - Describe the colors in your record.
    - What is the color of the ligand at pH 9?
    - What is the color of the chelate formed by the reaction of dye with Zn^{2+}?
    - What is the stoichiometry of the chelation reaction?

The strip contains a set of calibration standards for the determination of [Zn^{2+}] in any sample. The colored solutions are stable for about 3 hours, but may be photochemically faded by ultraviolet light. If you wish to store the color standards, place the strip onto a black background or cover with a piece of paper or a box.

14. Save the calibration standards for use in the following sections.

Section B. Zinc Determination in Unknown Samples

Goals:

(1) To be able to use the set of calibration standards prepared in the last section. (2) To carry out analyses of a variety of samples for the zinc concentration.

Experimental Steps:

1. Obtain the first unknown sample from your instructor.

   This first sample is an aqueous solution of zinc sulfate.

   - Be sure to record the sample number.

2. Use good transfer technique and the *same* microburet from the last section to deliver 1 drop of the unknown sample to a well in a clean 1 x 12 well strip.
3. Add 1 drop of Zincon dye solution and 2 drops of buffer pH 9. Stir.
4. Compare the unknown with the calibration standards.
   - Determine the zinc ion concentration in the unknown.
5. Carry out the determination of zinc ion concentration in the first unknown in triplicate.
   - Report the $[Zn^{2+}]$ (in mg L^{-1}) and the concentration of $ZnSO_4$ (in mol L^{-1}) in the unknown sample.
6. Obtain the second unknown sample.
   - Record your unknown sample number in your laboratory notebook.

   The second unknown sample is a solution prepared by dissolving a zinc dietary supplement tablet. The solution preparation was carried out as follows: A tablet containing zinc gluconate and excipients was weighed on an analytical balance and found to weigh 210 mg. The tablet was dissolved in dilute hydrochloric acid and diluted to exactly 1.00 L in a volumetric flask. Your second unknown sample consists of an aliquot (portion) of the solution from the volumetric flask.

7. Analyze the sample for zinc ion.
   - Report the $[Zn^{2+}]$ in mg L^{-1} and ppm and calculate the number of milligrams of zinc gluconate in the original tablet.
8. Save the calibration standards.

Section C. The Chelation of Zinc by Phytate and Other Naturally Occurring Substances

Goal:

To investigate the chemistry of zinc chelation by naturally occurring organic compounds.

Before You Begin:

You are going to be testing two types of food systems for potential chelating action: solutions and solids. Soluble chelating agents in solution are most conveniently tested by adding them to a solution containing zinc already chelated by Zincon. Insoluble chelating agents in solid form may be tested by adding a standard zinc solution to the solid and allowing the system to equilibrate. The resulting supernatant is filtered to remove any remaining solid particles and then analyzed for zinc as in the last section. Any zinc that binds to the solid food will be measured as a loss in zinc from the original standard solution. Your instructor will inform you of the potential chelating agents that are available to you. A partial list of available solutions is tea, juice from tofu, apple juice, lemon juice, and a solution of the dodecasodium salt of phytic acid. Each solution is in a labelled microburet.

Experimental Steps:

1. Obtain several of the available solutions of potential chelating agents.
2. Clean a 1 x 12 well strip.
3. Deliver 2 drops of standard zinc solution (10 mg L^{-1} Zn^{2+}) to each of 10 wells.
4. Deliver 2 drops of Zincon to each of the 10 wells. Stir.

   You now have produced the zinc chelate of Zincon.
5. Now try adding 1 drop of one of the solutions of potential chelating agents. Stir. Perhaps you can titrate the zinc out of the Zincon chelate.
6. Try some of the other solutions.
   - Describe the results of your experiments in your laboratory record.
7. In order to investigate the chelating effect of solid substances you need to clean a 24-well tray and obtain small amounts of 2 or 3 of the available solid foods. Place a small piece of each substance into its own well. For each analysis, follow Steps 8 through 13.
8. Deliver several drops of standard zinc solution (10 mg L^{-1} Zn^{2+}) to the well that contains the solid. Allow a few minutes for the heterogeneous reaction to occur.
9. While the reaction is occurring, prepare a micro filtration device: Place a 1/4" circle of filter paper into a small plastic cap. Use a slim straw to push it firmly into place. Place the cap over an empty well in the 1 x 12 well strip.
10. Use a clean microburet to remove a small volume of the solution from around the solid food.
11. Filter the solution through the paper circle by transferring a few drops at a time to the cap. Remove the cap.
12. Wash the microburet and use it to transfer 1 drop to a clean well.
13. Add 1 drop Zincon and 2 drops of buffer. Stir. Determine the zinc concentration by comparison with the set of calibration standards.
    - Record the zinc concentration you determined.
    - Calculate the zinc uptake by the solid in mg L^{-1} Zn^{2+}.

The zinc chelating agents in naturally occurring substances are almost always electron-rich, oxygen-containing ligands — e.g., tannins in tea, coffee, and fruit juices; phytate in soy and legume products; and citric, oxalic, and tartaric acids in spinach, grapes, and other types of fruit.

Section D. A Quantitative Instrumental Colorimeter for the Determination of Zinc

Goals:

(1) To construct an inexpensive instrumental colorimeter using a photoresistive light detector. (2) To prepare a light absorption calibration line for the quantitative determination of zinc.

Experimental Steps:

1. Construct a straw stand and clamp and place the stand into the well of a 96-well round-bottom tray.
2. Obtain a photoresistor (cadmium sulfide photocell) from Reagent Central. Do not bend the wire leads.
3. Cut off a 4 cm piece of slim straw and clip 2 small holes about 1 cm from the end (use an office punch).
4. Slip the photocell into the slim straw until the wire leads protrude through the holes.
5. Clamp the slim straw (and photocell) firmly in the straw clamp so that the photocell points vertically upwards.

   You have now made a quantitative light absorption (or emission) device that may be used as a colorimeter for the calibration of any series of light-absorbing standards.

6. Clip the leads from a multimeter to the wire leads from the photocell. It doesn't matter which lead (red or black) is clipped to which wire of the photocell.
7. Turn the multimeter on and switch the function control to measure resistance (Ω).

   The instrument should now look like:

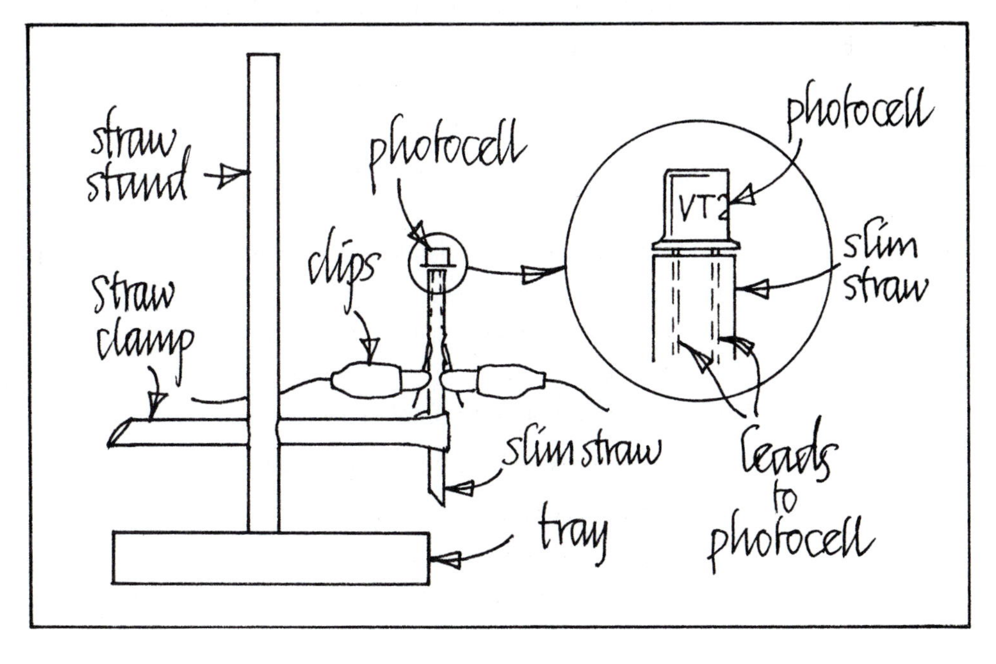

8. Place the instrument in a convenient position on the table. Once it is positioned, do not move the photocell (or stand) or you will change the amount of light falling on the photocell.

9. Pull out a thin-stem pipet to make a small-drop microburet.

10. Retrieve the 1 x 12 well strip containing the set of calibration standards that you made in Section A.

11. Suck up the solution in the first well into the microburet. Hold the microburet with the tip pointed vertically downward and tap the bulb so that the small volume of liquid is resting above the tip.

12. Hold the microburet tip about 0.5 cm directly above the photocell and deliver a free drop to the surface of the photocell, as shown below.

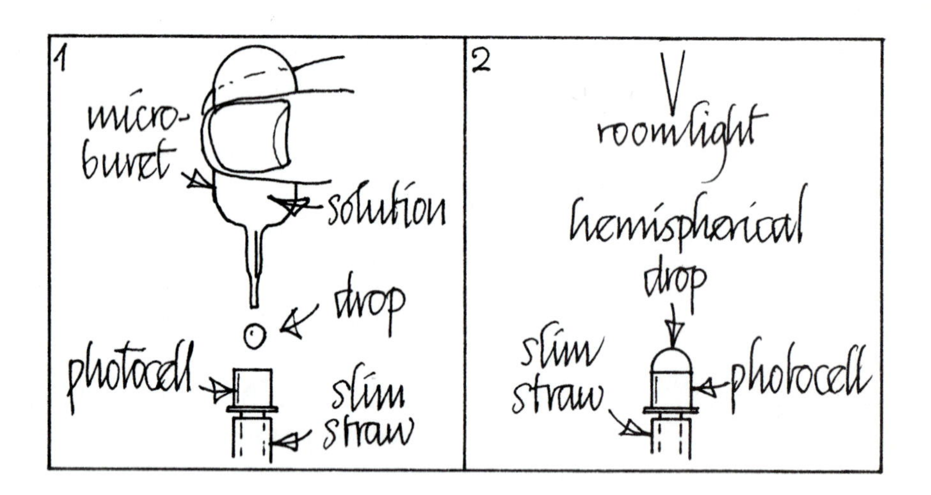

The drop will form a hemispherical cap over the glass surface of the photocell, and surface tension will keep it in place.

13. If you miss the photocell, or the drop is clinging to one side, use a cotton swab to remove the liquid and try again.
    - Read and record the resistance reading.

14. You may wish to remove the drop with a cotton swab and deliver another drop of the same solution to obtain a duplicate reading. Remove the drop with the swab (or a piece of microtowel).

15. Repeat Steps 11 through 14 for each of the solutions in the set of calibration standards.
    - Plot a graph of resistance versus zinc concentration. You calculated the zinc concentration in Step 5, Section A. Draw the best smooth line through the points. The curve may not be a straight line. That's OK!

You now have a light absorption calibration line with which you can determine the zinc concentration in any unknown sample. However, there are many caveats to consider. The conditions for the zinc determination of an unknown sample must obviously be identical with the conditions under which the calibration line was prepared. You must therefore

- Keep the photocell in the same position in order to ensure that the same amount of light is falling onto the system.
- Use the same experimental procedure for the preparation of the unknown sample solution as you did for the calibration standards — i.e., use the same large-drop microburet!
- Use the same small-drop microburet to deliver all the calibration standards and unknown sample to the photocell surface.
- Use the same multimeter and photocell for all the standards and the unknown sample.
- Recalibrate the instrument if any of the conditions change.

16. Obtain the unknown sample from your instructor. Prepare the sample as in Steps 1 through 3, Section B.
    - Use the light absorption calibration line to determine the zinc concentration of the unknown sample.

The quantitative colorimeter may be used to measure the concentration of any solution that absorbs light in the visible spectral range of 400 — 600 nm. Of course, an appropriate set of calibration standards must be constructed for each type of measurement.

Section E. Calculation of Your Daily Zinc Intake

Goal:

To use a table of zinc concentration in foods to calculate your daily nutritional zinc intake.

Experimental Steps:

1. Peruse the table of zinc content in foods presented in the Introduction section. Remember that the RDA is 15 mg per day for adults, with an additional 5 mg during pregnancy and 10 mg during lactation. An allowance of 10 mg is recommended for preadolescent children.*
    - Calculate your daily intake.

* Data from *Recommended Dietary Allowances*, 9th ed., National Academy of Sciences, Washington, D.C., 1980.

Appendix 1

Periodic Table of the Elements

IA	IIA											IIIA	IVA	VA	VIA	VIIA	VIIIA
1 **H** 1.008																	2 **He** 4.003
3 **Li** 6.941	4 **Be** 9.012											5 **B** 10.81	6 **C** 12.01	7 **N** 14.01	8 **O** 16.00	9 **F** 19.00	10 **Ne** 20.18
11 **Na** 22.99	12 **Mg** 24.31											13 **Al** 26.98	14 **Si** 28.09	15 **P** 30.97	16 **S** 32.06	17 **Cl** 35.45	18 **Ar** 39.95
19 **K** 39.10	20 **Ca** 40.08	21 **Sc** 44.96	22 **Ti** 47.90	23 **V** 50.94	24 **Cr** 52.00	25 **Mn** 54.94	26 **Fe** 55.85	27 **Co** 58.93	28 **Ni** 58.70	29 **Cu** 63.55	30 **Zn** 65.38	31 **Ga** 69.72	32 **Ge** 72.59	33 **As** 74.92	34 **Se** 78.96	35 **Br** 79.90	36 **Kr** 83.80
37 **Rb** 85.47	38 **Sr** 87.62	39 **Y** 88.91	40 **Zr** 91.22	41 **Nb** 92.91	42 **Mo** 95.94	43 **Tc** (98)	44 **Ru** 101.1	45 **Rh** 102.9	46 **Pd** 106.4	47 **Ag** 107.9	48 **Cd** 112.4	49 **In** 114.8	50 **Sn** 118.7	51 **Sb** 121.8	52 **Te** 127.6	53 **I** 126.9	54 **Xe** 131.3
55 **Cs** 132.9	56 **Ba** 137.3	57 **La*** 138.9	72 **Hf** 178.5	73 **Ta** 180.9	74 **W** 183.9	75 **Re** 186.2	76 **Os** 190.2	77 **Ir** 192.2	78 **Pt** 195.1	79 **Au** 197.0	80 **Hg** 200.6	81 **Tl** 204.4	82 **Pb** 207.2	83 **Bi** 209.0	84 **Po** (209)	85 **At** (210)	86 **Rn** (222)
87 **Fr** (223)	88 **Ra** (226.0)	89 **Ac**† (227)	104 **Rf**	105 **Ha**	106 **Unh**	107 **Uns**	108	109 **Une**									

58 **Ce** 140.1	59 **Pr** 140.9	60 **Nd** 144.2	61 **Pm** (145)	62 **Sm** 150.4	63 **Eu** 152.0	64 **Gd** 157.3	65 **Tb** 158.9	66 **Dy** 162.5	67 **Ho** 164.9	68 **Er** 167.3	69 **Tm** 168.9	70 **Yb** 173.0	71 **Lu** 175.0
90 **Th** 232.0	91 **Pa** (231)	92 **U** 238.0	93 **Np** (244)	94 **Pu** (242)	95 **Am** (243)	96 **Cm** (247)	97 **Bk** (247)	98 **Cf** (251)	99 **Es** (252)	100 **Fm** (257)	101 **Md** (258)	102 **No** (259)	103 **Lr** (260)

*Lanthanides

†Actinides

Appendix 2 Table of Atomic Masses

Element	Symbol	Atomic Number	Atomic Mass	Element	Symbol	Atomic Number	Atomic Mass	Element	Symbol	Atomic Number	Atomic Mass
Actinium	Ac	89	(227)†	Hafnium	Hf	72	178.5	Promethium	Pm	61	(145)
Aluminum	Al	13	26.98	Helium	He	2	4.003	Protactinium	Pa	91	(231)
Americium	Am	95	(243)	Holmium	Ho	67	164.9	Radium	Ra	88	226.0
Antimony	Sb	51	121.8	Hydrogen	H	1	1.008	Radon	Rn	86	(222)
Argon	Ar	18	39.95	Indium	In	49	114.8	Rhenium	Re	75	186.2
Arsenic	As	33	74.92	Iodine	I	53	126.9	Rhodium	Rh	45	102.9
Astatine	At	85	(210)	Iridium	Ir	77	192.2	Rubidium	Rb	37	85.47
Barium	Ba	56	137.3	Iron	Fe	26	55.85	Ruthenium	Ru	44	101.1
Berkelium	Bk	97	(247)	Krypton	Kr	36	83.80	Samarium	Sm	62	150.4
Beryllium	Be	4	9.012	Lanthanum	La	57	138.9	Scandium	Sc	21	44.96
Bismuth	Bi	83	209.0	Lawrencium	Lr	103	(260)	Selenium	Se	34	78.96
Boron	B	5	10.81	Lead	Pb	82	207.2	Silicon	Si	14	28.09
Bromine	Br	35	79.90	Lithium	Li	3	6.941	Silver	Ag	47	107.9
Cadmium	Cd	48	112.4	Lutetium	Lu	71	175.0	Sodium	Na	11	22.99
Calcium	Ca	20	40.08	Magnesium	Mg	12	24.31	Strontium	Sr	38	87.62
Californium	Cf	98	(251)	Manganese	Mn	25	54.94	Sulfur	S	16	32.06
Carbon	C	6	12.01	Mendelevium	Md	101	(258)	Tantalum	Ta	73	180.9
Cerium	Ce	58	140.1	Mercury	Hg	80	200.6	Technetium	Tc	43	(98)
Cesium	Cs	55	132.9	Molybdenum	Mo	42	95.94	Tellurium	Te	52	127.6
Chlorine	Cl	17	35.45	Neodymium	Nd	60	144.2	Terbium	Tb	65	158.9
Chromium	Cr	24	52.00	Neon	Ne	10	20.18	Thallium	Tl	81	204.4
Cobalt	Co	27	58.93	Neptunium	Np	93	(237)	Thorium	Th	90	232.0
Copper	Cu	29	63.55	Nickel	Ni	28	58.70	Thulium	Tm	69	168.9
Curium	Cm	96	(247)	Niobium	Nb	41	92.91	Tin	Sn	50	118.7
Dysprosium	Dy	66	162.5	Nitrogen	N	7	14.01	Titanium	Ti	22	47.90
Einsteinium	Es	99	(252)	Nobelium	No	102	(259)	Tungsten	W	74	183.9
Erbium	Er	68	167.3	Osmium	Os	76	190.2	Uranium	U	92	238.0
Europium	Eu	63	152.0	Oxygen	O	8	16.00	Vanadium	V	23	50.94
Fermium	Fm	100	(257)	Palladium	Pd	46	106.4	Xenon	Xe	54	131.3
Fluorine	F	9	19.00	Phosphorus	P	15	30.97	Ytterbium	Yb	70	173.0
Francium	Fr	87	(223)	Platinum	Pt	78	195.1	Yttrium	Y	39	88.91
Gadolinium	Gd	64	157.3	Plutonium	Pu	94	(244)	Zinc	Zn	30	65.38
Gallium	Ga	31	69.72	Polonium	Po	84	(209)	Zirconium	Zr	40	91.22
Germanium	Ge	32	72.59	Potassium	K	19	39.10				
Gold	Au	79	197.0	Praseodymium	Pr	59	140.9				

*The values given here are to four significant figures.
†A value given in parentheses denotes the mass of the longest-lived isotope.

Appendix 3

Table of Units and Metric Conversions

Mass

SI unit: kilogram (kg)

1 kilogram	= 1000 grams
	= 2.2046 pounds
1 pound	= 453.59 grams
	= 0.45359 kilogram
	= 16 ounces
1 ton	= 2000 pounds
	= 907.185 kilograms
1 metric ton	= 1000 kilograms
	= 2204.6 pounds
1 atomic mass unit	= 1.66056×10^{27} kilograms

Pressure

SI unit: pascal (Pa)

1 pascal	= $1\ N\ m^{-2}$
	= $1\ kg\ m^{-1}\ s^{-2}$
1 atmosphere	= 101.325 kilopascals
	= 760 torr (mmHg)
	= 14.70 pounds per square inch
1 bar	= 10^5 pascals

Length

SI unit: meter (m)

1 meter	= 1.0936 yards
1 centimeter	= 0.39370 inch
1 inch	= 2.54 centimeters (exactly)
1 kilometer	= 0.62137 mile
1 mile	= 5280 feet
	= 1.6093 kilometers
1 angstrom	= 10^{-10} meter
	= 100 picometers

Energy

SI unit: joule (J)

1 joule	= $1\ kg\ m^2\ s^{-2}$
	= 0.23901 calorie
	= 9.4781×10^{-4} btu (British thermal unit)
1 calorie	= 4.184 joules
	= 3.965×10^{-3} btu
1 btu	= 1055.06 joules
	= 252.2 calories

Volume

SI unit: cubic meter (m^3)

1 liter	= $10^{-3}\ m^3$
	= $1\ dm^3$
	= 1.0567 quarts
1 gallon	= 4 quarts
	= 8 pints
	= 3.7854 liters
1 quart	= 32 fluid ounces
	= 0.94633 liter

Temperature

SI unit: kelvin (K)

0 K	= −273.15°C
	= −459.67°F
K	= °C + 273.15
°C	= $\frac{5}{9}$(°F − 32)
°F	= $\frac{9}{5}$(°C) + 32

2 millimeters/Division 5th Accent

2 millimeters/Division 5th Accent

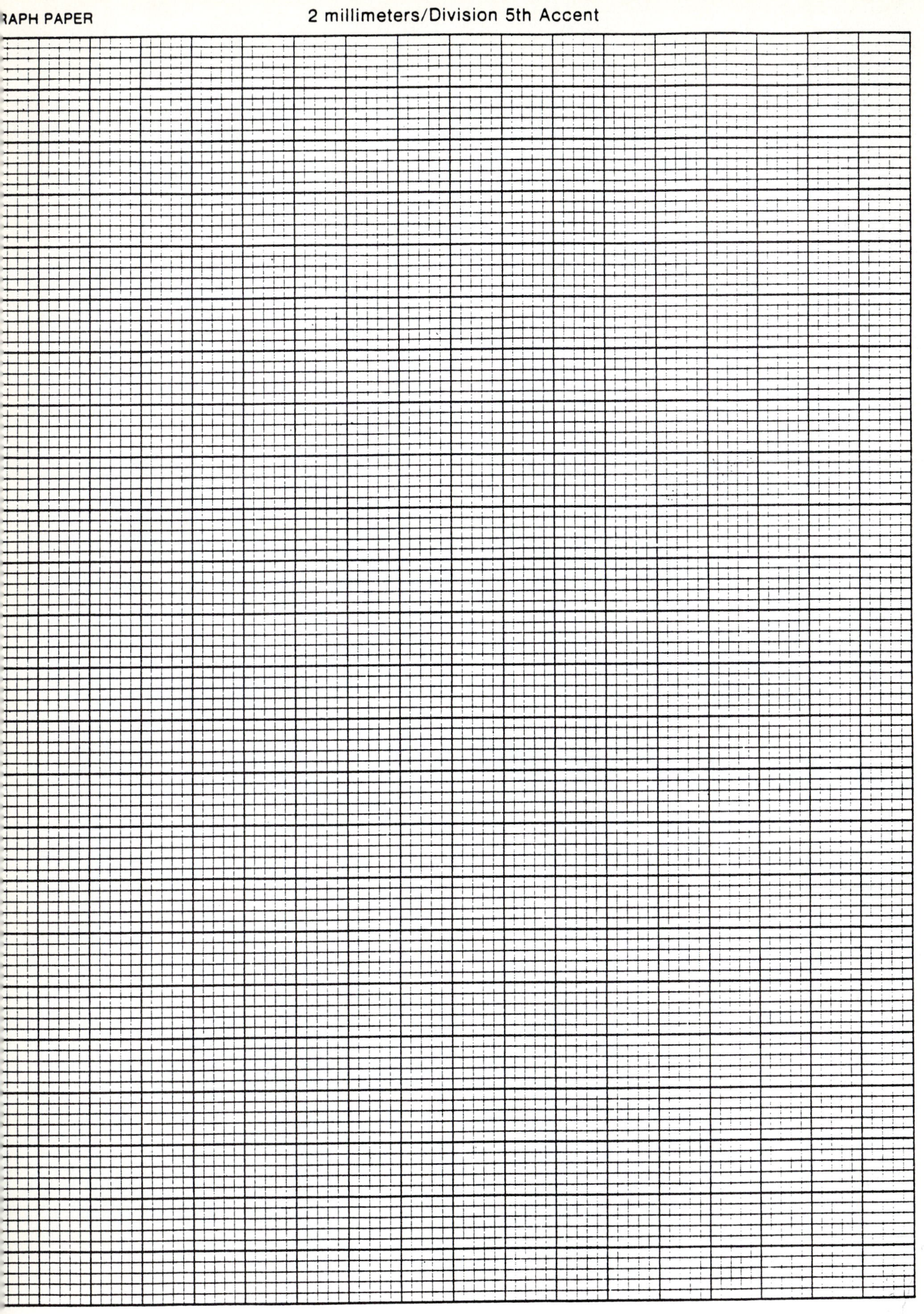

2 millimeters/Division 5th Accent

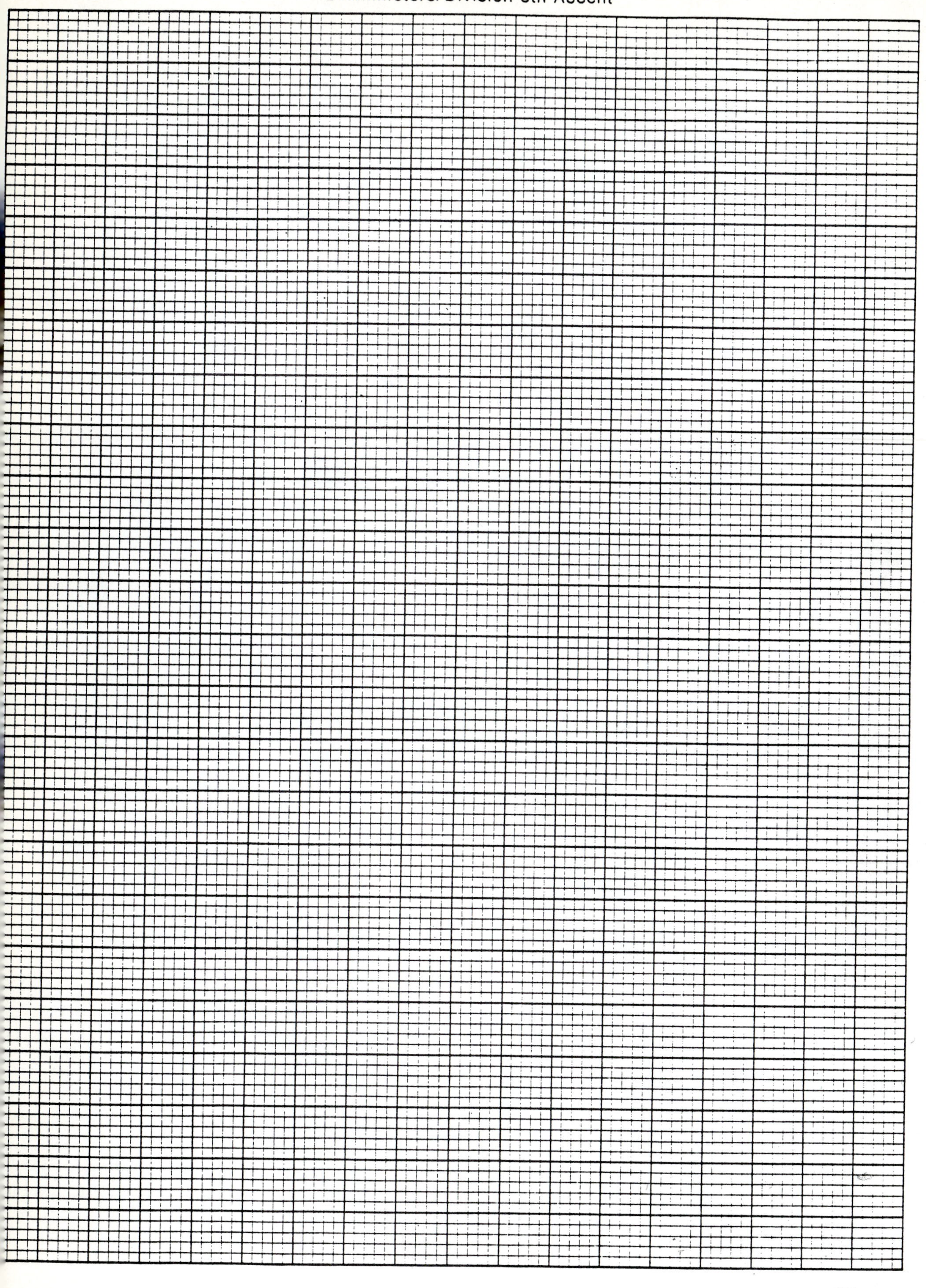